Manufacturing Industry: The Impact of Change

Second Edition

Manufacturing Industry: The Impact of Change

Second Edition

Michael Raw

Head of Geography, Bradford Grammar School

LANDMARK GEOGRAPHY

Collins Educational

An Imprint of HarperCollins*Publishers*

MADRAS COLLEGE LIBRARY

Contents

1 Manufacturing industry

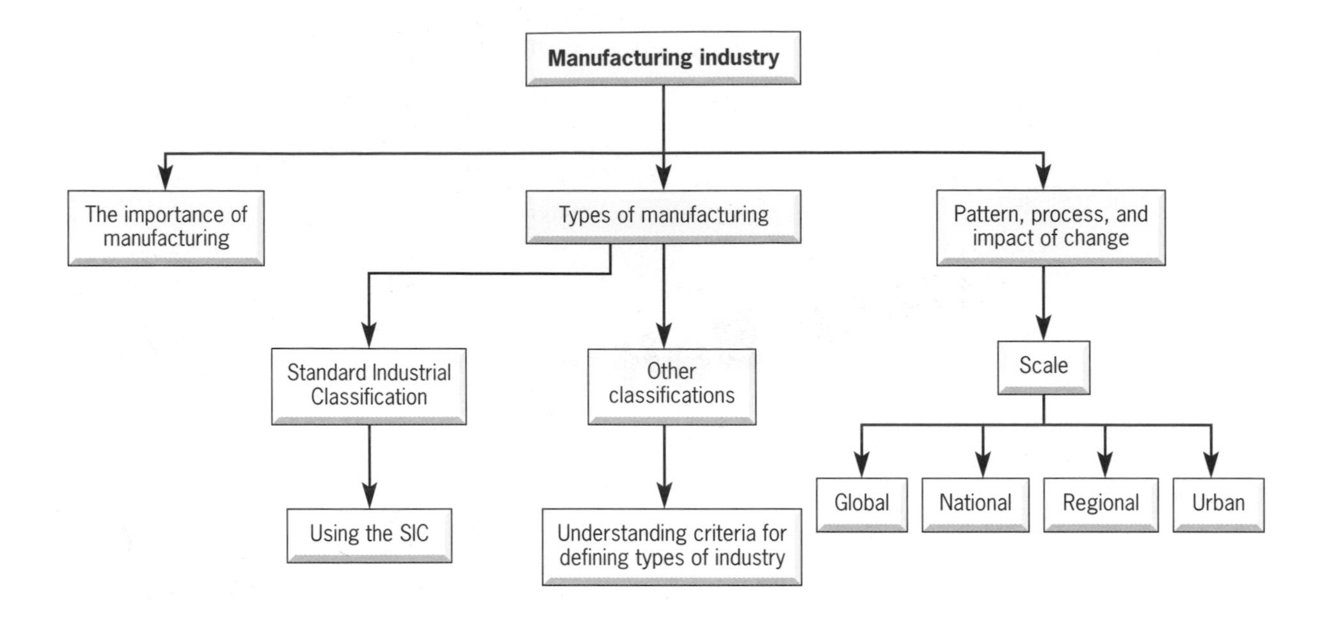

1.1 The importance of manufacturing

Manufacturing is defined as the production of goods by industrial processes. The newspaper headlines in Figure 1.1 remind us of the importance of manufacturing as an economic activity. They tell us that manufacturing is important for employment, for trade and for the creation of wealth. Manufacturing is also important for the impact it has on the environment. In other words, it is important because of the enormous influence it has on our quality of life.

How can we measure the importance of manufacturing? Figure 1.2 suggests one way: by the contribution that it makes to a country's wealth or **gross domestic product** (GDP). In the UK, manufacturing is responsible for about 22 per cent of GDP, in Japan it's around 28 per cent. In less-economically developed countries (LEDCs) manufacturing is far less important. Fairly typical are Malawi and Botswana in southern Africa. In Malawi, manufacturing contributes only 12 per cent to GDP; in neighbouring Botswana the figure is just 6 per cent.

An alternative measure of manufacturing's importance is provided by employment figures. Manufacturing employs 4 million people in the UK, which is around 18 per cent of the workforce. In contrast, the proportion in Botswana is estimated at 11 per cent.

Table 1.1 The United Kingdom's Standard Industrial Classification

Sections	Activities
A–C	Agriculture, forestry, fishing, mining and quarrying
D–F	Manufacturing industries, public utilities and construction
G–Q	Services

1.2 Types of manufacturing

Standard Industrial Classification

Almost without thinking we label industries according to what they make – e.g. the chemical industry, the brewing industry, the clothing industry, etc. Indeed this is the criterion used in the UK's **Standard Industrial Classification** (SIC). The SIC, revised in 1992, consists of ten major divisions, of which three are occupied by manufacturing (Table 1.1).

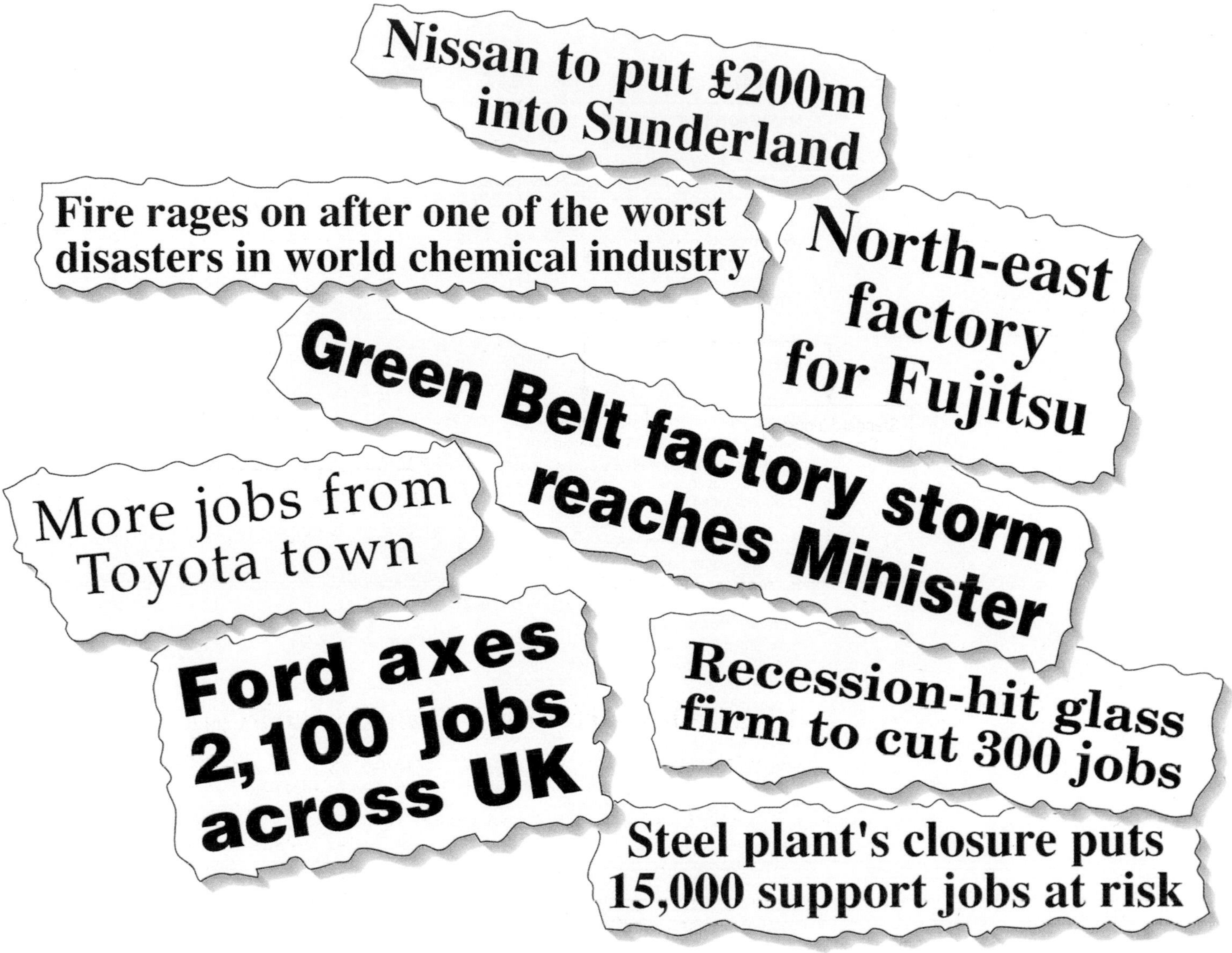

Nissan to put £200m into Sunderland
Fire rages on after one of the worst disasters in world chemical industry
North-east factory for Fujitsu
Green Belt factory storm reaches Minister
More jobs from Toyota town
Ford axes 2,100 jobs across UK
Recession-hit glass firm to cut 300 jobs
Steel plant's closure puts 15,000 support jobs at risk

Figure 1.1 Manufacturing industry hits the headlines

40
30
20
10
0
Percentage of GDP
United Kingdom
Japan
United States
1970 1980 1990 1995
1970 1980 1990 1995
1970 1980 1990 1995

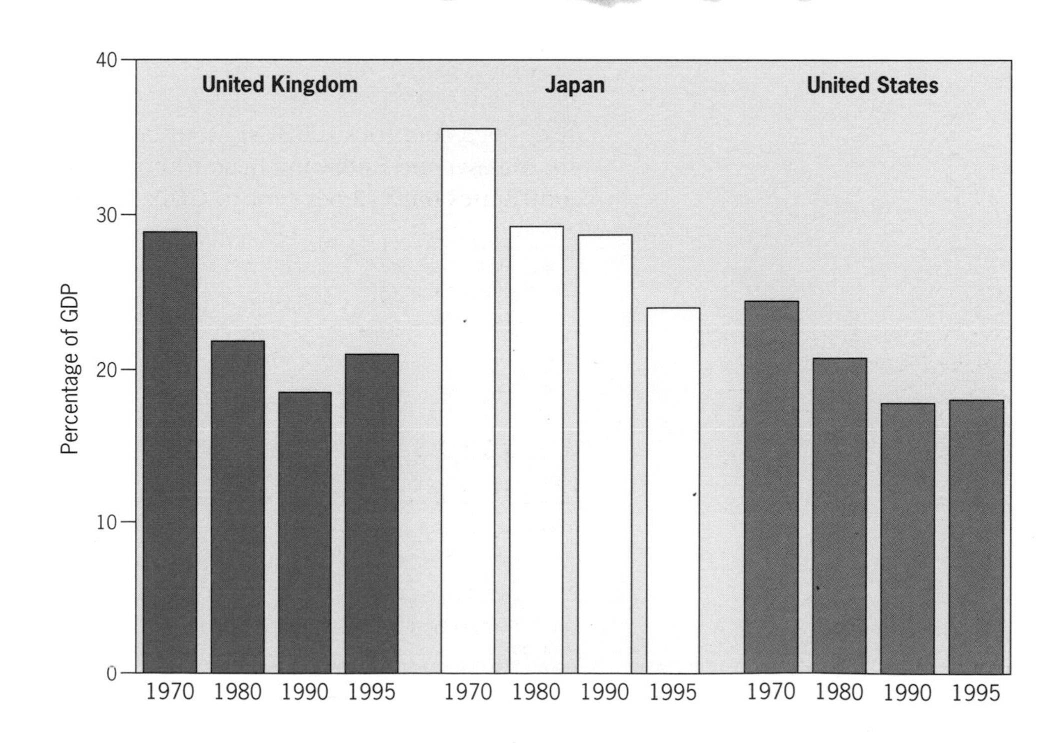

Figure 1.2 Manufacturing as a percentage of GDP

Table 1.2 Employees in employment: industry: production industries

GREAT BRITAIN SIC 1992	Section, sub-section, group or class	June 1999				
		Male		Female		
		Full-time	Part-time	Full-time	Part-time	All
MANUFACTURING	**D**	**2,727.7**	**66.5**	**878.4**	**205.7**	**3,878.4**
Manufacture of food products; beverages and tobacco	DA	260.0	21.9	119.3	54.9	456.0
of food	15.1–15.8	220.2	20.9	104.5	52.2	397.8
of beverages tobacco	15.9/16	39.7	1.1	14.8	2.6	58.3
Manufacture of textiles textile products	DB	110.0	3.7	115.3	27.0	256.0
of textiles	17	83.0	2.2	53.9	9.2	148.3
of made-up textile articles	17.4	11.1	1.1	15.3	3.2	30.7
of textiles, excl. made-up textiles	Rest of 17	71.9	1.0	38.6	6.0	117.6
of wearing apparel; dressing of fur	18	26.9	1.6	61.3	17.9	107.7
Manufacture of leather leather products including footwear	DC	13.8	0.2	10.5	1.7	26.2
of leather and leather goods	19.1/19.2	5.8	0.1	3.2	0.9	10.0
of footwear	19.3	8.0	0.1	7.3	0.8	16.2
Manufacture of wood, wood products	DD (20)	65.1	1.6	8.0	6.4	81.1
Manufacture of pulp, paper, paper products; publishing, printing	DE	281.4	7.4	138.5	32.7	460.1
of pulp, paper, paper products	21	77.8	0.4	23.1	3.1	104.4
of corrugated paper, paperboard, sacks, bags, cartons, boxes, cases and other containers	21.21	30.8	0.1	10.5	1.7	43.1
of pulp, paper, sanitary goods, stationery, wallpaper and paper products n.e.c.	Rest of 21	46.9	0.3	12.7	1.4	61.3
Publishing, printing, reproduction of recorded media	22	203.6	7.1	115.4	29.6	355.7
printing, service activities related to printing	22.2	135.8	2.2	53.8	10.0	201.8
publishing, reproduction of recorded media	Rest of 22	67.8	4.8	61.6	19.6	153.9
Manufacturing of coke, refined petroleum products, nuclear fuel	DF (23)	20.0	2.3	3.4	0.8	26.4
of refined petroleum products	23.2	10.5	2.3	1.4	0.5	14.7
Manufacture of chemicals, chemical products, man-made fibres	DG (24)	167.6	1.6	65.9	8.8	243.9
Manufacture of rubber and plastic products	DH (25)	165.2	2.4	46.0	14.7	228.4
Manufacture of other non-metallic mineral products	DI (26)	107.4	0.9	23.9	4.0	136.2
Manufacture of basic metals and fabricated metal products	DJ	428.7	8.5	73.0	13.1	523.3
of basic metals	27	103.9	0.7	11.0	1.6	117.2
of fabricated metal products, except machinery	28	324.8	7.8	62.0	11.5	406.1
Manufacture of machinery eqpt. n.e.c.	DK (29)	304.0	2.2	54.1	10.1	370.3
Manufacture of electrical optical equipment	DL	341.8	5.4	133.4	17.8	498.4
of office machinery, computers	30	31.6	0.3	11.7	1.8	45.4
of electrical machinery n.e.c.	31	116.1	1.4	43.6	6.6	167.8
of electric motors, etc.; control apparatus and insulated cable	31.1–31.3	68.6	0.7	26.3	3.5	99.1
of accumulators, primary cells, batteries, lighting eqpt., electrical eqpt. n.e.c.	31.4–31.6	47.5	0.7	17.4	3.1	68.7
of radio, TV, communication eqpt.	32	79.7	1.3	37.2	3.6	121.8
of electronic components	32.1	29.5	0.4	13.7	1.1	44.7
of radio, TV, telephone apparatus; sound and video recorders etc.	32.2–32.3	50.2	0.9	23.5	2.4	77.1
of medical, precision, optical equipment and watches	33	114.4	2.3	40.8	5.9	163.4
Manufacture of transport equipment	DM	324.3	2.4	38.0	3.7	368.4
of motor vehicles, trailers	34	188.1	1.5	23.4	2.2	215.2
of other transport eqpt.	35	136.3	0.9	14.6	1.5	153.3
of aircraft and spacecraft	35.3	87.9	0.2	10.1	1.2	99.4
of other transport equipment except aircraft, spacecraft	Rest of 35	48.4	0.7	4.5	0.3	53.9
Manufacturing n.e.c.	DN	138.4	5.9	49.2	10.0	203.5
of furniture	36.1	81.8	3.6	30.1	5.7	121.3

Each division is then broken down into classes, which are further split into groups as shown in Table 1.2. The smallest classes are called 'activity headings'.

Other classifications

Although the most obvious way to classify industries is to group them according to what they make, many other criteria are available and are frequently used in this book. Some of them are listed in Table 1.3. You should note that not every criterion is relevant to every industry.

Table 1.3 Criteria for recognising types of industry

	Aluminium smelting	Biotechnology
Heavy	•	
Light		•
Large-scale	•	
Small-scale		•
Processing	•	
Assembly		
Capital-intensive	•	
Labour-intensive		•
Material-oriented		
Market-oriented		
Transnational	•	
National		•
Fordist		
Flexible		•

?

Working through the next two exercises should make you familiar with a number of basic terms and also help you to find your way around this book.

1 Define what is meant by each of the criteria listed in Table 1.3.

2 Make a copy of Table 1.3 and analyse the characteristics of the car, steel and aerospace industries. Use reference books and your own knowledge.

1.3 Pattern, process and the impact of change

There are three strands to the geography of manufacturing: spatial patterns of manufacturing activity; the processes which give rise to these patterns; and the impact of change. Each strand has two features in common: it usually occurs unevenly in space; and it occurs at different scales (Table 1.4). Both are constant themes throughout this book.

Table 1.4 Scale and the geography of manufacturing

Scale	Pattern	Process	Impact of change
Global	✓	✓	✓
National	✓	✓	✓
Regional	✓	✓	✓
Urban	✓	✓	✓

?

Employment figures based on the SIC can be obtained from the Census of Economic Activity which should be available in a good reference library.

3 Record the employment figures by SIC for your home region or connurbation.

4 What proportion of the working population is employed in manufacturing? How does this compare with the national average? If there is a difference, try to explain it.

5a Using the employment classes from the SIC, construct bar charts for manufacturing for your town and for the UK (Table 1.2).
b Describe the structure of manufacturing shown by the two graphs and comment on any differences.

Global scale

At the global scale the most striking spatial pattern is the concentration of manufacturing industry in the economically developed world, especially in Europe, North America and Japan. However, this pattern is constantly changing. In some countries (notably on the Pacific Rim of Asia) rapid industrialisation has taken place in the past 30 years (Chapter 2). This region, which includes Japan, one of the world's most successful industrial nations, is already challenging the industrial supremacy of the West. Alongside this trend there has been a shift in the scale of manufacturing. Increasingly manufacturing is dominated by large transnational firms that are geared to global markets and global patterns of production (Chapter 8).

We can also see at the global scale how manufacturing has adversely affected the natural environment. Industrial pollution has contributed to the two great environmental issues of the late twentieth century: global warming and the thinning of the earth's ozone layer.

National scale

Spatial patterns of manufacturing activity are just as uneven at the national scale, where it is the regions that are affected. The processes responsible for this are historic (Chapters 3 and 4). They are often tied up with the availability of resources (particularly access to coal) and have their roots in the nineteenth century. For over half a century the economic tide has been against these regions. The impact of change in areas like North-East England, Lorraine, South Belgium and Pittsburgh has been severe, creating high unemployment, social decline and stressed physical environments (Chapters 10 and 12).

If the nineteenth-century industrial regions have been the losers in the manufacturing shift, then the clear winners have been the 'sun-belt' regions of Europe and North America. Attracted by a favourable climate and a high-quality environment, fast-growing high-tech industries sought locations in regions like southern California and the Côte d'Azur in France (Chapter 5).

Regional scale

The main feature of industrial patterns within regions is the distinction between conurbations and large cities, and small towns and rural areas. Although industry is still strongly concentrated in large urban areas, there has been an urban–rural manufacturing shift in favour of small towns and rural areas. This trend is found in more economically developed countries (MEDCs) and has been a consistent feature of the past 25 years (Chapter 11).

Urban scale

Even in cities, manufacturing industry has always been unevenly distributed. For the first half of the twentieth century industry was closely identified with inner-city areas. Since then massive decentralisation to the outer suburbs and beyond has occurred (Chapter 11). In the inner city the impact on the economic and social life of residents has been wholly negative, leading to urban decay, long-term unemployment and social deprivation. The physical environment has suffered too, with widespread dereliction. Neither has the impact in the outer suburbs always been beneficial. Industrial decentralisation has contributed to urban sprawl, and this has led to conflict for land on the city's edge between industry, farming and recreation (Chapter 13).

Finally, it is at the urban scale that manufacturing industry is often an uneasy neighbour with housing. Pollution, and the risk of accidents, may heavily affect surrounding residential areas, and give rise to important environmental issues (Chapter 12).

Figure 1.3 The diversity of urban industrial environments in the UK.

Summary

- Manufacturing is the production of goods by industrial processes.
- The importance of manufacturing industry can be gauged by the wealth it creates, by its contribution to employment and by its environmental impact.
- Manufacturing industries are grouped into different classes. The most formal grouping, based on what an industry makes, is the UK Standard Industrial Classification (SIC).
- Geographers group industries using a wide range of other criteria. These criteria include the following distinctions: light and heavy, large-scale and small-scale, processing and assembly, capital-intensive and labour-intensive, market-oriented and material-oriented, transnational and national, Fordist and flexible.
- There are three main strands to the study of the geography of manufacturing: pattern, process and change. Each strand can be studied at different scales: global, national, regional and urban.

2 Patterns, processes and change

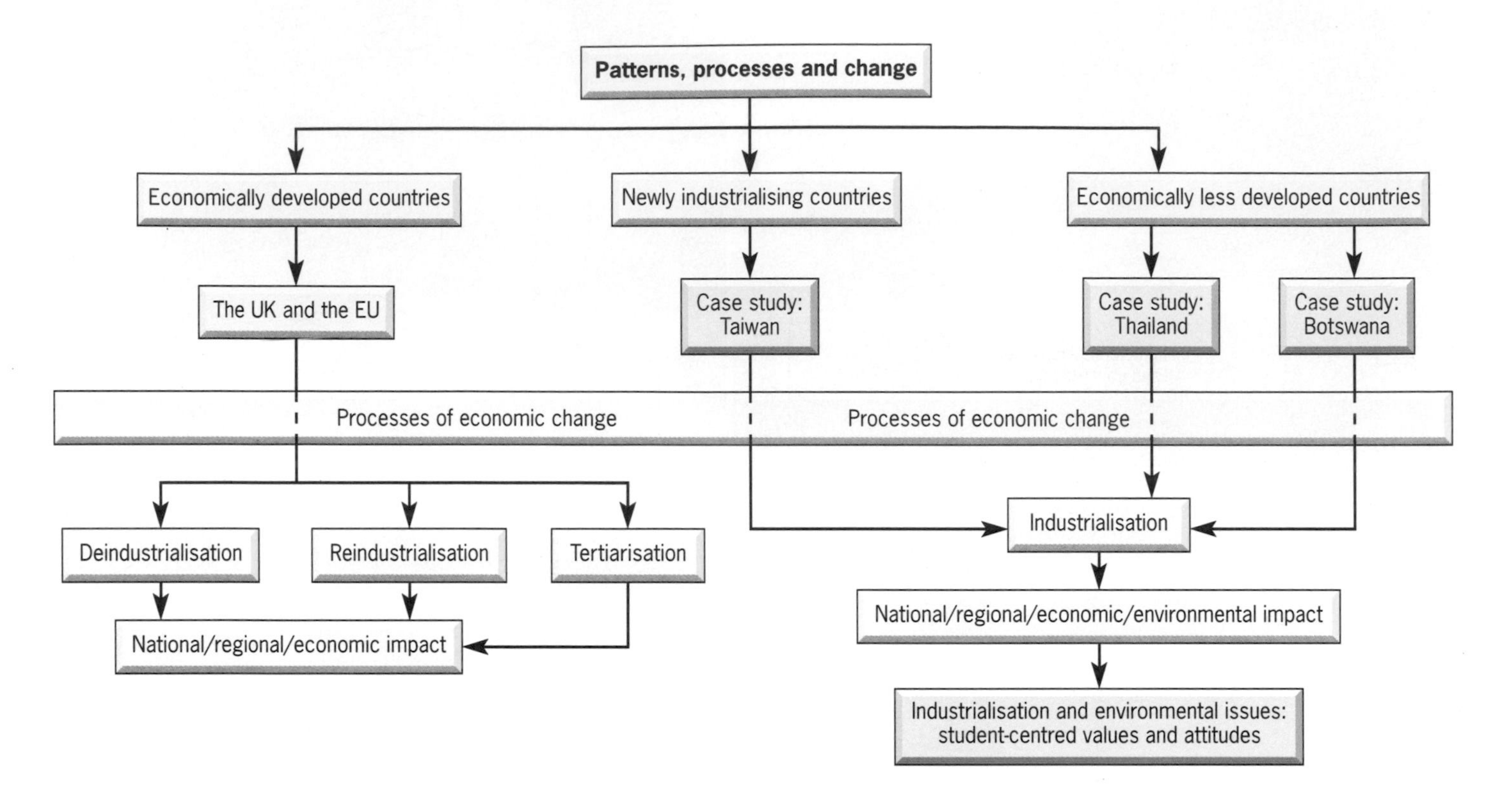

2.1 Introduction

In this chapter we shall look at some recent trends in manufacturing at a global scale. To do this we can group countries into three broad classes: more economically developed (MEDCs), less economically developed (LEDCs) and **newly industrialising countries** (NICs). You should, however, appreciate that enormous differences exist within the economically less-developed world. Generalising about such a diverse area is very difficult.

2.2 Change in the economically developed world: the UK and the EU

Compared with 30 years ago, manufacturing in the economically developed world has a smaller proportion of the workforce, contributes a smaller share to national wealth and has a very different structure. These changes relate to three processes: **deindustrialisation**; **reindustrialisation** and **tertiarisation**.

Deindustrialisation

Deindustrialisation takes place when a country's manufacturing base shrinks. It is associated with a steep decline in manufacturing employment, and a fall in manufacturing's contribution to GDP. Both the UK and the EU suffered large-scale deindustrialisation in the recession between 1975 and 1985 (Figs 2.1, 2.2, 2.3). It was most severe in the older, heavier industries such as iron and steel, chemicals, shipbuilding and textiles (Table 2.1). For example, the West Yorkshire woollen industry, which had low productivity and out-dated plant, lost two-thirds of its workforce between 1970 and 1986.

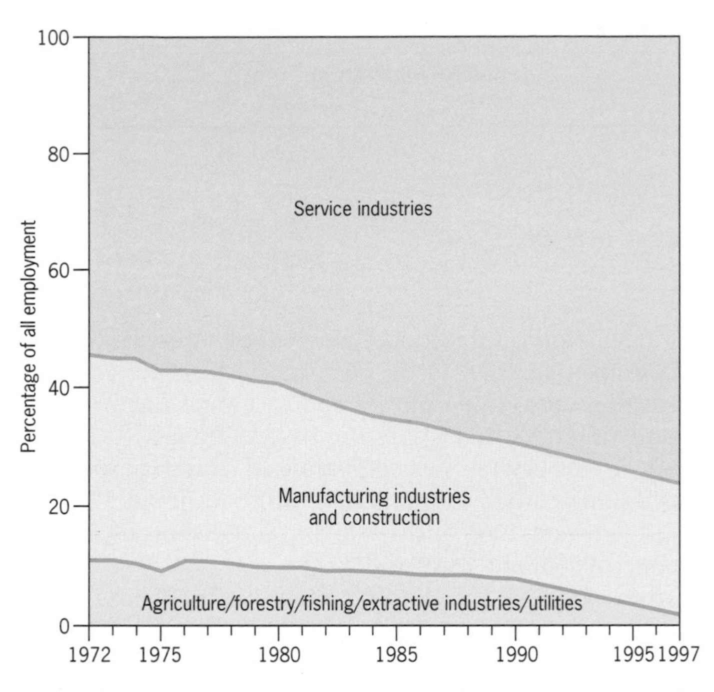

Figure 2.1 Changing relative importance of employment sectors in the UK, 1972–97

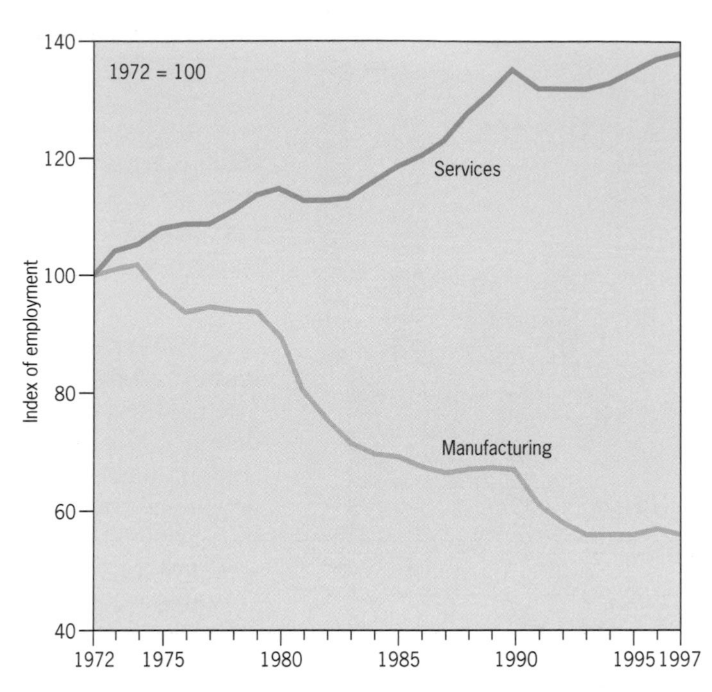

Figure 2.2 Changing employment patterns in manufacturing and services in the UK, 1972–97

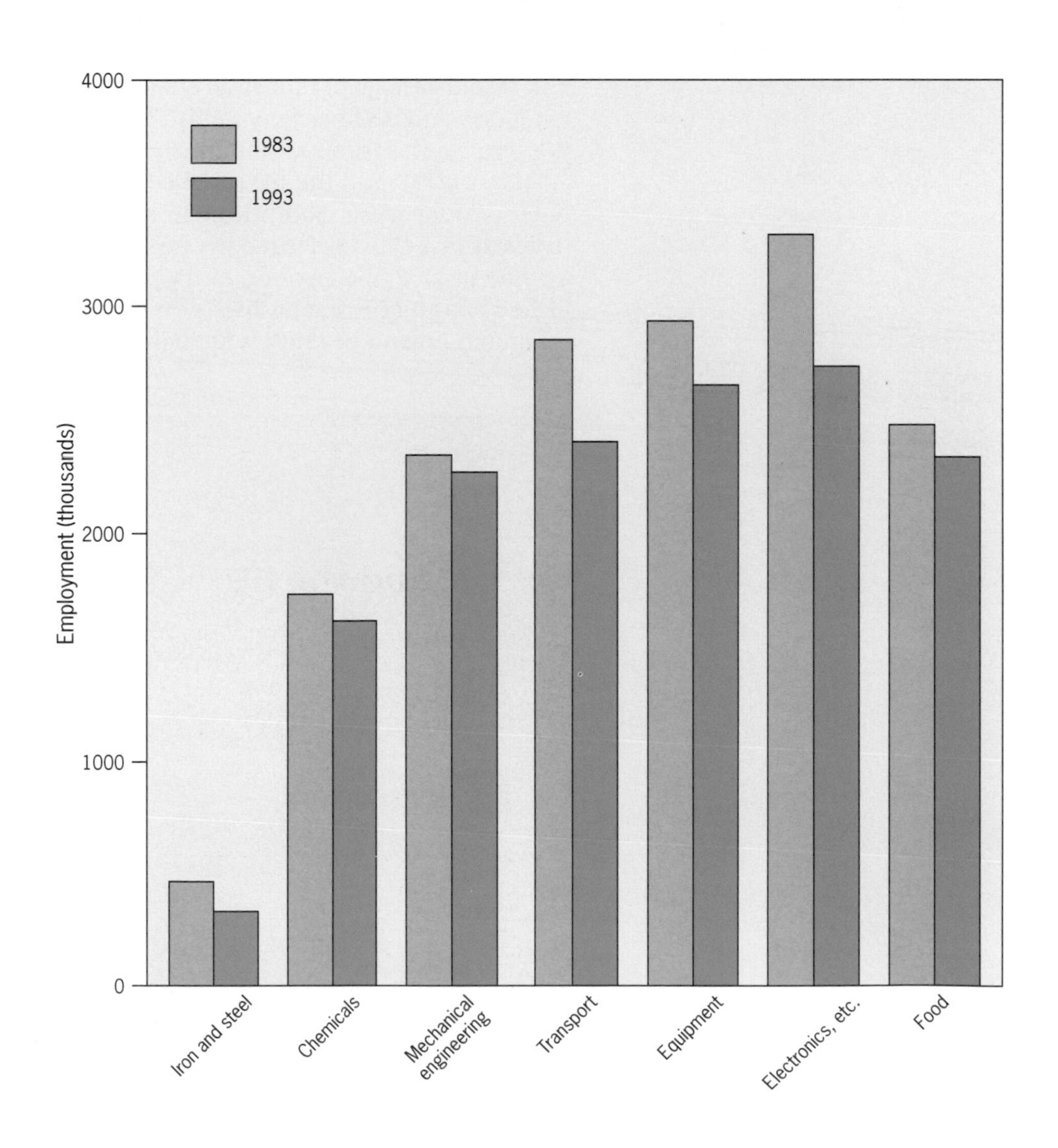

Figure 2.3 Employment change by manufacturing sector: EU, 1983–93

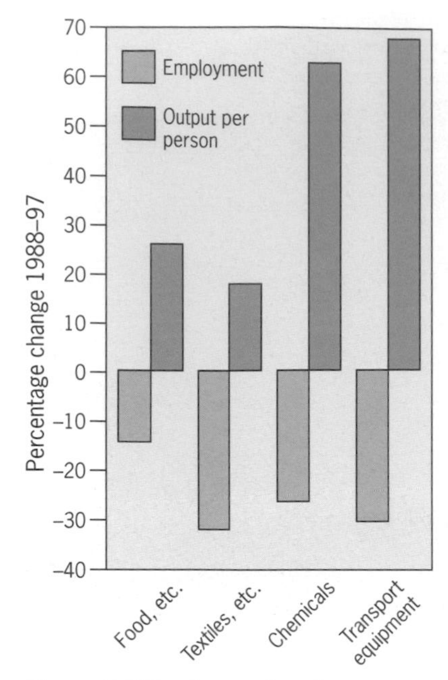

Figure 2.5 Employment and productivity changes in selected industries in the UK, 1988–97

Table 2.1 Employment change in selected industries in the UK, 1981–97

	Employment ('000s)		
	1981	**1997**	**% change**
Chemicals and synthetic fibres	383	240	–37.3
Motor vehicles and components	361	220	–39.0
Food, drink, tobacco	664	424	–36.1
Textiles, footwear, clothing	614	323	–47.4

Regions that were most dependent on traditional heavy industries were hardest hit by deindustrialisation (Fig. 2.4). In the UK, the collapse of Sheffield's steel industry left thousands unemployed and vast areas of dereliction in the Lower Don Valley. Wearside was the UK's leading shipbuilding centre in 1980. Yet by 1990 its last shipyard had closed. In other developed countries it was a similar story. Employment in the Ruhr steel industry in Germany halved between 1961 and 1988, and in Pittsburgh (USA) steel lost 130 000 jobs between 1965 and 1985.

Where traditional industries did survive, they were often more efficient and achieved higher output levels than before (Fig. 2.5). However, they always employed fewer workers. Moreover, new industries and services were often unwilling to move to these areas. Redundant miners did not necessarily make good car workers.

Reindustrialisation

The recent history of industrial change in the UK and the EU is not just about job losses and decline. New industries have emerged to replace the old ones. We refer to this trend as reindustrialisation. How has this happened?

First, the UK and the EU received massive investment by foreign firms, especially Japanese, South Korean and Taiwanese. This **foreign direct investment** (FDI) occurred because firms were eager to set up business in Europe prior to the creation of the Single Market in 1992. Second, thousands of new manufacturing businesses were formed in the 1980s. And third, there was rapid growth of knowledge-based, high-technology industries.

Figure 2.4 Derelict slipways at Sunderland – the end of a great shipbuilding era

?

To answer the following questions, read through the text from the beginning of the chapter, and then study Figures 2.1, 2.2, 2.3 and 2.5.

1 Draw an annotated diagram to show the links between deindustrialisation, reindustrialisation and tertiarisation in MEDCs.

2 How did the UK's employment structure change between 1972 and 1997? Give reasons for the changes shown by Figures 2.1 and 2.2.

3 Study Figure 2.5.
a State briefly the changes in employment and productivity.
b Suggest why output per worker increases as employment declines.

4a Describe the pattern of change in Figure 2.3.
b Explain why some types of industry contracted faster than others.

Figure 2.6 Foreign inward investment. High-tech firms at Livingston in Scotland.

In the 1980s, the UK attracted a greater share of Japanese FDI than any other EU country. By 1997, the UK still accounted for half of all Japanese FDI in the EU. Sixty per cent of this investment has been in electronics and motor vehicles. The bulk has gone to old industrial regions such as South Wales and central Scotland
(Fig. 2.6).

Figure 2.7 Grenoble, a modern business centre in a high-amenity area

?

5 Make a newspaper search for articles on FDI in the UK and the EU. Make a list of the investments, using the following headings: firm, nationality, economic activity, location, value of investment, employment creative, reasons for investment.

6 Look carefully at the location shown in Figure 2.7 and suggest reasons why it is an attractive environment for modern manufacturing industries.

Figure 2.8 Producer services have become firmly established in London Docklands

The recent growth of small manufacturing enterprises (SMEs) in regions such as Emilia–Romagna in Italy and East Anglia has led to **diffuse industrialisation**. In Emilia–Romagna, diffuse industrialisation has centred around SMEs in mature industries such as clothing, furniture and leather goods. Production processes are simple and firms are export-oriented. A nexus of firms work closely together supplying each other with materials, semi-finished products and components. This type of organisation allows firms to respond flexibly to consumer demand. Similar diffuse industrialisation has occurred in high-tech industries. New and established firms have located in relatively unindustrialised regions such as Rhône–Alpes in France and Bavaria in Germany. These regions provide the sort of high-amenity environment needed to attract well-qualified managers, scientists and technicians.

Tertiarisation

Service industries now employ nearly 17 million people in the UK, and account for 65 per cent of GDP. This tendency for services to take up a growing share of economic activity is known as tertiarisation.

The shift to services resulted from a complex set of factors. In the first place it is a response to rising incomes. As people become better-off, they spend a larger proportion of their income on personal, financial and leisure services. Tertiarisation is also prompted by demographic factors. An ageing population requires more resources devoted to health care. New technology also creates new service activities. A good example is the growth of cash cards and mobile phones, which were almost unknown 20 years ago.

Tertiarisation is also linked to deindustrialisation. As employment in manufacturing has fallen, redundant factory workers have often had little option but to find work in the tertiary sector.

In the 'sunbelt' of the EU – Mediterranean France, Italy, Greece and southern Spain – tourism has helped the tertiary sector to grow. Meanwhile, the expansion of **producer services** directed towards manufacturing industry (e.g. advertising, legal services, management consultancy, market research), has been strongest in major international cities like London, Paris, New York and Tokyo. Today, these cities are centres of telecommunications and information, and play a key role in financing the global economy. Unfortunately the growth of producer services has been weak in Europe's nineteenth-century manufacturing cities. Cities like Liverpool, Liège and Marseille are less and less relevant to the needs of the **post-industrial economy**.

2.3 Change in the economically less-developed world: industrialisation

We saw in the last section that, in MEDCs, tertiary activities are now more important than manufacturing. We described these countries as post-industrial. In LEDCs the situation is very different. There, rapid economic change is linked to industrialisation and to an increase in the proportion of the workforce in manufacturing.

Rightly or wrongly, many people in LEDCs think that industrialisation is the fastest way to economic growth and social change. In the next few pages we shall see how industrialisation has transformed Taiwan into a wealthy country. We shall also look at the progress of industrialisation in Thailand and Botswana, and assess how likely they are to join Taiwan as newly industrialising countries (NICs).

Taiwan: a 'tiger' burning bright

?

7 Study Figure 2.9 and give three reasons why Taiwan's geography has been of little help to its industrial development.

Figure 2.9 Taiwan

Figure 2.10 The mountainous nature of Taiwan inhibits industrial development

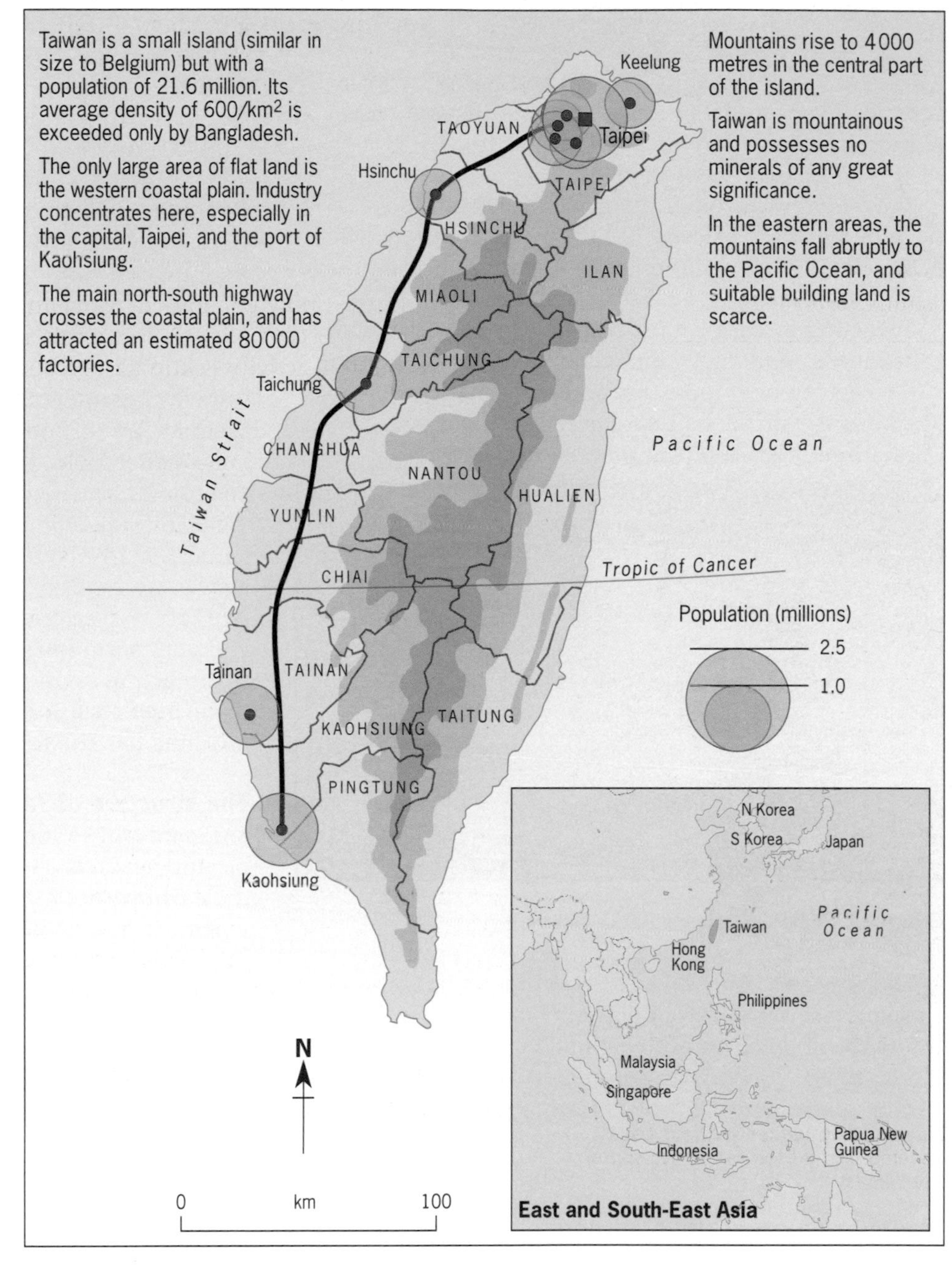

In 1960 Taiwan was just another poor Asian country. The economy was based on an undeveloped agricultural sector. Forty years later its GNP per capita had risen from $120 to over $14,000 with its economy geared to capital- and technology-intensive industry. How was this incredible growth achieved?

Early industrial development

Three factors assisted Taiwan's early industrial growth between 1945 and 1970: first, Japanese colonial rule, which ended in 1945 and modernised the country's

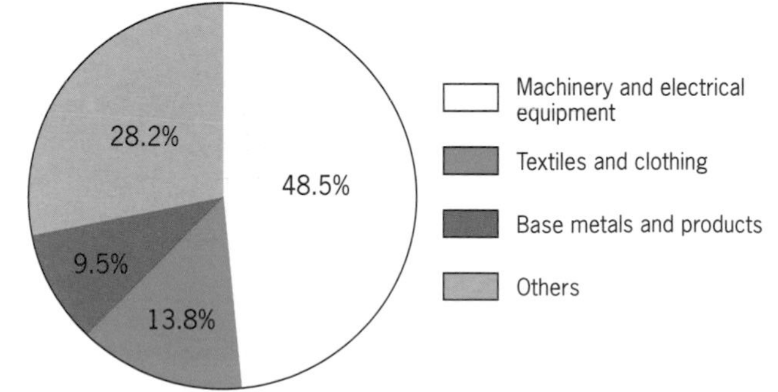

Figure 2.11 Taiwan: major exports, 1997

Table 2.2 Taiwan's changing GNP by economic sector (% share)

	1982	1987	1992	1997
Agriculture	7.7	5.3	3.6	2.7
Industry	44.3	46.7	39.9	34.9
Services	48.0	48.0	56.5	62.4

infrastructure, extending its road and rail networks; second, a substantial amount of economic and technical aid has been given by the USA; and third, there is a supply of relatively cheap labour.

Taiwan went through the usual stages of industrial development for a newly industrialising country (NIC):

- developing its own industries to cut imports and save on hard currency (i.e. import substitution);
- attracting foreign export-oriented firms (particularly Japanese) and developing its own export industries;
- developing new industrial sectors which were less labour intensive, had higher added value, and required more sophisticated technology.

Figure 2.12 Hsinchu Science Park, Taiwan

The Taiwan government directed industrialisation through a series of development plans, though the influence of government was small compared to other Asian 'tigers' such as Singapore and South Korea. Recent plans have set targets for the sustainable development of industry, as well as rates of economic growth and 'balanced development'.

Recent industrial development

Taiwan's industry is dominated by small to medium-sized manufacturing enterprises (SMEs). Some 98.5 per cent of Taiwanese companies are SMEs and they account for three-quarters of all employment. Industry is diverse, with electronics, textiles and shoes, petrochemicals and machinery prominent.

8 Describe and explain the changes in Taiwan's GNP by economic sector between 1982 and 1997 (Table 2.2).

Labour

On the face of it, Taiwan is an unlikely success story. Apart from physical obstacles to industrialisation (Fig. 2.9), the country's domestic market is small and its labour expensive. However, expensive labour is not a problem if the workforce is highly educated and skilled. Taiwan's universities produce nearly 50,000 new engineers a year, more than a quarter of all graduates. In addition, many more young people go to study at American universities.

Overseas investment

Between 1990 and 1998, 80,000 Taiwanese businesses relocated offshore, an investment worth $80 billion. About half moved to China and half to other South-East Asian countries. Most industries moving offshore were low skilled and labour intensive (e.g. toys, footwear, garments etc.). Wages in mainland China and Vietnam are just one-twentieth of those in Taiwan and land is affordable. Today more than half of Taiwan's electronics production is overseas.

Since 1990, Dongguan, an industrial region in China's Pearl River basin, has attracted 4,000 Taiwanese factories which employ 2 million migrant workers. Typical industries include the assembly of computers, clothing and shoes. Many clothing firms manufacture items that will be sold by leading foreign companies such Nike and Reebok. Taiwan's investment in China has reached an estimated $38 billion.

The structure of Taiwan industry

SMEs in Taiwan form a web of subcontractors in the production chain. In electronics (the leading sector) most companies manufacture parts for computers, videos, CDs, etc. that are eventually sold under international brand names such as Compaq and Dell. Indeed, three-quarters of Taiwan's electronics production is eventually sold under someone else's brand name.

Industries based in Taiwan are increasingly capital-intensive (petrochemicals, machinery) and technology-intensive (electronics). In 1998, these industries accounted for 71 per cent of total manufacturing output, compared to 48 per cent in 1986. Taiwan is one of the world's major semi-conductor suppliers and is also the world's leading supplier of motherboards, monitors, personal computers, mice, keyboards, scanners and laser disk drives. The core of Taiwan's high-tech industry is the government-established Hsinchu Science Park near Taipei (Fig. 2.12). Known as Silicon Valley East, it accounts for one-third of Taiwan's manufacturing exports (Table 2.3). A second science park is currently being built near the southern city of Tainan.

Figure 2.13 Air pollution in Taipei is a consequence of uncontrolled industrial growth in Taiwan

Table 2.3 Hsinchu Science Park, Taiwan

Industry	Companies	Employees
Integrated circuits	51	18,297
Computers and peripherals	39	9,868
Telecommunications	30	3,928
Optoelectronics	25	2,585
Precision machinery	16	1,015
Biotechnology	9	215
Total	170	35,908

The price of economic success

The speed and success of Taiwan's industrialisation has caused severe environmental degradation. However there is no doubt that environmental problems have been made worse by the country's geography (Table 2.4). Taiwan is a small country with a large population and a mountainous relief. Most of the population and industrial activity is centred in the western alluvial plains, which comprise less than half the island's total area.

Table 2.4 Environmental pressures in Taiwan and selected MEDCs

	Taiwan	USA	Japan	Germany	UK
Pop density (km^2)	589	27	332	228	238
Motor vehicle density (no/km^2)	367	230	211	99	102
Factory density (no/km^2)	4.32	0.04	1.14	0.12	0.6

Despite government restrictions in the 1990s, air pollution from factories and car exhausts remains a serious concern. Emissions of SO_2 and NO_2 cause air pollution at levels that are injurious to health – on 6 per cent of days a level of pollution is three times higher than the MEDC average.

In Taiwan, factories have routinely discharged liquid toxic waste into rivers and solid waste into landfill sites, resulting in contaminated food chains and soils.

A growing environmental awareness has led to mass protests and demonstrations. During the 1990s, the environmental movement successfully blocked the building of a fourth nuclear power station, and halted investments in cement factories in the north, and petrochemicals plants in the south of the island.

9 From the late 1980s, Taiwan invested heavily in other countries in East and South-East Asia. Describe these industries and explain why they relocated overseas.

10 Taiwan, with over 150 000 factories, has an industrial structure dominated by small and medium-sized firms. Explain why Taiwan's dependence on small to medium-sized firms might be both an economic strength and an economic weakness.

11 Essay: With reference to Taiwan, Thailand and Botswana (in the next section), discuss the view that successful industrialisation depends on natural resources (minerals, energy supplies, good-quality farmland, etc.).

Values and beliefs

Throughout East Asia, the environment has taken second place to industrialisation. The pollution problem in Taiwan is nowhere near as bad as in China, Hong Kong and the Philippines. Yet of Taiwan's 22 million people, only 600,000 are served by sewers, and most of the country's factories pour untreated effluent into rivers and emit toxic fumes into the air. This raises the question:

Is environmental protection a luxury that LEDCs cannot afford?

To answer this question you need to decide what your values and beliefs are. You will then be in a position to state your opinion or attitude on the whole issue of industrialisation and its impact on the environment.

Values are what you desire to be true (e.g. you may value economic development more than environmental protection). Beliefs are what you think is true (e.g. you may believe that only through economic development can sufficient wealth be created to care for the environment). By analysing your own values and beliefs you can clarify your own attitude towards issues, and by looking at other people's values and attitudes you can better appreciate their views.

Study Table 2.5.

12 State what you think are the likely values and beliefs of the so-called 'accommodators'. This group occupies the middle ground and tries to accommodate both industrialisation and environmental protection.

13 Clarify your own attitude towards this issue. Then answer the question, setting out your values and beliefs, making sure that they are consistent with your attitude.

Table 2.5 The issue of industrialisation and environmental protection

	Extreme economist	Accommodator	Extreme environmentalist
VALUES	Economic development and the raising of people's material living standards must always have priority.	?	Economic development is only acceptable if it has no adverse effects on the environment.
BELIEFS	Wealth has to be generated in the first place if people are to enjoy the environment.	?	Industrial development causes irreversible damage to the environment, including loss of species, wildlife and habitat.
	The environment is a distraction which would raise costs, limiting a country's ability to compete internationally.	?	The quality of life is not measurable just in terms of income and material possessions.
	Damage to the environment is localised and its overall impact is exaggerated.	?	We have no right to destroy the environment. We have a duty to protect it for future generations.
	As people's material needs are met it will be possible to give conservation a higher priority.	?	Industrial development benefits governments and large industrial corporations rather than the people. It is ordinary people who suffer the consequences of environmental destruction and pollution.
ATTITUDE	In favour of industrialisation.	?	Against industrialisation.

Thailand: the next Asian NIC?

Between 1985 and 1995 Thailand had the world's fastest growing economy. This astonishing economic success led to confident forecasts about the future. Many experts believed that by the end of the twentieth century Thailand would join Taiwan, South Korea, Singapore and Hong Kong as Asia's fifth NIC.

Table 2.6 Thailand's changing economy (% GDP)

	1960	1990	1993
Agriculture	39.8	14.2	10.0
Industry	12.5	25.2	29.0

Thailand's progress towards take-off was brought about by its success in attracting foreign direct investment, particularly from Japan and Taiwan. They favoured Thailand for several reasons, not least for its stable government, which encouraged foreign investment. Thailand also had a competitive edge over other East Asian countries because of its low costs of land, building and labour. Wages, for instance, were just one-sixth of Taiwan's, and a third of those in Malaysia. The country's other major asset is its large market. With a population of 60 million, and rising incomes, Thailand remains an attractive prospect for many foreign companies.

Three stages of industrialisation

The path taken by industrialisation is very similar to Asia's NICs. First-stage industries were labour-intensive and based on local agricultural or mineral products (Fig. 2.15). They included industries such as textiles and clothing. They were succeeded by second-stage, **import-substitution** industries, producing goods like cars and electrical equipment, which would otherwise have to be imported. In the main, import-substitution industries were **'screwdriver' industries** assembling imported components.

Thailand has now reached the key third stage. The shift to more capital- and less labour-intensive industries has begun. Several large petrochemical

Figure 2.14 Thailand

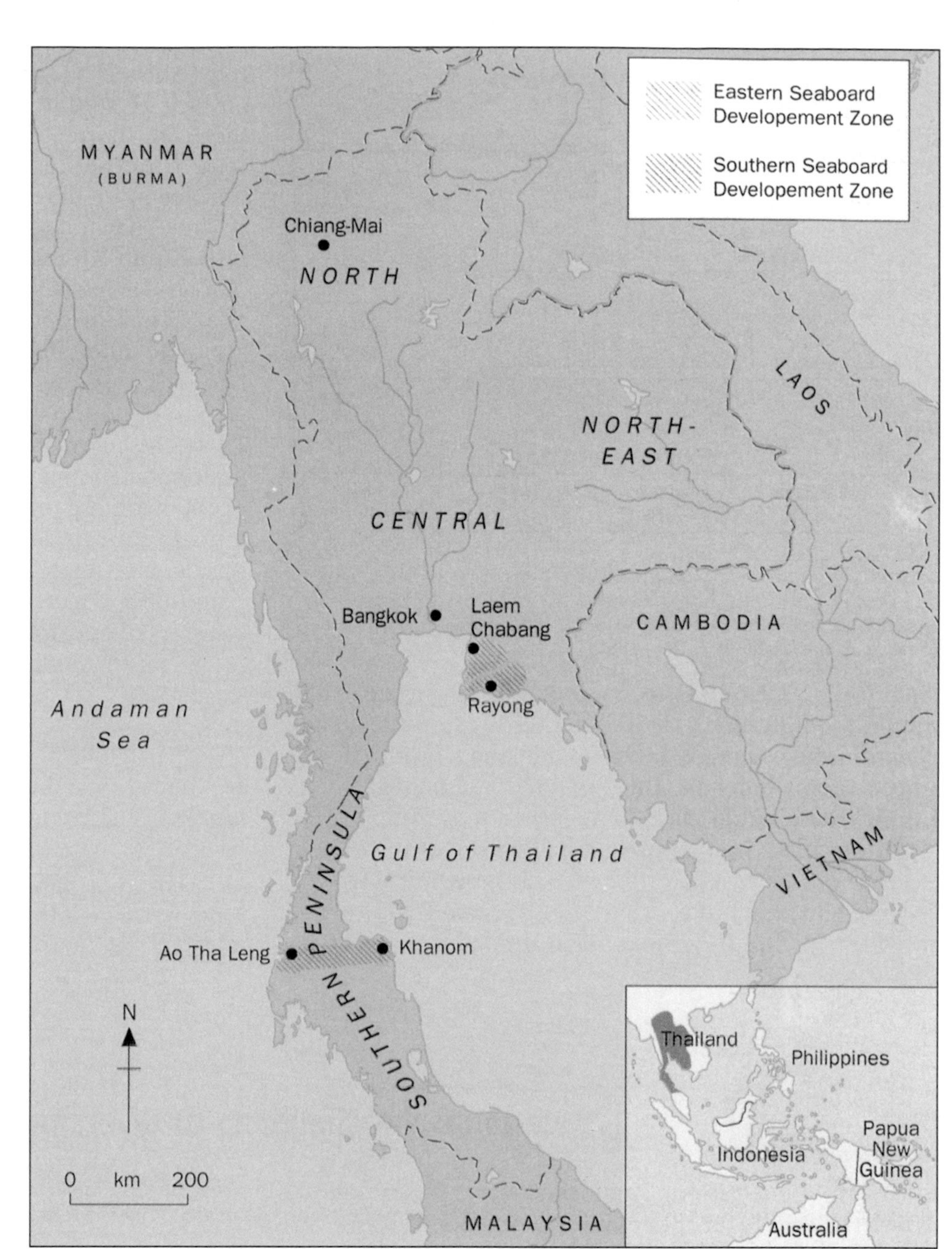

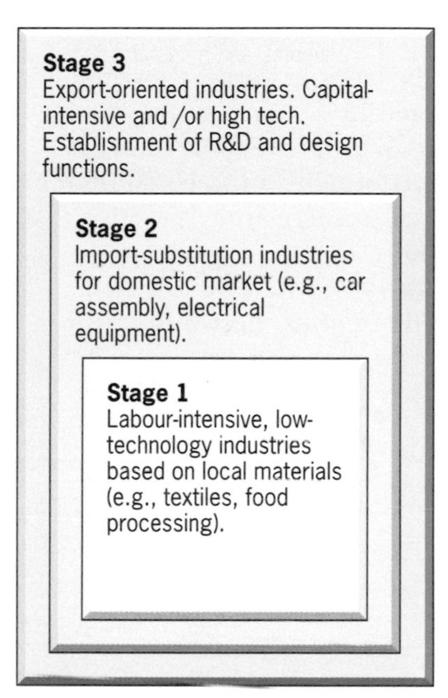

Above
Figure 2.15 The changing pattern of manufacturing with industrialisation and economic expansion

Below
Figure 2.16 Thailand's trading partners and destination of exports (*Source: Financial Times*, 5 Dec. 1996)

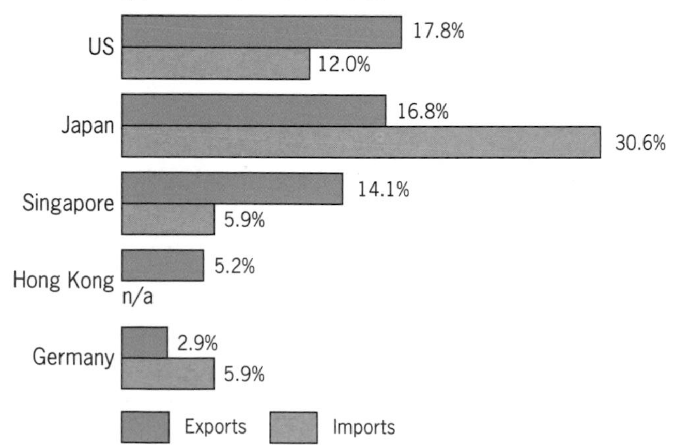

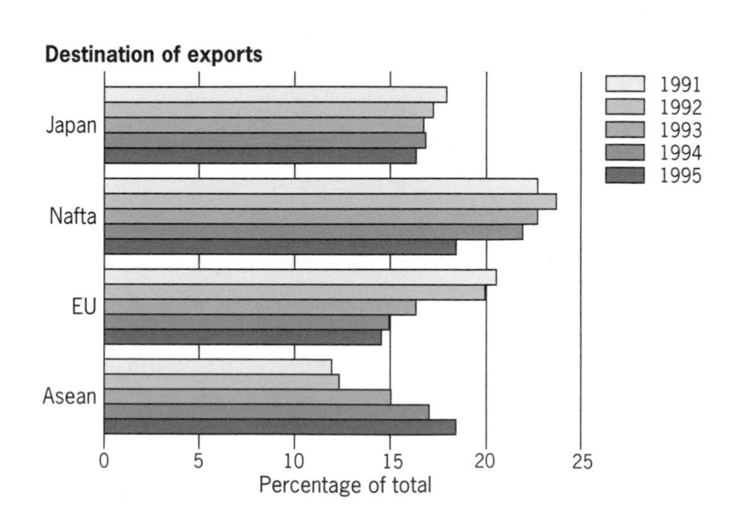

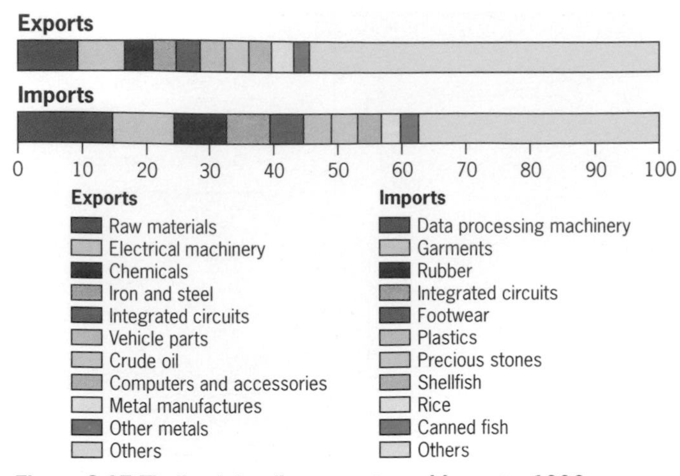

Figure 2.17 Thailand: leading exports and imports, 1996

projects are under construction south-east of Bangkok on the Gulf of Thailand. Electronics is also growing rapidly (Fig. 2.17). It is technology-intensive, and should promote R&D, design and testing activities. Electronics has already become Thailand's second largest export industry.

Both of these industries (petrochemicals and electronics) are export-oriented. This is significant, because successful industrialisation by Asia's NICs was based on an export-led manufacturing boom.

Problems to overcome

Thailand's rapid industrialisation came to an abrupt halt in 1997. A huge foreign debt, overvalued currency and inefficient industries led to an economic crisis. The International Monetary Fund (IMF) and the Pacific Rim countries put together a rescue package. In return for strict financial measures (which for two or three years would cause the Thai economy to contract), the government received a loan of $16 bn.

But even before the 1997 crisis Thailand faced a number of problems. The country's economic success led to a steady rise in wages. Thus by the mid-1990s labour costs in Thailand were three times higher than in China. As a result, labour-intensive industries such as textiles and footwear began to relocate elsewhere in Asia. High wage costs together with an inadequate infrastructure led to a sharp drop in export growth. Moreover several huge investments in steel, oil refining and petrochemicals proved unsuccessful. The drive to become an exporter of more advanced products received a further blow in 1997 when the US TNC, Texas Instruments, pulled out of an electronics joint venture in Thailand.

14 Why does successful industrialisation involve a shift from import-substitution to export-oriented industries?

15 Study the location of Thailand's two development zones (Figure 2.14). Suggest three possible reasons for their location.

Botswana: the challenge of diversification

Botswana is one of four landlocked states in southern Africa (Table 2.7). Its population is tiny (only 1.48 million) yet in area (565 000 km^2) it is equal to Thailand. Like most LEDCs, Botswana is weakly urbanised: 80 per cent of the population lives in rural areas, and works in subsistence farming. The rest are mainly concentrated in the capital city, Gaborone, and the larger centres of Francistown, Lobatse and Selebi-Phikwe (Fig. 2.18).

16 What is the average population density of Botswana? How does it compare with Taiwan?

17 Study Figure 2.18.
a Describe the distribution of settlement in Botswana.
b Why are the central and southern areas of the country sparsely populated?

Table 2.7 The landlocked states of southern Africa

	Population (millions)	GNP per capita (US$)	Foreign exchange reserves (US$m)
Botswana	1.48	2941	4820
Lesotho	1.90	660	410
Malawi	7.75	179	120
Swaziland	0.89	1660	149

Given its small population, landlocked position and location in sub-Saharan Africa, you might expect Botswana to be desperately poor. In fact, for much of the 1980s Botswana had one of the fastest growing economies in Africa. Even so, its success is relative: it is estimated that half of its population lives below the poverty line.

Botswana's economic success can be summed up in one word: diamonds. Measured in terms of value, it is

18 Study the information in Table 2.7 and comment on:
a Botswana's economic situation.
b How the country compares with the other states in southern Africa.

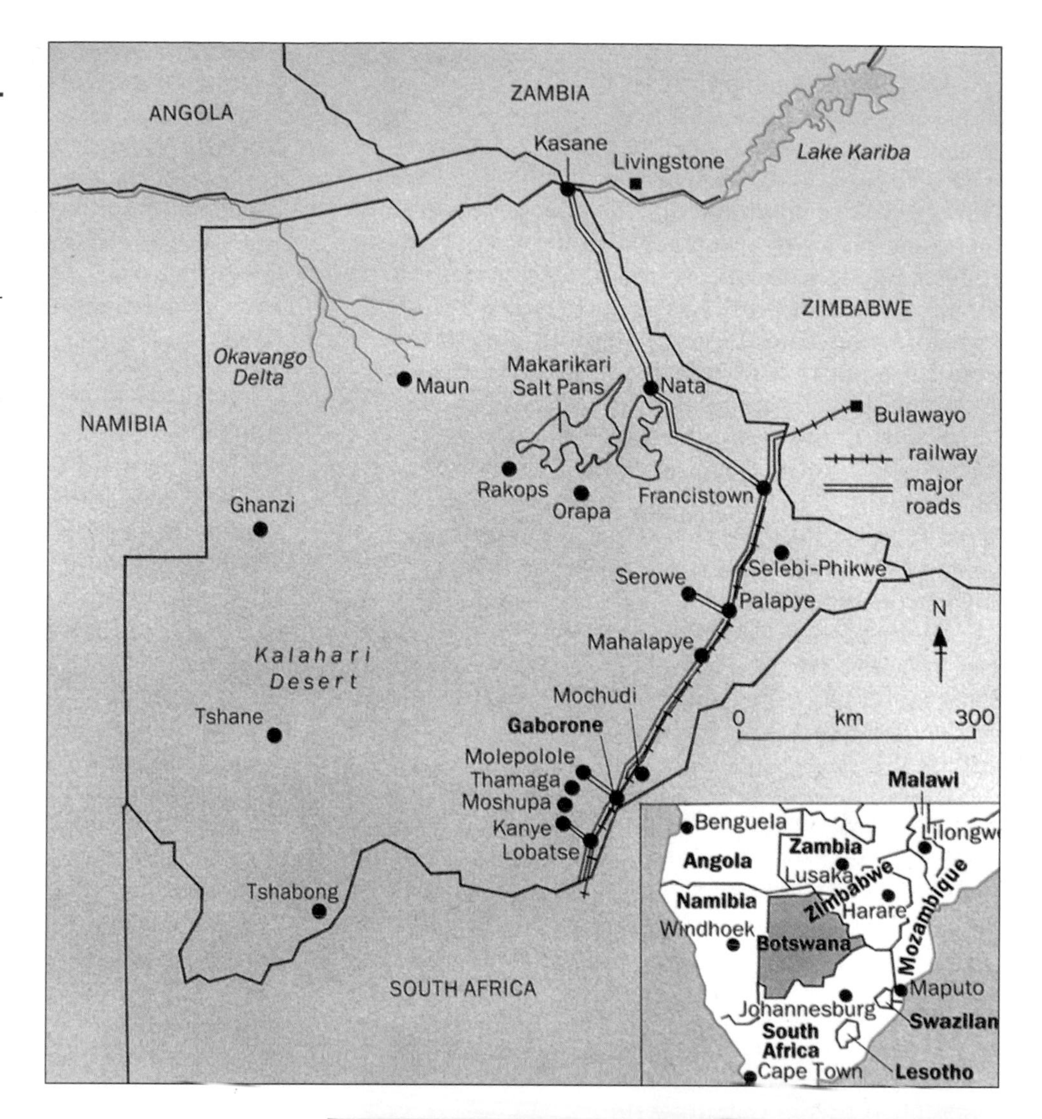

Figure 2.18 Botswana

the world's second largest diamond producer. Diamonds, together with nickel and copper, account for nearly 90 per cent of its exports (Figs 2.19, 2.20) and over half the government's income. Diamond mining is managed by De Beers of South Africa. Mining creates wealth, but few jobs: only 10 000 jobs for a population of 1.48 million. Its dominance means that other sectors of the economy are poorly developed. Manufacturing contributes only 6 per cent of GDP, and agriculture, which employs four out of every five people, a mere 4 per cent.

Searching for 'new engines of growth'

Highly specialised economies, based on a narrow range of mineral and agricultural exports, are common in the economically developing world. Such economies are

Figure 2.19 Botswana: value of exports

Figure 2.20 The open-cast pit at the Jwaneng diamond mine in Botswana

fragile. Botswana is no exception. Hence the government's attempts to broaden the economic base and search for what it calls 'new engines of growth'.

Unfortunately, **diversification** is not easy. One of the best options would be to expand the mining sector. Botswana has known reserves of gold, semi-precious stones, potash, soda ash and chromite. Recently the salt pans of the Makarikari (Fig. 2.18) have attracted huge investment (including a rail link to Francistown) for the extraction of soda ash. This project is a joint venture between the Botswanan government and a group of South African companies.

The agricultural sector offers few opportunities for diversification. Because of low rainfall, only 25 per cent of the land area is suitable for agriculture. Of this, less than 5 per cent can support arable farming. Drought is a recurrent problem, but irrigation is too costly to be a realistic option.

Industrialisation has been a priority in the government's development plans. The emphasis has been on industries like meat processing, based on local primary products; import substitution industries; and export-oriented industries.

Inward investment is encouraged by a stable, democratic government free from corruption. Various incentives are offered to foreign investors through the Botswana Development Corporation. They include loans and equity capital. Also, by investing in infrastructure (roads, water, energy supplies, etc.), education and health care, the government hopes to create the sort of environment to attract foreign investment. There have been some notable successes: Hyundai, the South Korean TNC, has set up two car assembly plants in Gaborone since 1990.

Obstacles to industrialisation

While Botswana has achieved some success in broadening its economy, major problems limit industrialisation. These problems relate to: infrastructure, population growth, unemployment, labour, the domestic market, dependence on neighbouring countries and competition.

Infrastructure

In spite of government spending in this area (including a new international airport at Gaborone in 1984 and the completion of the Trans-Kalahari highway in 1998), Botswana's infrastructure is still not yet good enough. Any large country with a small scattered population will have difficulty in building an adequate road network. In Botswana there are only 18,500 kilometres of roads, and less than a quarter are bitumen-surfaced. This leaves huge areas remote and isolated. In addition there is only one railway (apart from local lines serving mining industries); electricity is only available in the main towns; and telecommunications are poorly developed.

Population growth and unemployment

Population growth rates have fallen rapidly in the last 30 years, thanks to a steep decline in fertility. Nevertheless, the current growth rate of 3.1 per cent a year is still high. With a youthful population, large numbers of young people enter the job market each year. At the moment the formal economy can absorb fewer than half of them. In the 1990s AIDS began to have a significant impact. Life expectancy fell from 63 to 52 years between 1990 and 1996. Current estimates suggest that 13 per cent of the population are HIV positive. The implications of the AIDS epidemic for Botswana's future economic and demographic growth are considerable.

Landlocked situation

Botswana has no direct outlet to the sea and depends on shipments through South Africa for most of its foreign trade. This dependence on another country is a potential weakness.

Skilled labour

Industrialisation is held back by shortages of managers, scientists and technicians. Botswana has to rely heavily on the skills of immigrants from developed countries.

Domestic market

Industrialisation is helped where there is a large domestic market for consumer goods. With a total population of less than 1.5 million, the Botswanan market is very small. TNCs locating in southern Africa can easily supply Botswana from neighbouring South Africa or Zimbabwe.

Competition

The most basic obstacle to industrialisation is competition from Botswana's much larger and more powerful neighbours, South Africa and Zimbabwe. Both countries, with larger domestic markets, better infrastructure and larger pools of skilled labour, have a strong competitive advantage in manufacturing over Botswana.

?

19 Draw up a table using the following headings: importance of manufacturing; diversity; types of industry; natural resources; labour; inward investment; obstacles and prospects.

a Summarise in your table the main features of the economies of Thailand and Botswana.

b Which country in your opinion has the better prospects of achieving industrialisation and economic take-off? Briefly justify your decision.

20 Essay: Many less-developed countries see industrialisation as the only route to economic development and prosperity. Using the information on Taiwan, Thailand and Botswana in this chapter, state your attitude towards this issue, and support it by outlining the values and beliefs you hold.

Summary

- Industrial change at a global scale can be studied in the context of MEDCs, LEDCs and NICs.
- The relative importance of manufacturing industry in MEDCs countries has declined in the last 30 years.
- The relative decline of manufacturing in MEDCs is explained by the processes of deindustrialisation and tertiarisation.
- The effects of deindustrialisation have been most severe in older, heavier manufacturing industries such as steel and textiles. Regions dependent on these industries have been hardest hit by deindustrialisation.
- Reindustrialisation based on small businesses and new technologies has often been most successful outside regions of traditional heavy industries.
- Tertiarisation and the growth of producer services has been most successful in prosperous metropolitan regions such as London, Paris and Frankfurt.
- Industrialisation based on export-oriented industries has brought rapid economic development to a number of East Asian countries, including South Korea, Taiwan, Singapore and Hong Kong. These countries are known as newly industrialising countries.
- Industrialisation starts with low-technology industries based on local materials. These are succeeded by import-substitution industries, and finally by export-oriented industries.
- Rapid industrialisation has often resulted in environmental damage and widespread pollution in East Asia.
- Major obstacles to industrialisation face most LEDCs. These often include inadequate infrastructure, skills shortages, rapid population growth, small domestic markets and competition from other countries.

3 Materials, energy and transport

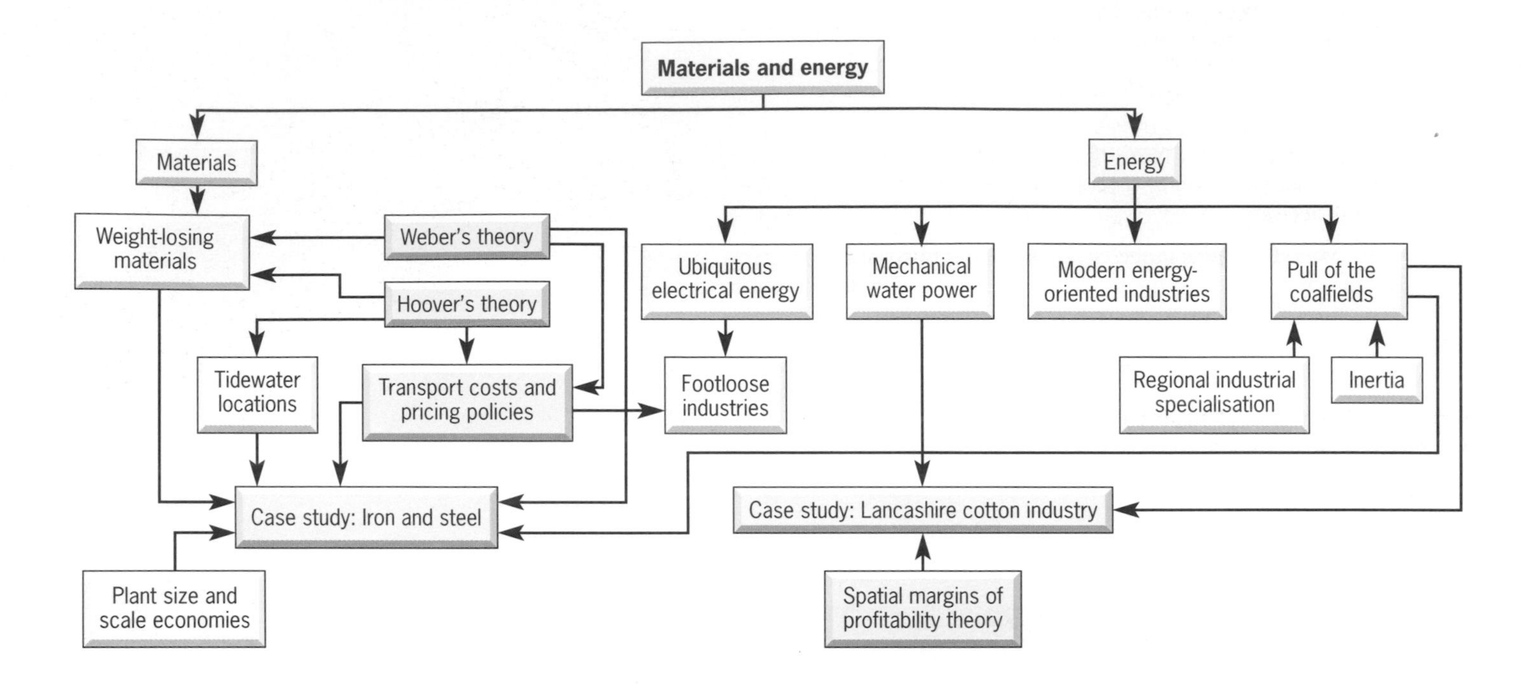

Figure 3.1 Sugar beet refinery at Bury St Edmunds, Suffolk

Table 3.1 Weber's material index and locational patterns (1955)

Material index	<1.0	1.0–2.0	>2.0
Number of industries:			
Located at materials	2	17	3
Not located at materials	16	14	1

3.1 Introduction

In this chapter, with the help of location theory, we shall consider the changing importance of materials, energy, transport and plant size in the location of industry. These factors, which often attract firms to material-oriented locations, are illustrated by case studies of the steel and textile industries.

3.2 Materials

Weight-losing materials

Today, few manufacturing industries use raw materials directly. Those that do, such as metal smelting and oil refining, are known as **processing industries**. Because most raw materials lose weight in manufacturing, processing industries often locate close to materials and energy supplies. Weber's theory of industrial location (see Weber's theory on pages 29–30) explains how a firm can reduce its transport costs by choosing a material-oriented location.

Sugar beet, for example, the raw material of the sugar-refining industry (Fig. 3.1), loses 90 per cent of its weight in manufacture. In the UK, this is sufficient to attract sugar refineries to the main beet-growing areas in eastern England (Figs 3.2, 3.3). Timber undergoes a similar weight loss in its conversion to wood pulp. Thus in Sweden, most pulp mills are found close to forested areas, especially along the Bothnian coast, in Norrland (Figs 3.4, 3.5).

Tidewater locations

Nowadays, most developed countries depend heavily on imported materials. Seaports and seaport terminals give access to these materials, and offer attractive sites for industries like oil refining, chemicals, non-ferrous metal refining, flour milling and food processing (Figs 3.6, 3.7).

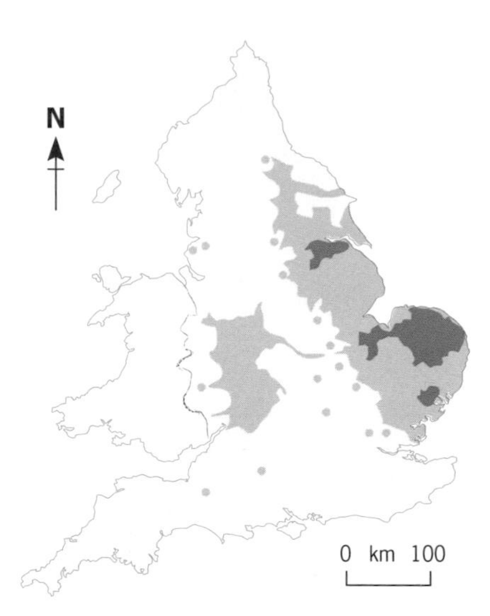

Figure 3.2 Distribution of sugar beet growing in England and Wales

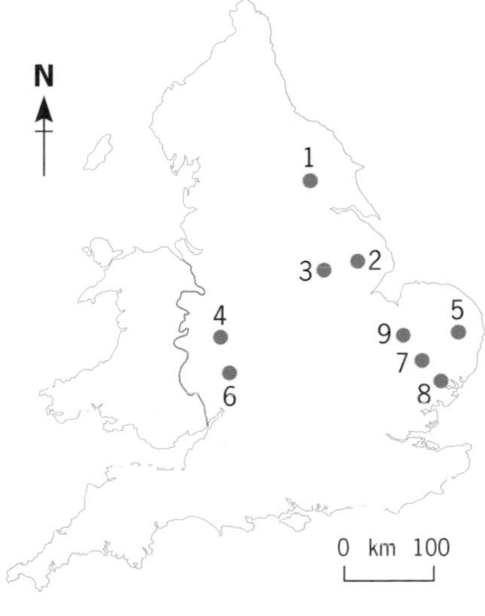

1 York
2 Bardney
3 Newark
4 Allscott
5 Cantley
6 Kidderminster
7 Bury St Edmunds
8 Ipswich
9 Wissington

Figure 3.3 Location of sugar beet refineries in England and Wales

1 Read through Weber's location theory (pages 29–30). What is the value of the material index for the sugar-refining industry? Does the material index correctly predict its location?

2 Study Table 3.1. How well would you say that Weber's material index predicted the location of industry in 1955?

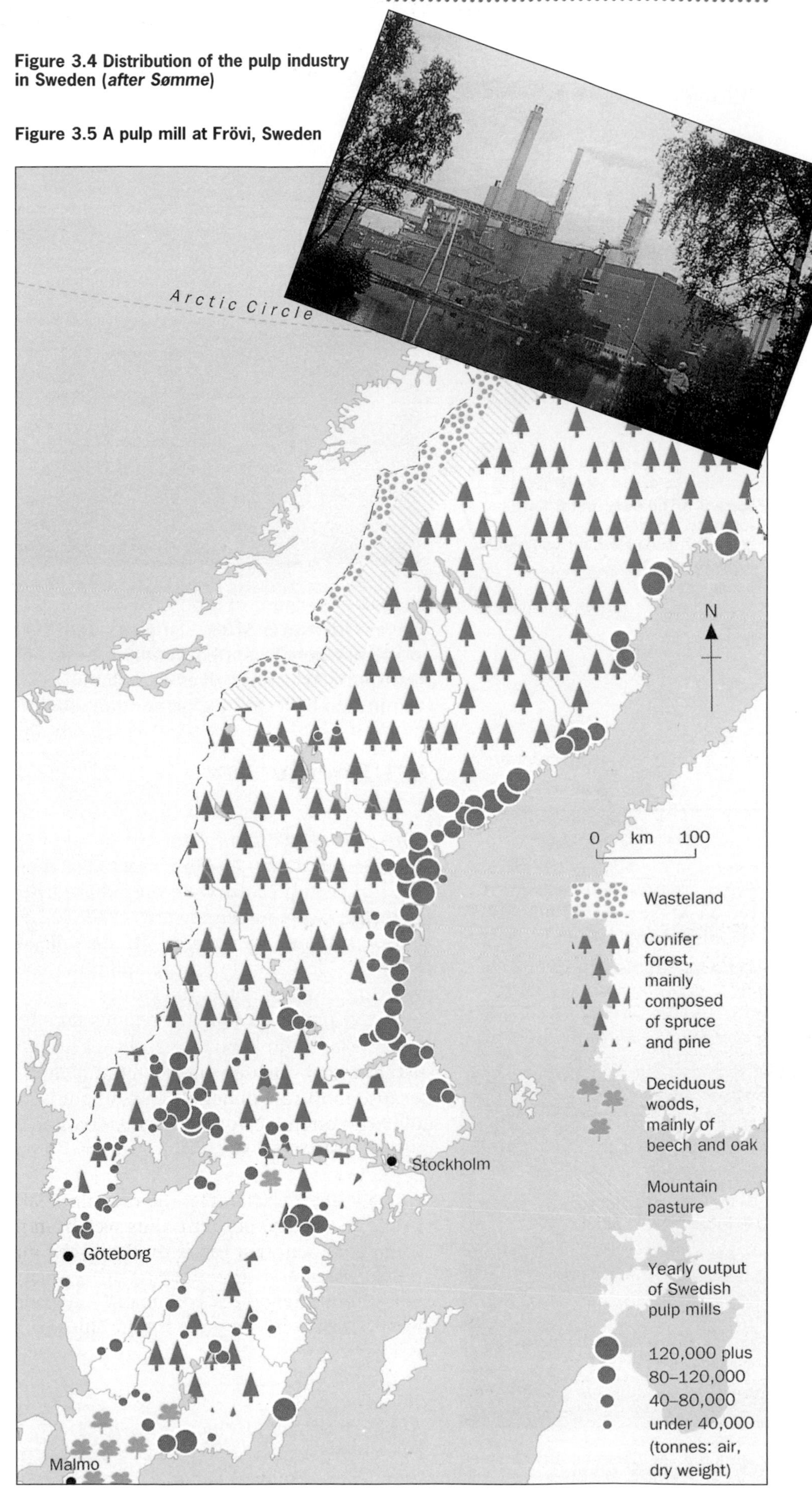

Figure 3.4 Distribution of the pulp industry in Sweden (*after Sømme*)

Figure 3.5 A pulp mill at Frövi, Sweden

Figure 3.6 The deep-water terminal at Richmond, Virginia
Figure 3.7 Kobe Harbour container terminal, Japan

Hoover's theory of industrial location tells us why processing industries often prefer **tidewater sites** (see pages 30–1). These sites are **break-of-bulk points**, where imported materials must be transferred from sea to rail or road transport. As this involves extra handling charges, transport costs can be minimised by locating a plant at a break-of-bulk point.

3.3 Energy

The pull of the coalfields

Energy, more than any other factor, shaped the industrial location patterns of the nineteenth century. Because the transport of coal was so expensive, industries were pulled towards coalfield sites. As a result coalfield regions, such as South Wales, the Sambre–Meuse valley in Belgium, the Ruhr in Germany, Nord-Pas-de-Calais in France and other coalfields, became the main centres of industry in the nineteenth century.

Surprisingly, the coalfield regions remain important centres of industry today. Coal is no longer a significant locational factor, but many **acquired advantages**, such as skilled labour and linkages between firms, are. Moreover, fixed investment in factory buildings and infrastructures such as roads and railways has tended to keep industry in the places where it started. We refer to this as **industrial inertia**.

?

3 Look at Figure 3.8. What features in the photograph are likely to contribute to industrial inertia and the survival of industry into the twenty-first century?

Figure 3.8 The Corus (ex-British Steel) plant at Sheffield, Yorkshire

Energy available everywhere

In the twentieth century coal gradually gave way to electricity as the main source of energy for industry. The distribution of electricity through transmission grids has made energy available everywhere (known as ubiquitous energy) in economically developed countries, and it is now of little importance as a locational factor. This is one reason why most modern industries are **footloose** or less constrained than in the past in their choice of location.

Modern energy-oriented industries

However, for a few energy-hungry industries, energy supplies remain a significant locational factor. Examples of such energy-oriented industries

Figure 3.9 Norsk Hydro's aluminium reduction plant at Hoyanger, Norway

include metal smelting and electro-chemicals. Southern Norway, with more than thirty plants, has the largest concentration of electro-metallurgical and electro-chemical industries in Western Europe (Fig. 3.9). They make a range of products, from aluminium, zinc and ferro-alloys, to fertilisers and ammonia. The attraction of southern Norway is cheap hydro-electric power (HEP). Fjords also provide sheltered, deep-water anchorages for the import of raw materials and the export of finished products.

One energy-oriented industry – aluminium smelting – has experienced a global shift in location in the past 30 years. Rising energy costs in the developed world have caused the industry to move to areas such as Latin America where production costs are relatively low. This trend was exemplified by Norsk Hydro's decision (1998) to build a primary aluminium smelter in Trinidad and Tobago by 2002. The new plant is gas-fired with a capacity of 475,000 tonnes a year. The main advantages of production in Trinidad and Tobago are: low energy and material costs thanks to local gas and bauxite resources; and access to the North American market.

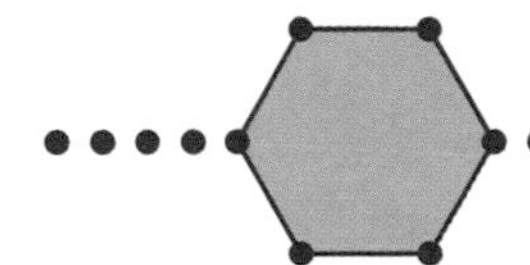

Weber's theory of industrial location

Alfred Weber's theory of industrial location was first published in 1909. Although highly simplified, his theory has proved useful in helping us to understand the modern location of heavy industries, and industrial location patterns in the nineteenth century when transport costs were much higher than today.

At its simplest Weber's theory answers the following question: Where should a firm using two raw materials to make a single product for a single market locate? Weber's answer was that the best or optimal location was where transport costs were as low as possible. To find this location he calculated transport costs by multiplying weight by distance (in units of tonne/km);

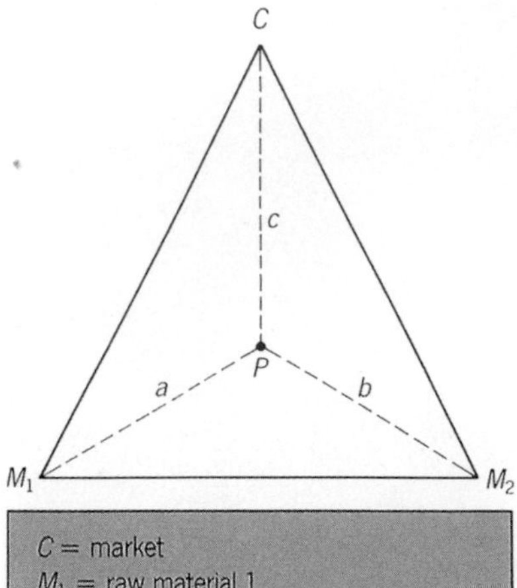

C = market
M_1 = raw material 1
M_2 = raw material 2
P = point of minimum transport cost
$a/b/c$ = distance between P and material sources and market

Weight of materials needed per unit of production:
M_1 = x tonnes
M_2 = y tonnes

Finished product weighs z tonnes

Least-cost location (P) is the point which minimises $x_a + y_b + z_c$

Figure 3.10 Weber's locational triangle

Figure 3.11 Weber's analysis of the effect of a cheap labour location using isodapanes (after Smith, 1981) Copyright © 1971 D.M. Smith, reprinted by permission of John Wiley & Sons, Inc.

assumed there was perfect competition between firms (no firm is able to influence price levels by its own actions); and that decisions were made by **economic man**, who acted on **perfect knowledge** and was dedicated to minimising costs.

Weber determined the point of minimum transport cost (P) by using his **locational triangle** (Fig. 3.10). In the triangle M_1 and M_2 are sources of raw materials and C is the market. The minimum transport cost location always lies within the triangle. It is found by drawing isolines of total transport costs (Fig. 3.11), known as **isodapanes**, which increase in value with distance from P.

In some circumstances a source of cheap labour (or savings made when firms locate together – **agglomeration economies**) can deflect a firm from the location of minimum transport costs. This happens when the savings in labour costs (or agglomeration economies) at a location exceed the additional transport costs incurred there.

In Figure 3.11 imagine that a source of cheap labour (L) results in a saving of £3 per unit of production. To determine whether a firm should locate there, Weber drew the £3 isodapane (which he called the critical isodapane) around P. From this, he devised a simple rule: a firm would locate at a source of cheap labour (L) if L lay inside the **critical isodapane**. You can see that, in this example, a location at L saves £3 on labour, but costs less than £3 in additional transport charges. Thus L, rather than P, is the optimal location.

Weber also devised a **material index** to show whether a firm would be **market-** or **material-oriented**. He divided materials into **localised** and **ubiquitous**. Ubiquitous materials are found everywhere, so they have no influence on industrial location. Thus the material index is the weight of localised materials divided by the weight of the finished product.

$$\text{Material index} = \frac{\text{weight of localised materials}}{\text{weight of the finished product}}$$

If the material index is more than 1 it means that the materials lose weight in the manufacturing process. Industries such as pulp and paper and iron and steel have high material indexes. In order to minimise their transport costs they should locate at materials. However, industries based either on pure or weight-gaining materials, with material indexes of 1 or less, should in theory locate at the market.

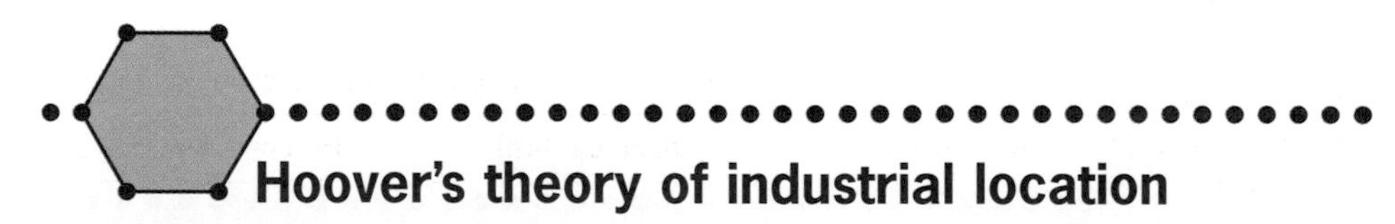

Hoover's theory of industrial location

Like Weber, Hoover defined the **optimal location** as one of least transport cost. To find this location Hoover assumed the most simple situation: a firm using a single material to make a single product, which is sold in a single market. A decision-maker is an economic man, who has perfect knowledge and a single goal: cost minimisation.

Hoover's theory is an advance on Weber's because it recognises the complexity of transport costs. Transport costs consist of two elements: **terminal costs** and **haulage costs**. Terminal costs include charges for handling freight, storage, docking in seaports and so on. They are fixed, regardless of the length of a journey. But haulage costs, which cover wages paid to drivers and crew, fuel used in transport, and insurance, vary according to the length of a journey.

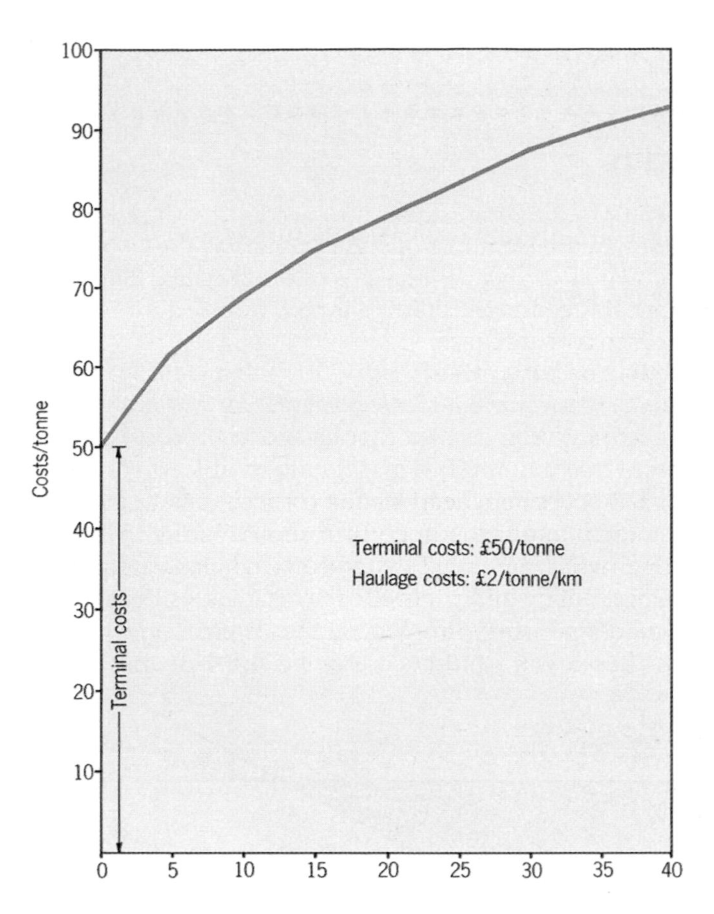

Figure 3.12 The tapering effect of freight rates

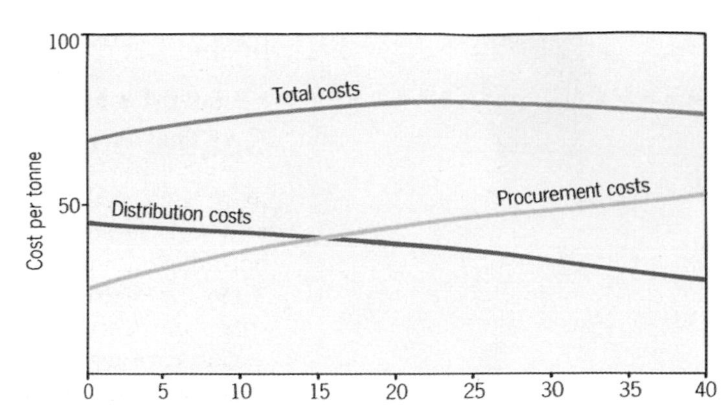

Figure 3.13 Hoover's analysis of freight rates: distribution costs less than procurement costs

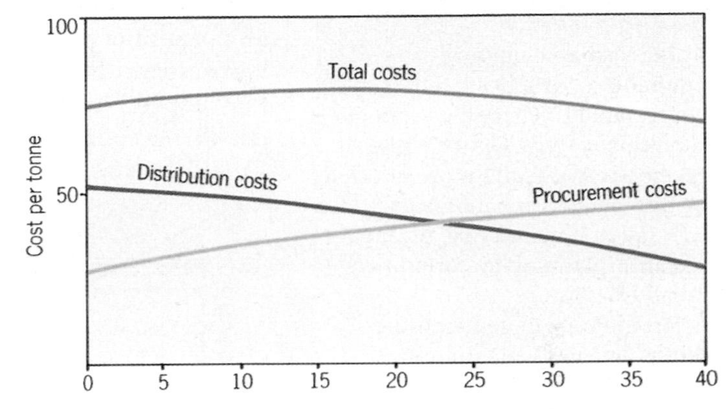

Figure 3.14 Hoover's analysis of freight rates: distribution costs greater than procurement costs

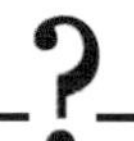

4 Sketch a graph to show how transport costs vary with distance according to Weber and Hoover.

5 Suggest why factory locations between materials and markets are ruled out as points of least transport cost in Hoover's theory.

6 Test for yourself the idea that a location between materials and markets can be the point of least transport cost.
a Plot the data in Table 3.2 as a graph (similar to Fig. 3.13). To do this, first, plot the procurement costs, from materials to the market. Then plot the distribution costs, from the market to materials. If you add the values for two graphs for each distance you will get the total transport cost curve.
b In this example explain where a firm should locate to minimise its transport costs.

The effect is to cause average transport costs (per tonne-kilometre) to fall with distance. Thus on long hauls, the fixed terminal costs can be spread over a larger distance, to give lower costs (per tonne-kilometre) than on short hauls.

If we plot the cost per tonne of a journey against distance (Fig. 3.12), we can see an initial steep rise corresponding to the fixed terminal costs. This is followed by a gently tapering curve, as haulage costs take effect, and average costs per tonne begin to fall with distance.

Table 3.2 Transport costs and industrial location

Distance (km)	Procurement costs	Distribution costs
0	50	50
5	61.5	61.5
10	68	68
15	72.8	72.8
20	76.8	76.8
25	80.3 91.8 **break of bulk**	80.3 91.8
30	98.3	98.3
35	103.1	103.1
40	107.1	107.1
45	110.6	110.6
50	115.8	115.8

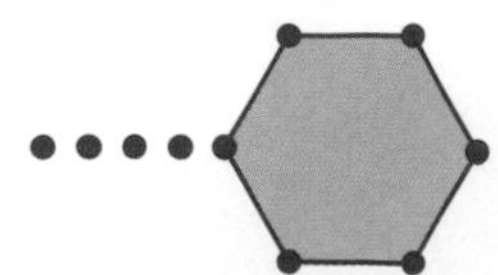

Spatial profit margins

D. N. Smith's theory of spatial profit margins takes into account both revenue and costs. It delimits an area where a firm can operate profitably, the boundaries of which are known as the spatial margins of profitability (Fig. 3.15). In Figure 3.15 the spatial margins are defined by the intersection of the space–cost curve with the revenue–cost curve.

Costs are made up of production costs (wages, rents, etc.) and transport costs. In Figure 3.15 production costs are the same everywhere, while transport costs increase with distance from O. You can see that location O is the optimal location: it is both the point of maximum profit and the point of least cost.

Unlike other theories, spatial profit margins does not assume that a firm automatically chooses the optimal location. It merely states that a firm must locate somewhere within the spatial margins of profitability. The exact location may depend on the knowledge, ability and goals of decision-makers (a more detailed account of these and other decision-making factors is given in Chapter 7). Either way, the location chosen will almost certainly be **sub-optimal**.

7 The spatial margins of profitability for a firm are not fixed, but respond to changing economic conditions. Study Figure 3.15 and say how these margins are affected by the following changes:

a A fall in demand (e.g. during a recession) causing uniform prices to fall to 40 units.

b An increase in demand (e.g. during an economic upturn) causing uniform prices to rise to 60 units.

c A lowering of transport costs (e.g. improved road network, lower energy costs, etc.) by 10 units.

d A fall in production costs (e.g. automation) by 10 units.

e An increase in production costs (labour disputes, material price rises, etc.) by 10 units.

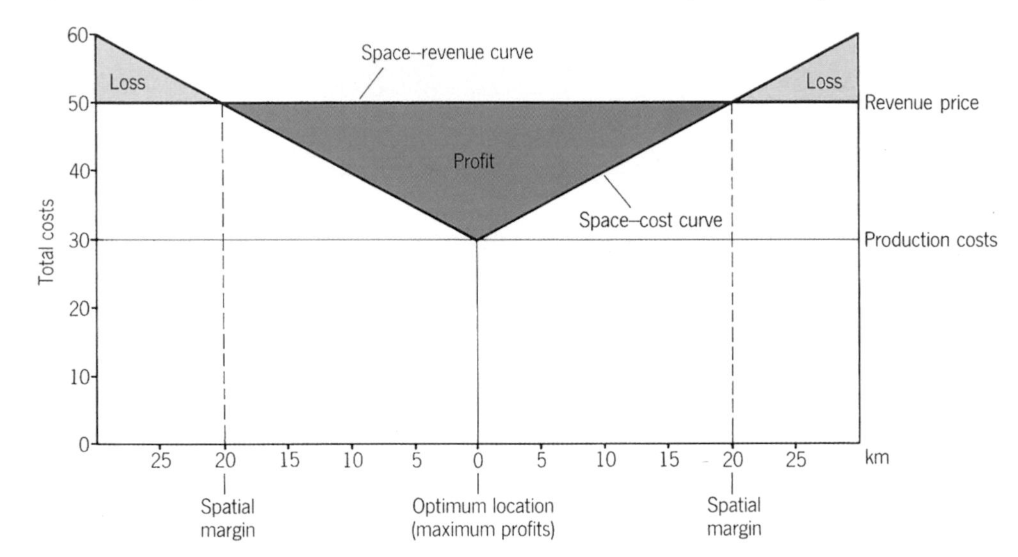

Figure 3.15 Simple model of the spatial margins of profitability

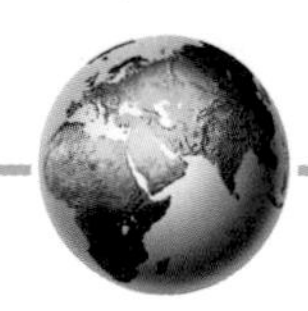

The changing location of the Lancashire cotton industry, 1780–1840

Phases of growth

The Lancashire cotton industry was transformed by technological change between 1780 and 1840 from a cottage to a factory-based industry. In the process it became the leading sector in Britain's industrial revolution. This transformation, based on new sources of energy and new machinery, falls into three overlapping phases:

1 A pre-industrial domestic hand-loom and hand-spinning phase which remained dominant until the end of the eighteenth century.

2 Factory-based production, using mechanical water power, which flourished briefly in the late eighteenth and early nineteenth centuries.

3 Factory-based production, dependent on steam power and coal, which became increasingly dominant after 1810.

The theory of spatial profit margins provides us with a framework in which to view the changing geography of the cotton industry (Figs 3.16–3.18). In all three phases, the industry was organised from Manchester. Manchester was the main source of capital and materials, and the principal market. Smaller centres such as Bolton, Rochdale and Blackburn operated at a local level as part of a hierarchy of industrial market centres.

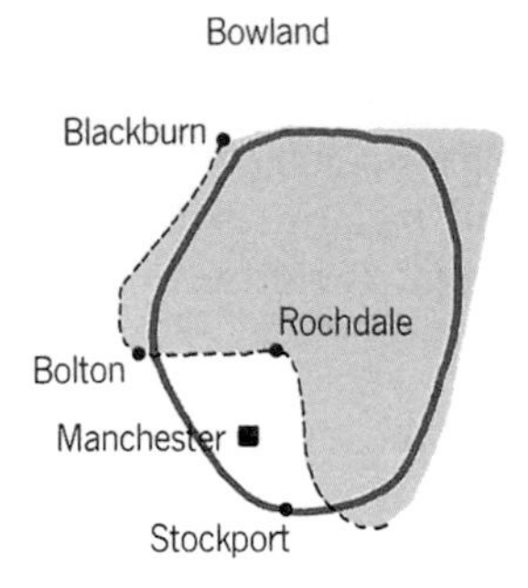

Figure 3.16 Pre-industrial

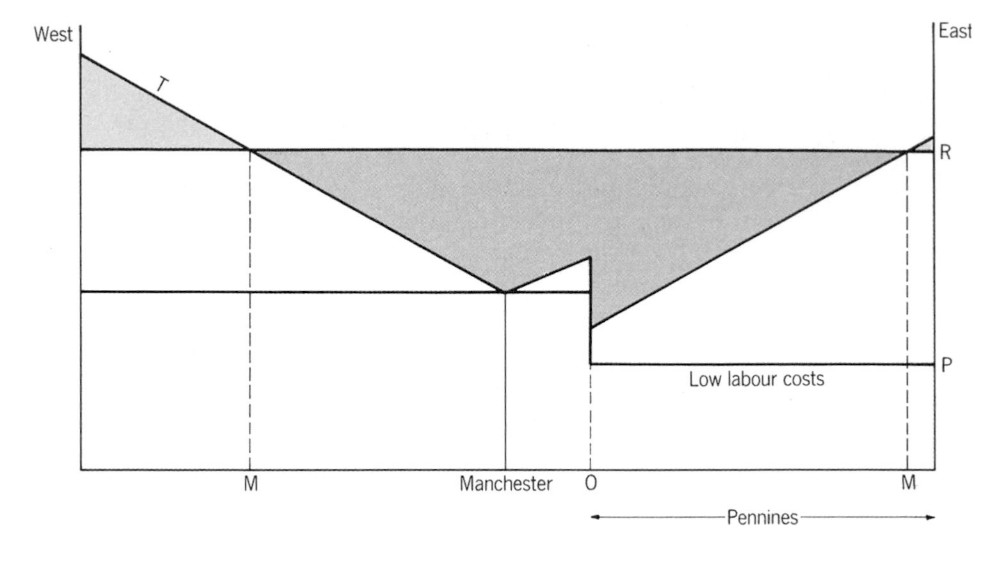

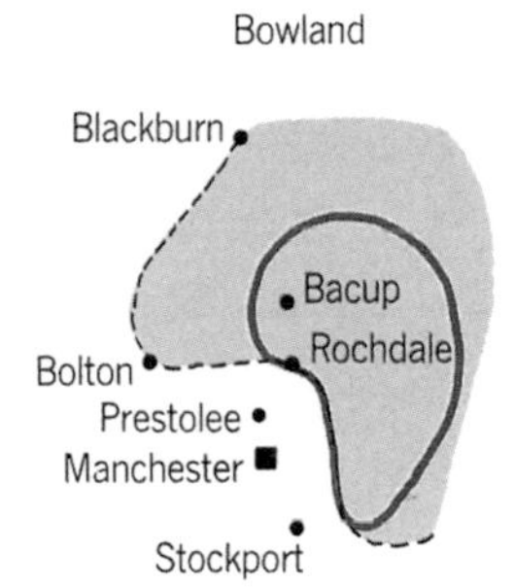

Figure 3.17 Early industrial

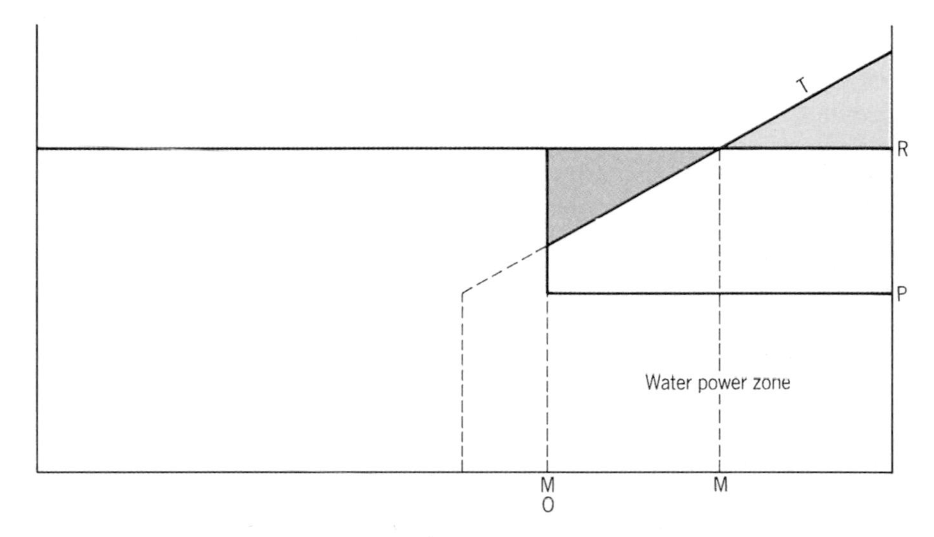

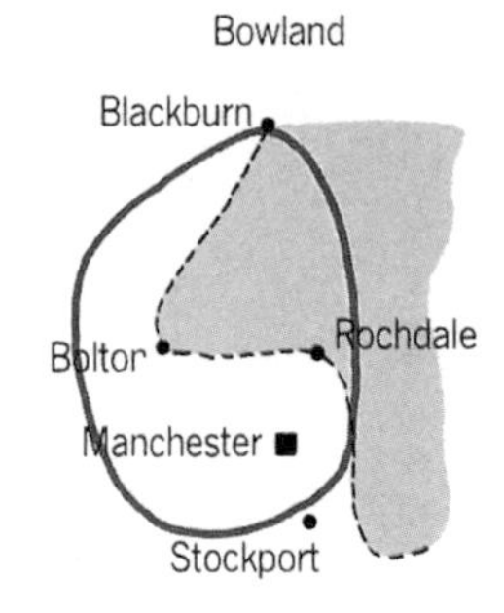

Figure 3.18 Late industrial

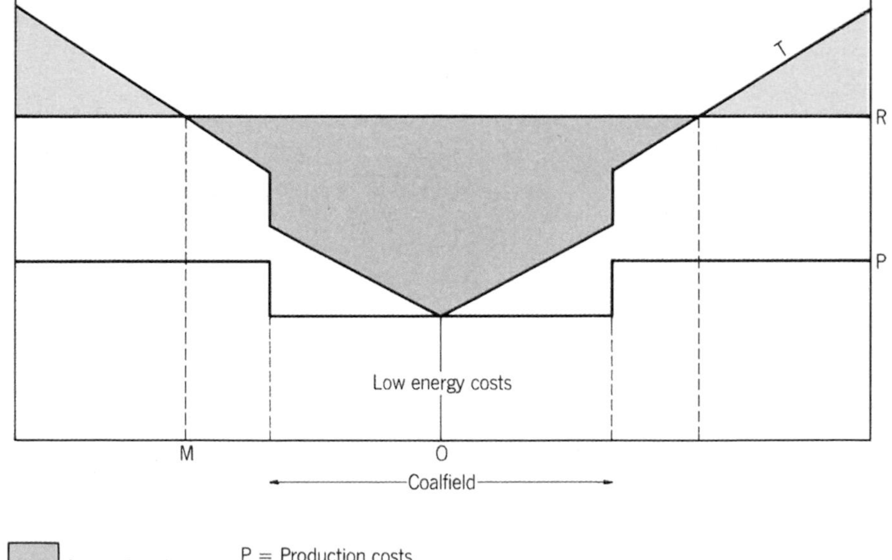

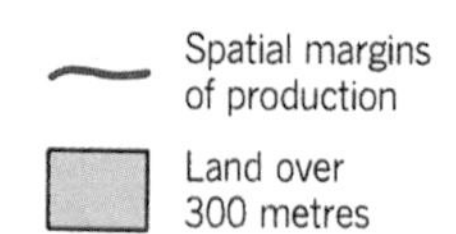

Figures 3.16–3.18 Spatial profit margins and the location of the Lancashire cotton industry, 1780–1840

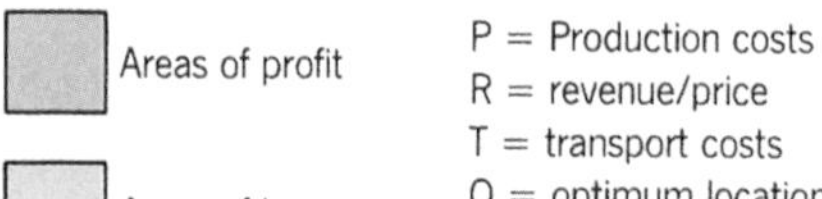

P = Production costs
R = revenue/price
T = transport costs
O = optimum location (maximum profits)
M = spatial margins of profit

The pre-industrial phase

In this phase (Fig. 3.19), labour costs were the main locational consideration. Entrepreneurs found surplus labour in the Pennine uplands to the north and east of Manchester. Employment opportunities were strictly limited in this region. The harsh physical environment meant that agriculture could not provide full-time work for everyone, and this problem had been made worse by rapid population growth during the eighteenth century. Figure 3.16 shows how cheap labour pushed the spatial margins of production deep into the uplands. The geographical limits to production were set by rising transport costs as distance from Manchester increased.

The early factory system

The early factory system (Fig. 3.20), centred on cotton spinning, was based on inventions such as the water frame and mule. They allowed a huge expansion of output which demanded new sources of energy. Everywhere water-powered mills sprang up on the fast-flowing streams draining the Pennines. Rivers like the Irwell became focal points for the industry. Falling nearly 300 metres between Bacup in Rossendale and Prestolee on the outskirts of Manchester, the Irwell had over 300 water-powered mills by 1835.

The immobility of water power meant that the industry was strongly **energy-oriented** during this second phase. However, the attractiveness of a site was also determined by its distance from Manchester. Thus, while the Pennines within a radius of 20 kilometres of Manchester were within the spatial margins of profitability (Fig. 3.17), more distant uplands, such as the Forest of Bowland, were never industrialised.

The mature factory system

The final phase involved factory-based production for both spinning and weaving (Fig. 3.21). With the introduction of steam power, activity shifted to the coalfield and was organised on a much larger scale. Again, costs increased from Manchester, but now the whole of the Lancashire coalfield lay within the spatial margins of profitability (Fig. 3.18). Even so, the cost of transporting coal before the development of the railways remained expensive. For example, coal mined at Wigan in 1840 doubled its pithead price when transported just 15 miles by canal to Preston. Thus for a long time the spatial margins of profitability extended only a short distance beyond the coalfield.

?

8 Essay: Describe how chance geographical circumstances led to growth and continuity of production in the Lancashire cotton industry between 1780 and 1840.

Figure 3.19 Pre-industrial spinning and weaving involved the whole family

Figure 3.20 The development of the power loom transferred the industry from cottage to factory

Figure 3.21 Manchester, about 1850

3.4 Transport

Transport costs and transport type

The structure of transport costs is complex. Hoover showed that costs per tonne-kilometre were lower for longer journeys than for shorter journeys. This was because the terminal costs, which are fixed, could be spread over more kilometres on longer journeys. (If you want a fuller explanation of Hoover's ideas, look at pages 30–31.) In addition to terminal costs, transport charges also include line or haulage costs. These are principally the costs of fuel and wages for the crew.

?

Each form of transport has a different mix of terminal and line costs. Study Figure 3.22 and answer the following:

9 Which form of transport has:
a the highest terminal costs?
b the highest line costs?

10 Terminal costs are higher for sea and canal transport than for trucks. Study the photograph of the seaport at Kobe in Figure 3.7 and try to explain why.

11 Explain why line costs are higher for trucks than for terminal charges.

12 Which form of transport is cheapest for:
a short journeys?
b long journeys?

Transport costs and type of freight

Transport costs also vary with the type of freight carried. Raw materials like iron ore and coal have lower rates per tonne than manufactured products such as steel and cars. This is because they often require less care and special handling, have lower insurance rates and can be transported in bulk. Bulk transport leads to the lowering of costs through economies of scale. Thus it is usually cheaper to ship raw materials than finished goods over long distances, and locate production near the market. At the global scale this trend is evident in the location of industries like steel and oil refining.

Transport and the location of heavy industry

Transport has a limited influence on the location of most manufacturing industries today. In the UK, for three out of four manufacturing firms, transport accounts for less than 3 per cent of gross output. In section 3.3 we saw that the widespread availability of electricity in economically developed countries had helped to make industry more footloose. As the real cost of transport has declined in the twentieth century, this trend has been reinforced, giving firms much greater freedom in their choice of location.

However, transport costs remain important for a number of processing industries, which use heavy, bulky, low-value raw materials (Table 3.3). For these industries, the theories of Weber and Hoover, based on locations of least transport cost, remain relevant.

Table 3.3 Manufacturing industries with above-average transport costs in the UK (transport costs as percentage of net output)

Bricks, pottery, glass, cement	14.6
Food, drink, tobacco	13.7
Petroleum products	8.8
Timber, furniture	7.5
Chemicals and allied industries	6.5
Metal manufactures	5.9

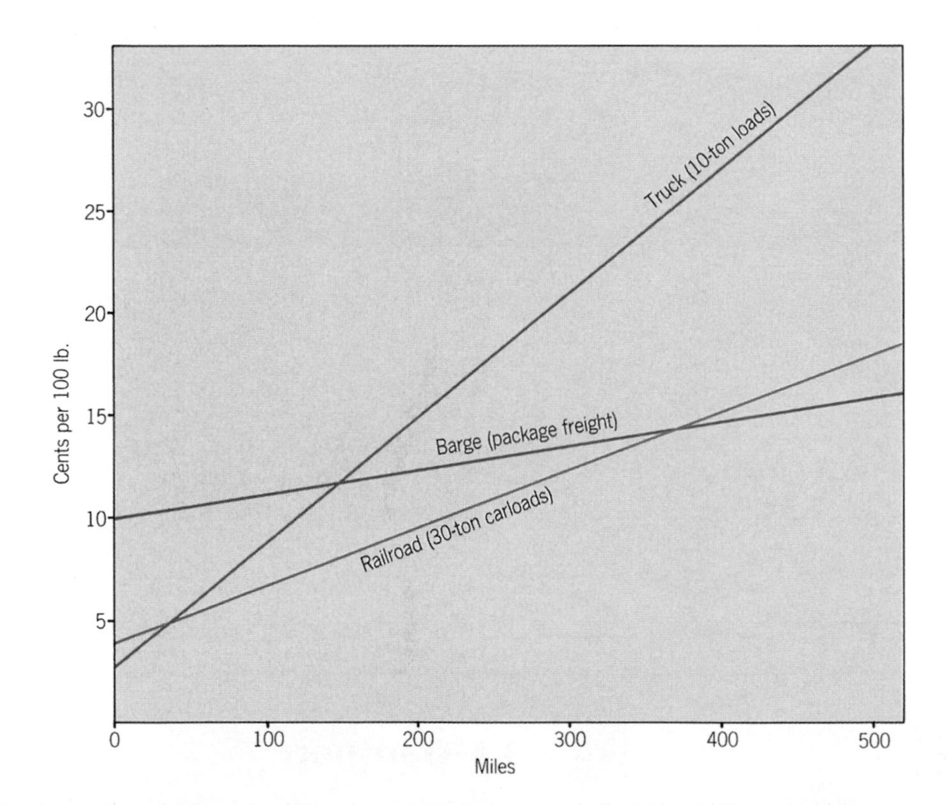

Figure 3.22 Freight rates and transport media: mileage-cost scales development for movements of commodities in carloads or equivalent in the lower Mississippi Valley area, 1939–40 (*Source:* Hoover, 1948)

?

13 Table 3.3 shows that both food and metal industries have above-average transport costs. Compare these two industries in Figure 3.23.
a State which industry is likely to have the higher material index and explain why.
b State which industry is likely to be material-oriented and which market-oriented. Explain your answer.
c State which industry is likely to be more footloose and explain why.

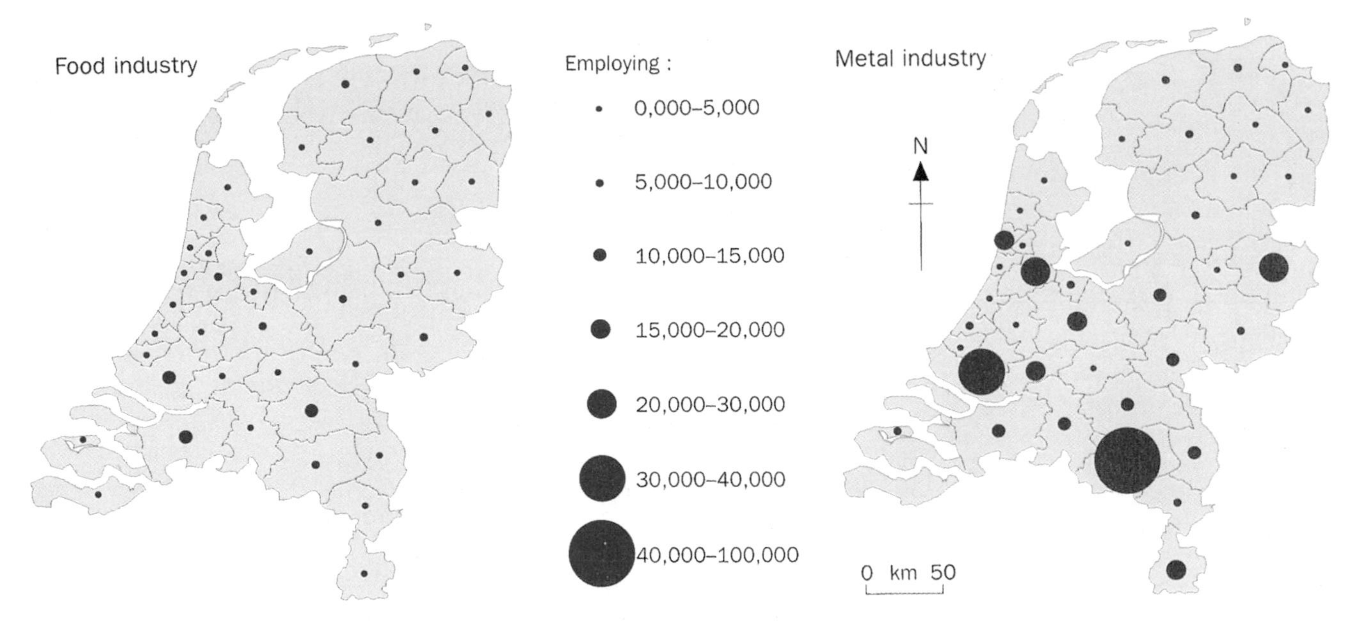

Figure 3.23 Distribution of the food and metal industries in the Netherlands

Transport and pricing policies

Transport costs per tonne-kilometre fall with distance. However, this so-called 'tapering effect', rather than being smooth, follows a step-like trend (Fig. 3.24). Instead of quoting a unique rate for every destination, transport hauliers make a uniform charge within broad geographical zones (a similar policy is often adopted by public transport systems, such as London's Underground). Thus when a journey crosses the boundary of a price zone, there is a sudden increase in cost.

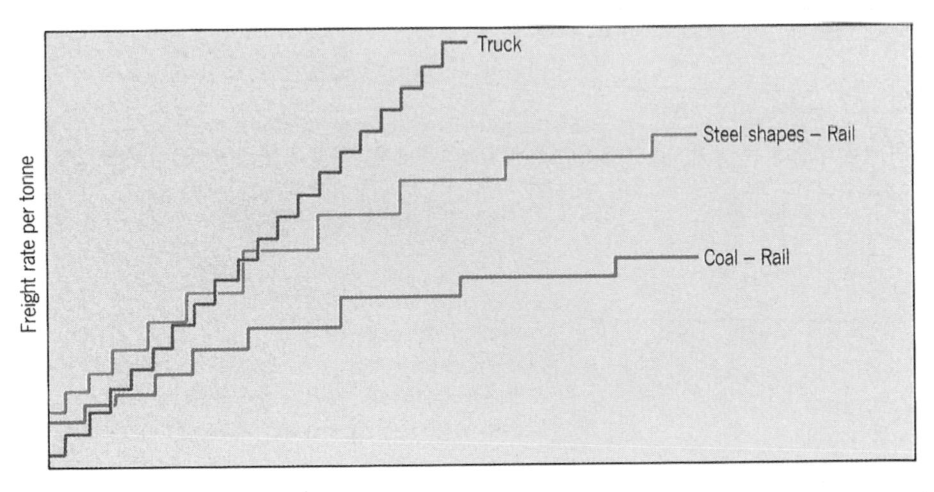

Figure 3.24 Freight-rate structures (*Source*: McCarty and Lindberg, 1966)

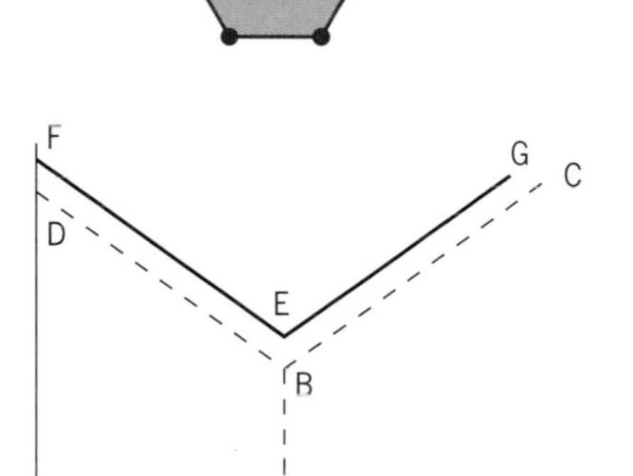

Figure 3.25 F.o.b. pricing

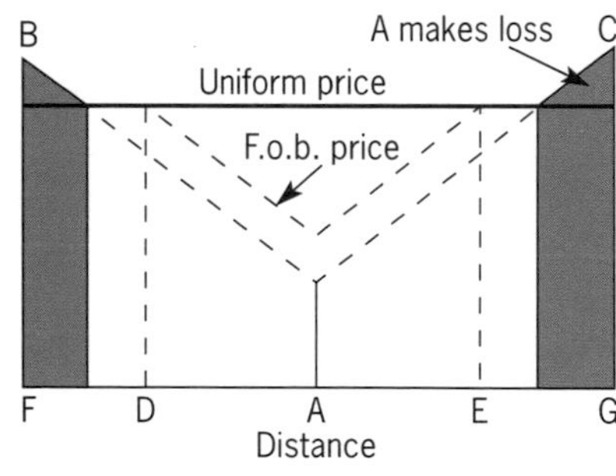

ABC = total costs for production centre A

DAE = zone where customers pay above f.o.b. prices and subsidise customers in EG and FD

Zone of loss

Figure 3.26 Uniform delivered pricing

F.o.b. pricing and uniform delivered pricing

Most locational theories assume that the price of a good reflects two things: its production costs, and its cost of transport to market. This is free-on-board (F.o.b.) pricing (Fig. 3.25). Under this system customers close to a plant enjoy lower prices than those further away.

In fact f.o.b. pricing is far less common than uniform delivered pricing (Fig. 3.26). **Uniform delivered prices** are set at a standard rate within regions or even entire countries. There is, therefore, no direct saving to customers locating close to the point of production. Customers close to a plant subsidise the transport costs of customers further away. Even so, uniform pricing does not give a firm the freedom to locate anywhere. Ideally, a firm should choose a location central to its market to keep its prices as low as possible, and maximise its sales.

?

14 What effect is f.o.b. pricing likely to have on the location of industries?

15 Read the paragraph on uniform delivered pricing. Suggest why transport costs are unlikely to be responsible for the agglomeration of manufacturing industry.

Basing-point pricing

Basing-point pricing is rarely used today, though early this century it was found in many US industries. A base point is fixed at a certain location, and delivered prices are calculated by adding production costs to the transport costs from the base point (Fig. 3.27). This is irrespective of whether a customer buys the good from the base point or not. Where this system operates, all sellers quote an identical delivered price to one customer. The classic example was the Pittsburgh Plus system in the US steel industry. It operated between 1900 and 1924 and, as you would expect, was highly unpopular with steelmakers outside Pittsburgh. Its purpose was to protect Pittsburgh's position as the country's leading steel centre. It also had the effect of encouraging steel-using industries to locate close to Pittsburgh.

The diagram shows two production centres, P and B. P is the base point, and P–C is the basing–point price. ACD is the delivered price from P. B has lower production costs (BE) and could quote a delivered price FEG. Under a simple f.o.b. system, B would serve the market as far as H. But under the basing–point system all customers must pay the delivered price ACD. Thus a customer at I would normally purchase from B, at a price IJ. Instead, the basing-point system forces I to pay IK, even if goods are purchased from B. Such a system discriminates against B and reinforces P's position in the market.

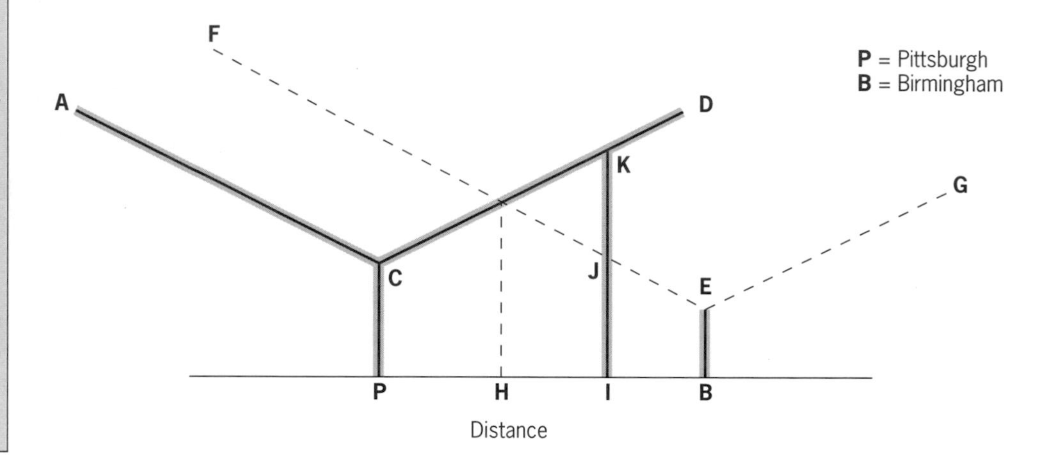

Figure 3.27 Basing-point pricing

3.5 Plant size

Economies of scale

Plant size can have an important effect on unit costs (the cost of manufacturing a product). Large plants usually have lower unit costs than small ones. Because these savings arise from within a plant, they are known as **internal economies of scale**. Three factors explain internal economies:

1 The existence of indivisibilities, i.e. the greater the number of units produced the smaller will be research and development costs associated with each unit.

2 Economies of specialisation. Increased size means that specialist machines and specialist staff (division of labour) can be used.

3 Economies arising from massed resources. Discounts may be obtained for buying materials in bulk.

These factors operate with different force in different industries. In footwear and iron castings few savings can be made by increasing the scale of

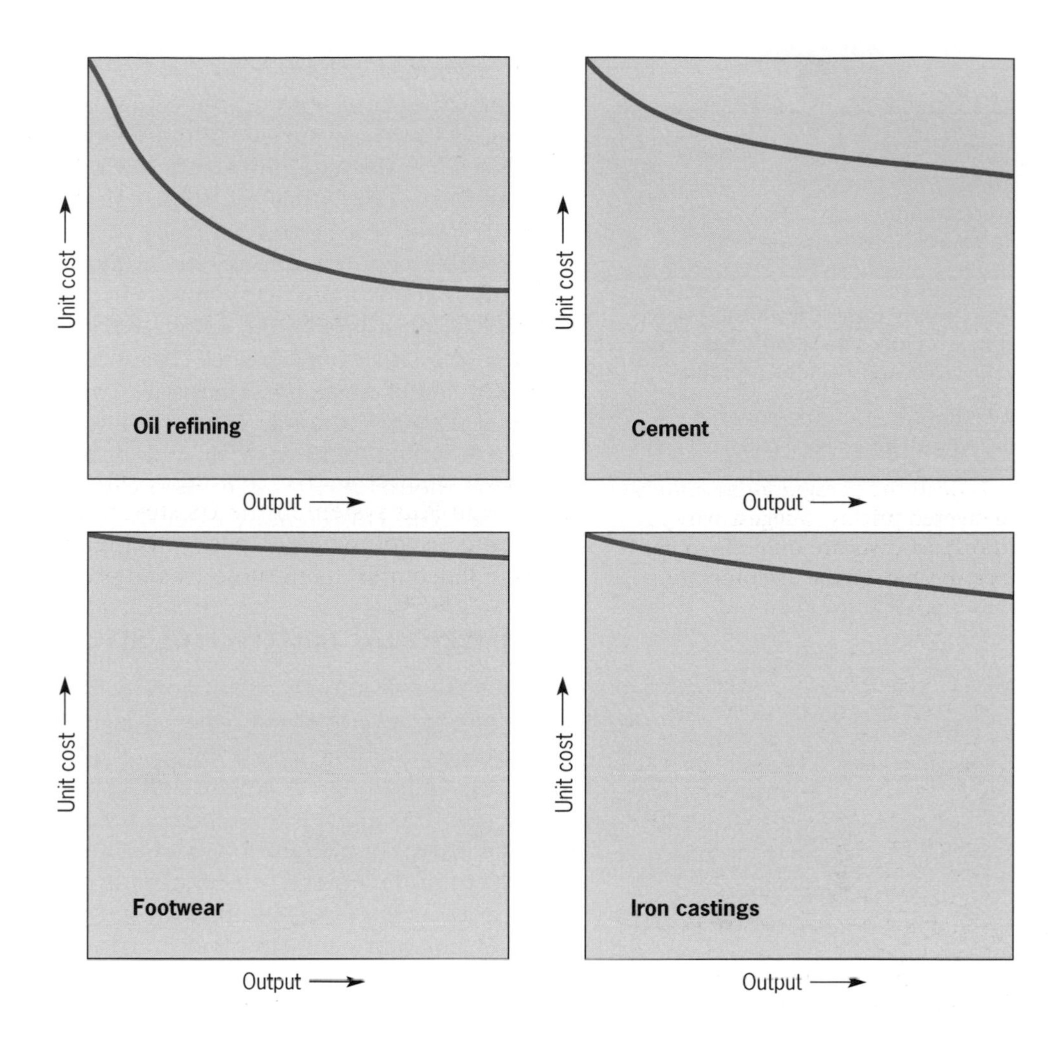

Figure 3.28 Economies of scale in selected industries

operations. But in oil refining and cement manufacture the savings are considerable (Fig. 3.28).

Unit costs fall with increasing plant size until the point of minimum cost is reached (Fig. 3.29). This point marks the **minimum efficient scale of production** (m.e.s.) and represents the optimal plant size.

Table 3.4 Minimum efficient scale of plants: selected industries in the UK

Industry	Plant size no. employees)
Ordnance and small arms	1109
Ofice machinery	470
Asbestos	465
Scientific and industrial instruments	102
Brewing and malting	41
Animal and poultry products	31

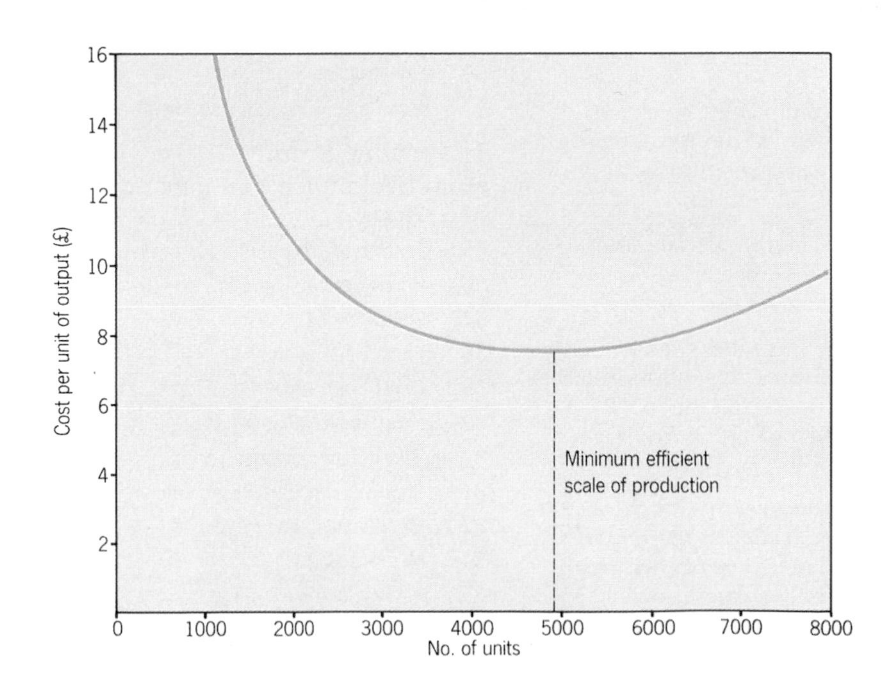

Figure 3.29 Relationship between plant size and unit costs

?

Table 3.5 shows manufacturing employment in three connurbations in 1991. The connurbations have varying degrees of industrial specialisation.

16 Calculate the specialisation index for the West Midlands, Tyne and Wear and Greater London.

17 Comment on the advantages and disadvantages of industrial specialisation.

Plant size and location

Plant size can place limits on the choice of location for a new plant. Usually very small plants can survive in most locations, but large plants may be limited by labour or space requirements or both.

Small towns offer a limited labour market and restrict the size to which a plant can grow. For instance, a study in South-West England showed that towns with a population of under 25 000 were limited to branch plants with fewer than 200 workers. Only towns with over 50 000 people were able to receive plants with over 500 employees. In northern Mexico, US branch plants locate in the larger border cities (Tijuana, Ciudad Juárez) where there is sufficient labour to support their scale of operation.

The minimum economic size for a modern integrated iron and steelworks is around 8 million tonnes a year. A plant of this size needs access to a workforce of several thousand. Thus a location close to a large urban centre is essential. Large steelworks also require extensive and flat sites of up to 10 km^2, which sets further constraints on their location.

3.6 Regional industrial specialisation

Regional specialisation is a common feature of the industrial geography of most economically developed countries. It is usually explained by **comparative advantage**. The basic idea is that some places, by reason of their location, have a cost advantage in certain industries (e.g. access of materials, skilled labour, etc.). As a result these places become highly specialised and depend on just one or two industries. Regional industrial specialisation was particularly prominent in the nineteenth century. It was at this time that Lancashire became pre-eminent in cotton textiles, Clydeside in shipbuilding, and South Wales in iron and steel. However, in the past 30 years deindustrialisation, globalisation and foreign direct investment have reduced specialisation and caused regional economies to converge.

Specialisation is a high-risk strategy: if a staple industry collapses it creates large-scale unemployment and years of painful economic adjustment

?

Visit the main reference library in your nearest town and look at the Census Reports on Economic Activity for the 1991 or 2001 census for the county in which you live.

18 Obtain the latest figures on unemployment from the *Labour Market Trends* magazine for a sample of 10 counties.

19 Calculate the specialisation index for manufacturing employment for each county.

20 Using Spearman's rank correlation (see Appendix Al) test the hypothesis that the more specialised a county is, the higher its level of unemployment.

21 Was the hypothesis accurate? If not, can you suggest reasons why? (NB: excessive dependence on an industry only creates problems when that industry starts to decline.)

Table 3.5 Standard occupational classification manufacturing industry

Standard Occupational Classification		% Manufacturing employment		
		West Midlands	Tyne & Wear	Greater London
50	Construction	8.63	11.37	18.29
51	Metal machining and instrument making	17.23	11.76	8.93
52	Electrical/electronic	6.61	10.60	13.60
53	Metal forming, welding	9.25	11.73	6.63
54	Vehicles	4.40	4.20	5.72
55	Textiles, garments etc.	5.04	5.76	6.26
56	Printing etc.	2.05	2.99	4.52
57	Woodworking	4.83	6.52	8.15
58	Food preparation	1.23	1.78	1.67
59	Other crafts etc.	3.72	5.39	4.6
80	Food, drink, tobacco	1.12	2.66	1.39
81	Textiles and tanneries	0.19	0.033	0.21
82	Chemicals, paper, plastics etc.	3.06	3.22	1.70
83	Metal making and treating	2.60	0.47	0.29
84	Metalworking	10.15	3.16	2.93
85	Assemblers/lineworkers	6.25	3.53	2.58
86	Other process operatives	8.27	6.85	6.14
89	Plant and machine operatives	5.38	7.66	6.4

(see section 12.2). The solution is diversification. By establishing a broad-based economy a region (or town) ensures that the decline of any one industry no longer spells disaster.

Specialisation index

The extent of specialisation can be measured by the **specialisation index** (SI). Its calculation is shown below:

$$SI = \sqrt{P_1^2 + P_2^2 + P_3^2 + P_n^2}$$

where P_1 is the percentage share of the first industry in an area, P_2 is the percentage share of the second industry in an area and so on. The higher the value of the index, the more specialised a place is. For instance, if there were only one industry in a place the index would be 100. With two industries, each employing 50 per cent of all industrial workers, the index would be 71; with four industries, each with a 25 per cent share of workers, the index would be 50 and so on.

Iron and steel

Iron and steel is the classic industry of Weber's locational theory. Few modern industries have locational patterns so closely linked to transport costs. The principal raw materials – iron ore, coking coal and limestone – are bulky, weight-losing and have low weight: value ratios. These materials have always had a strong influence on the location of steel production, and this is still true today.

Global production

World steel production increased massively in the period after 1945. In 1997 it was 794 million tonnes, more than five times greater than the 1948 figure. But growth has not been smooth. The slow-down of the world economy in the 1970s depressed steel production. As a **producer-goods industry**, steel is sensitive to demand from other industries such as motor vehicles, heavy engineering and construction. Falling demand in the 1970s and early 1980s hit these industries hard. The knock-on effects for steel were severe.

Steel production is dominated by the major industrial economies, especially China, the USA, Japan and the EU (Fig. 3.30). Some NICs, notably Taiwan and South Korea, have emerged as low-cost producers and have captured important export markets. With the exception of China, Brazil, India and Mexico, few of the world's poorer countries are major steelmakers. Given the scale of investment needed, and the importance of a domestic market, this is not surprising.

Organisation – big is beautiful

Low-cost production of steel is only possible in large firms that can achieve economies of scale. Hence steel is an industry of large firms. The largest firm in the world – POSCO of South Korea – would rank seventh in the league table of steel-producing countries. The importance of size was underlined by a spate of mergers between firms in the 1990s. In 1999 British Steel merged with the Dutch steelmaker Hoogvens to create the third largest steel producer in the world.

Although the global steel industry is dominated by a small number of very large firms, unlike the car industry, few are transnational organisations. In the past, governments have had a strong influence on

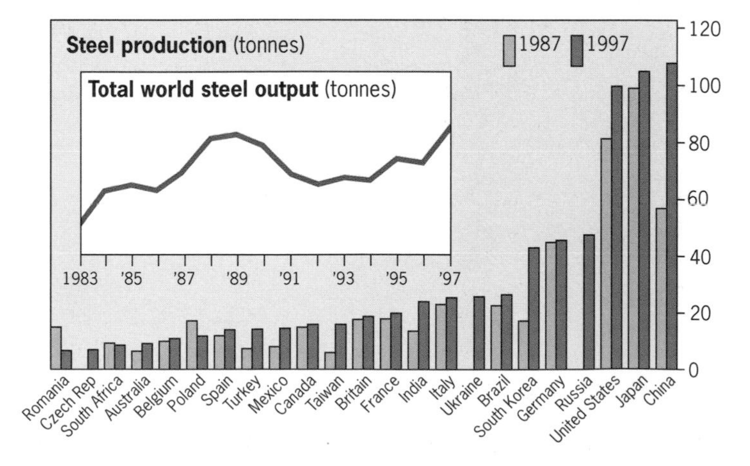

Figure 3.30 World's leading steel producers, 1997

steelmaking. There were two reasons for this: first, steel's importance to other industrial sectors; and second, the huge investment needed to develop and sustain a modern steel industry.

Political influences on steelmaking can operate on an international scale. The European Coal and Steel Community (ECSC), founded in 1951, was the first EU institution. From 1951 to 2002 the ECSC regulated steel production in the EU. Quotas allocated to member states kept production in line with demand. The phasing-out of the ECSC from 2002 will bring an end to direct EU intervention in the steel industry.

Technological change and location in the UK

In the last 300 years, the technology of iron- and steel-making has undergone enormous changes. Innovations in manufacturing methods, new materials and transport have greatly altered the geography of production.

Pre-industrial – a rural industry

Before the eighteenth century, iron-making was on a small scale. Its distribution was determined by the location of iron ore, water power and charcoal for smelting. Charcoal exerted the strongest influence. In Britain, iron-making was attracted to districts where timber was plentiful, such as the Weald of Sussex, and the Forest of Dean in Gloucestershire.

Industrial Revolution – the pull of coalfields and orefields

Dwindling supplies of timber led to the invention of smelting iron ore with coke. This technological breakthrough is attributed to Abraham Darby at Coalbrookdale in Shropshire, some time around 1709. Although it took many years to perfect the process, there is no doubt that Darby's invention revolutionised the iron industry. Output of pig iron increased from 17 000 tonnes in 1720 to 250 000 tonnes in 1805; by the end of the nineteenth century, production stood at more than 8 500 000 tonnes.

Technological change was swiftly followed by locational change. There was a movement to the coalfields (Fig. 3.31). Especially attractive were regions like the Black Country in the West Midlands, and Stoke-on-Trent in North Staffordshire, where coking coal and iron ore were found together. With 8 tonnes of coal needed to make 1 tonne of iron in 1828,

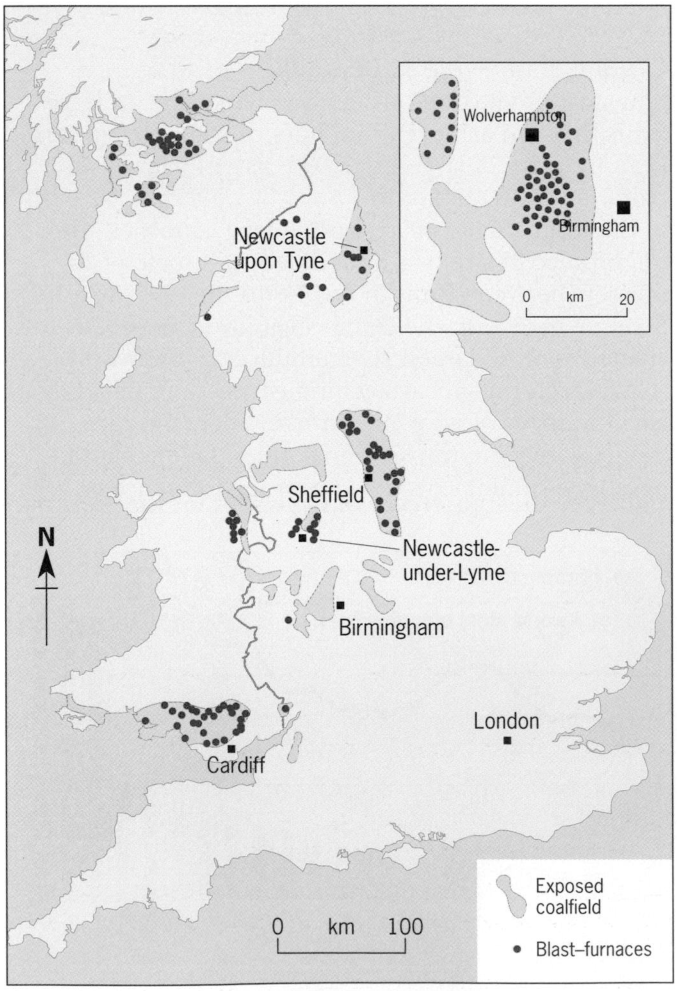

Figure 3.31 Distribution of ironworks in the UK, 1849

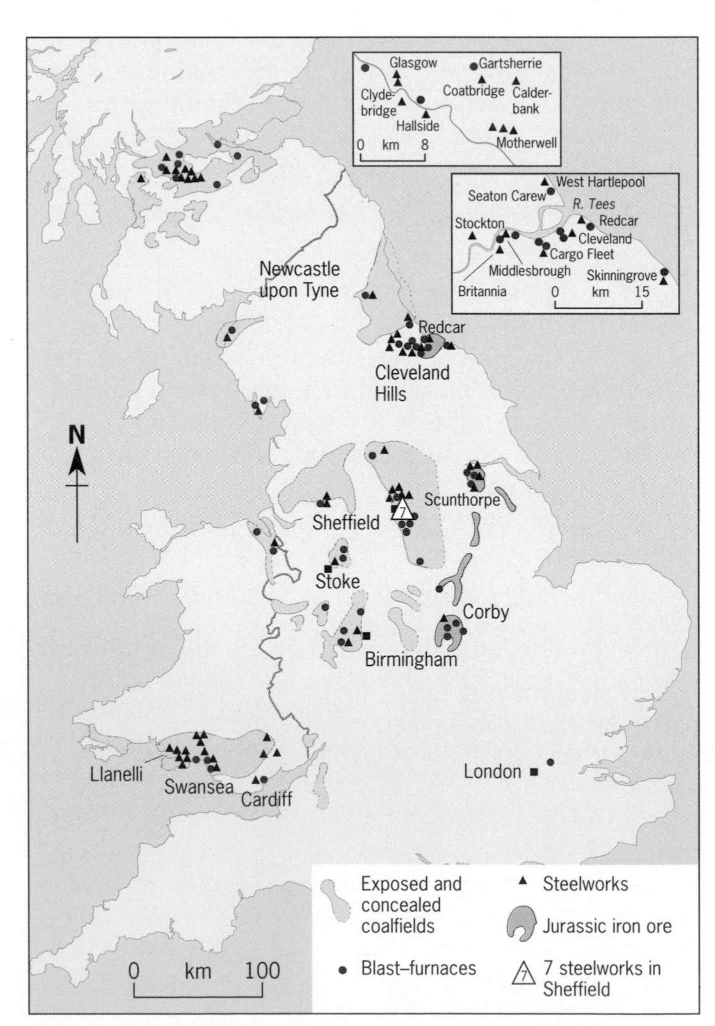

Figure 3.32 Distribution of iron and steelworks in the UK, 1937

coalfield locations were essential to keep transport costs down. However, with the development of the railways, and improvements in blast-furnace technology, the coalfields began to lose their dominance. By 1850, the ratio of coal to iron was down to 4:1; 50 years later it was just 2.5:1.

The invention of the Gilchrist–Thomas process in 1878 made it possible to smelt iron from phosphoric ores for the first time. This opened up the Jurassic orefields of Lincolnshire and the East Midlands. Because these ores were 'lean' (maximum 35 per cent iron content) new plants were drawn to orefield locations. Scunthorpe, established in 1890, was situated close to the phosphoric ores of Lincoln Edge. Forty-five years later a steelworks was built at Corby in Northamptonshire using local phosphoric iron ore (Fig. 3.32).

Post-1945 – the move to tidewater

Modern steel industries, relying heavily on imported materials, show a strong preference for tidewater locations (coastal sites). The UK steel industry, for example, imports more than 98 per cent of its iron ore. This is partly due to domestic supplies running out, and to new steel-making technologies which require higher-quality ores. Imported ores have an iron content (around 60 per cent) which is three times greater than domestic ores.

However, the main reason for the shift to imported materials is cost. Imported materials are cheaper because they can be shipped in huge consignments by bulk carrier vessels. The effect has been to pull steel production to coastal sites near to deep-water terminals. The advantages that domestic raw materials had have all but disappeared. Now any country with a deep-water port can assemble the basic materials at costs competitive with the most efficient steel producers in the world.

Coastal sites have one further advantage: they offer large areas of cheap, flat land. This is crucial because lower costs and profitability are tied up with the scale of production. In the period after 1945, steelworks increased steadily in size, until by 1980 their minimum economic scale had reached 8 million tonnes a year, though few plants outside Japan were so large.

New steel-making technologies (Fig. 3.33), in particular the basic-oxygen process and continuous casting (CONCAST), have favoured large-scale production. The basic-oxygen process replaced open-hearth steel-making in the 1960s and 1970s and greatly speeded-up the steel-making cycle. Whereas it took 6–7 hours to convert iron to steel using the open-hearth system, the basic-oxygen method reduced this to just 45 minutes. Thus basic-oxygen steel-making increased the demand for molten iron, which meant larger blast furnaces and larger steelworks.

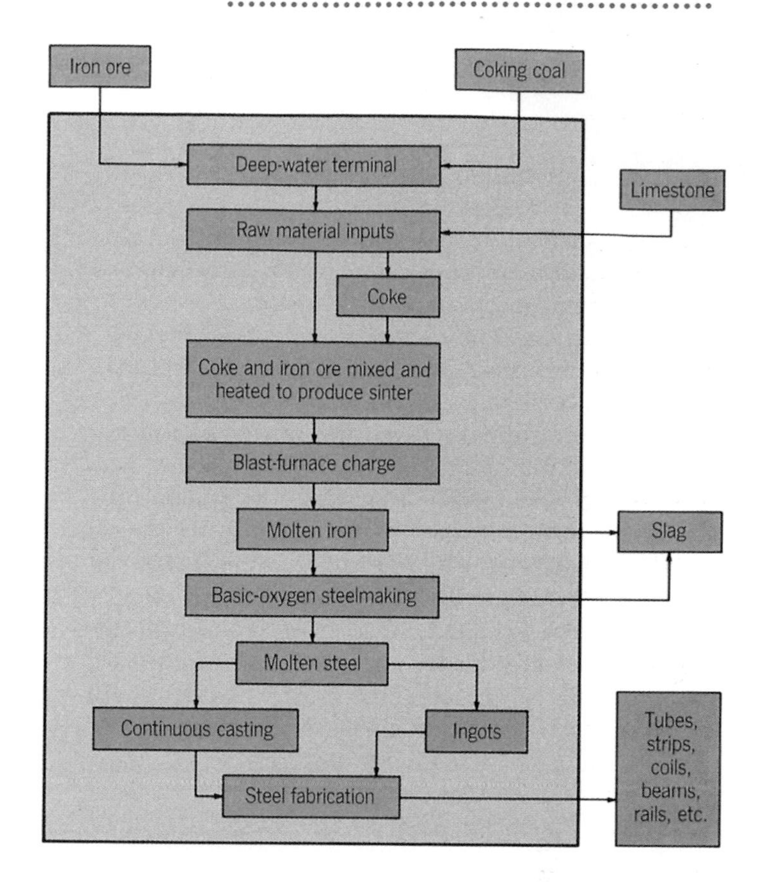

Figure 3.33 A modern integrated iron- and steelworks

CONCAST is the direct production of steel slabs from molten steel. Most EC steel is now produced in this way. It increases efficiency and reduces energy costs, as steel ingots no longer have to be re-heated for conversion to slabs before rolling. Also, by bringing steelmaking and casting together, concast has encouraged integration and economies of scale in production.

Not only have individual plants increased their output, but **vertical integration** of production is now standard. This means that iron smelting, steel-making and the fabrication of steel products all take place together on the same site. In locational terms this again reinforces the importance of large sites for the industry.

22 Explain how Weber's theory (pages 29–30) can help us to understand the changing location of Britain's iron and steel industry since 1600.

23 How does Hoover's theory (pages 30–1) help to explain the location of the modern steel industry?

24 'Market-oriented at a global scale, but material-oriented at a national scale.' Explain this statement with reference to the iron and steel industry.

The UK steel industry in the 1990s

Volume steel production in the UK is concentrated at four integrated sites: Redcar–Lackenby, Port Talbot, Llanwern and Scunthorpe (Fig. 3.34). These sites reflect the long and complicated history of the industry. Only Redcar–Lackenby and Port Talbot, with their own deep-water terminals, could be regarded as optimal. Llanwern (Newport) occupies a coastal site, but has no attached ore terminal. Ore is transported by rail from Port Talbot, 50 kilometres away. Scunthorpe, originally sited close to phosphoric ores, relies on foreign ores imported through Immingham in North Lincolnshire. Although Llanwern and Scunthorpe are satisfactory locations for modern steel-making, neither is ideal.

Special steels are made at Sheffield and Rotherham in South Yorkshire using steel scrap and electric arc furnaces. This is a nineteenth-century coalfield location which survives partly due to the firm being disinclined to move because of its investment in fixed capital.

?

25 Compare the photographs of steelworks in Figures 3.35 and 3.36.

a What are the disadvantages of the site in Figure 3.35 for modern steelmaking? Suggest possible reasons for the initial location of the steelworks in Figure 3.35.

b List the advantages for modern steelmaking of the site in Figure 3.36. Explain why locations like the one in Figure 3.36 are so attractive today.

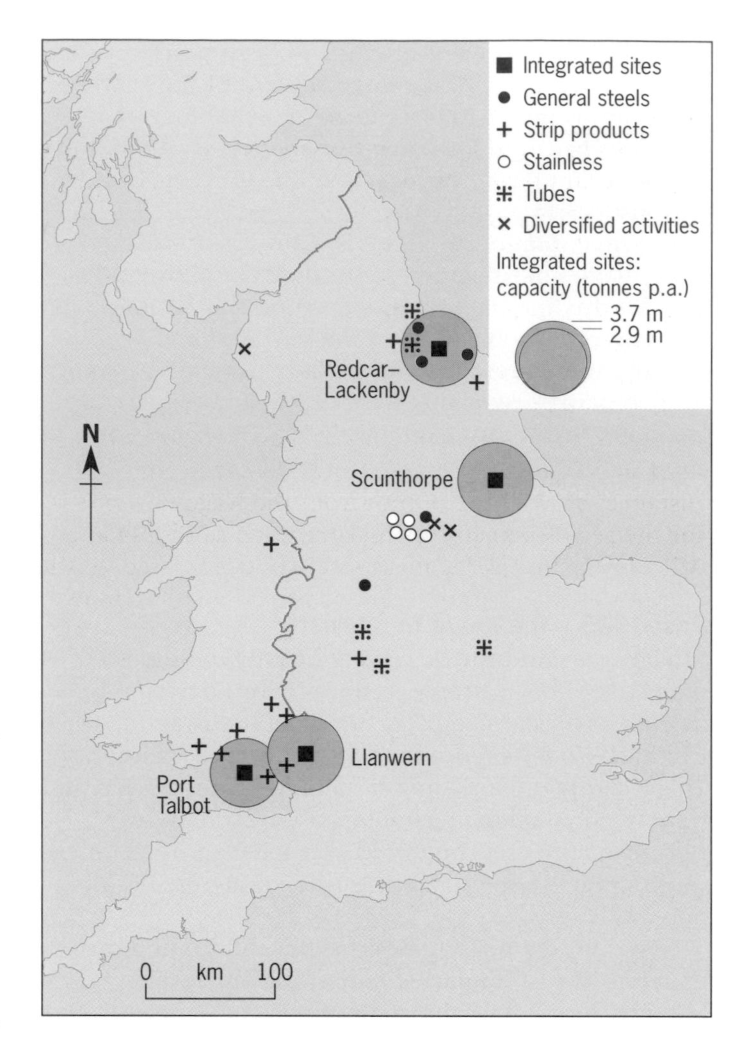

Figure 3.34 British Steel's production sites

Figure 3.35 British Steel's former steelworks at Ebbw Vale

Figure 3.36 British Steel's modern Lackenby steelworks

Japan's steel industry

Largely because of the destruction of other sites during the Second World War, Japan was the first steelmaker to opt for large-scale, vertically integrated production at coastal sites. Today, all Japan's major steel plants are located on the Pacific coast. With Japan importing 99 per cent of its iron ore, and 94 per cent of coal requirements, steel sites are strongly material-oriented. Coastal sites also assist exports, which are one-third of total output.

While Japan's steelworks occupy material-oriented sites, their situation is market-oriented. Every major industrial concentration (Tokyo, Osaka, Nagoya) has its own steel industry geared to local demand. For instance, most of the output from the Nagoya works is for the region's major car makers, such as Toyota and Mitsubishi. Just as Japanese steelworks are large, so the companies that operate them are equally large: three (Nippon Steel, NKK and Sumitomo) are ranked in the world's top twenty steel companies.

Nippon Steel Corporation

Nippon Steel is the world's second largest steel company. It operates seven volume steel-making sites in Japan (Fig. 3.37), which had a combined output of 25.1 million tonnes of steel in 1998. Its integrated Kimitsu works near Tokyo illustrates many of the features of modern steel-making. It occupies a reclaimed coastal site of more than 10 km^2, and almost all of its raw materials come from overseas. Iron ore is imported from Australia and Brazil, and coking coal from Australia, the USA and Canada. These materials are unloaded at its own deepwater terminal which can handle bulk carriers of up to 250,000 tonnes. Output from its four blast furnaces and five basic-oxygen furnaces was 8.44 million tonnes in 1998. Finished steel products made on site include heavy plates, beams, tubes and wire rods. Eighty per cent of production is destined either for export or the nearby Tokyo market (Fig. 3.38).

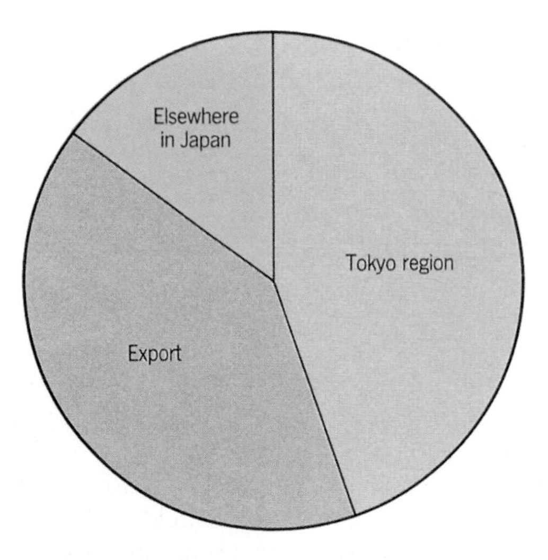

Figure 3.38 Markets for steel production from the Kimitsu works

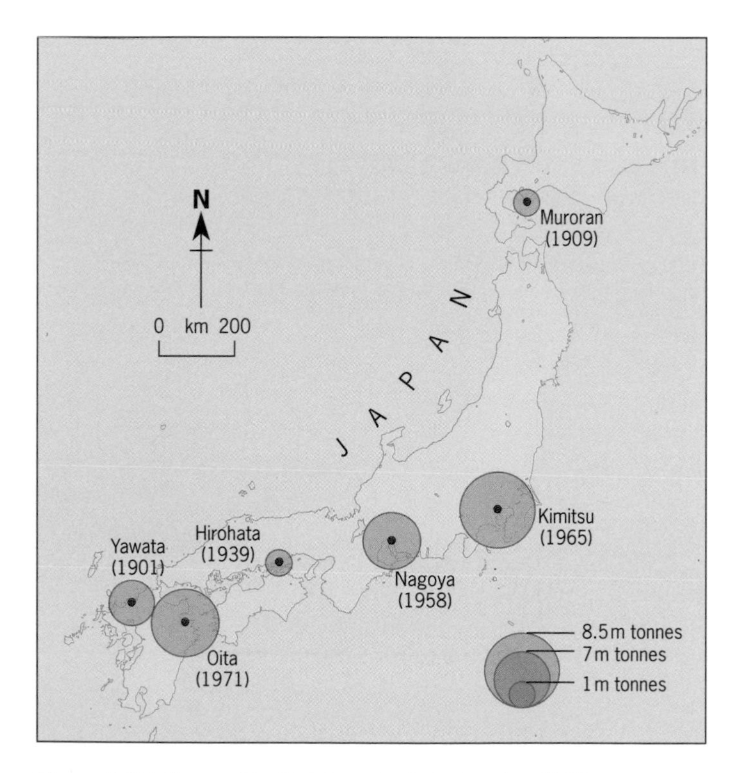

Figure 3.37 Nippon Steel's integrated steelworks, 1998

In contrast, Kamaishi, established in 1857, represents an early phase in the growth of the steel industry, when location was determined by domestic raw materials. In this case iron ore was available locally, and coal was shipped from Hokkaido. A direct parallel is found in nineteenth-century locations in Britain, Germany and the USA.

Nippon Steel closed its last remaining blast furnace at Kamaishi in 1988. The works suffered the disadvantages of a cramped site, small size and remoteness from the main markets further south. Rationalisation of production in the face of competition from South Korea, and the increase in the value of the yen, made Kamaishi an obvious target. Recently Nippon Steel's policy has been to reduce its iron and steel capacity and concentrate large-scale production at its most cost-efficient works, at Kimitsu, Nagoya and Oita.

?

26 Draw an annotated diagram to show the influences on Japan's steel industry.

Figure 3.39 British Steel's blast furnace on Teesside, the largest of its kind in Europe

Summary

- Manufacturing industries using raw materials which lose weight in processing (e.g. sugar refining, wood pulp) are often material-oriented.
- If an industry has a material index of more than 1, it is likely to be material-oriented.
- The coalfields shaped the industrial geography of nineteenth-century Europe.
- Despite the decline of coal as a source of fuel, coalfields have remained important centres of industry in the twentieth century owing to inertia and acquired advantages.
- In economically developed countries, electrical energy is ubiquitous: this development has contributed towards the footlooseness of many modern industries.
- Both Weber and Hoover identify the optimal location for a factory as the point of least transport costs.
- Hoover's analysis of transport costs recognises the distinction between terminal and haulage costs.
- In contrast to Weber, Hoover's theory shows that intermediate locations between materials and markets are rarely least-transport-cost locations.
- Hoover's analysis explains why industries locate at break-of-bulk points such as ports and deep-water terminals.
- The theory of spatial profit margins says that firms will always locate within the spatial profit margins, though not necessarily at the point of maximum profit.
- The changing locational patterns in the Lancashire cotton industry 1780–1840 can be explained by changing sources of energy.
- Transport costs are affected not only by distance, but by transport media, the nature of the freight carried and by pricing policies.
- Unit costs fall with increasing plant size. These savings are known as economies of scale.
- Global steel production is dominated by MEDCs.
- Economies of scale are vital to the steel industry and explain the organisation of production in very large plants.
- The locational patterns of the steel industry have experienced several shifts in the last 200 years in response to technological change.

4 Markets

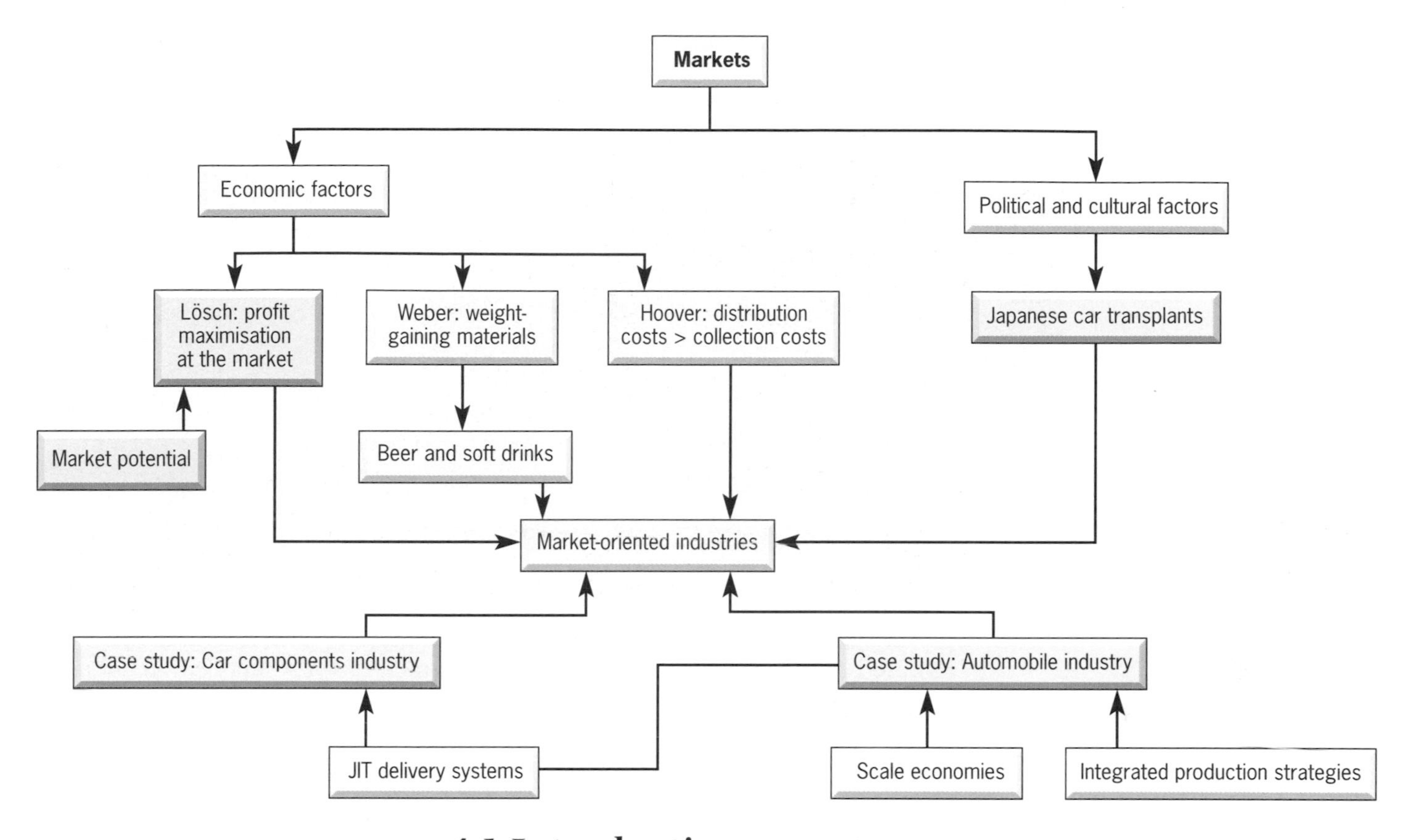

4.1 Introduction

Markets are places where manufactured goods are sold. They vary in scale from continental-sized areas to countries, regions and towns. In this chapter we shall investigate the importance of markets to the location of manufacturing firms.

4.2 Markets and locational theory

Lösch's theory (see page 50) said that firms should locate at markets. He reasoned that the market was where sales, and therefore profits, were maximised. Weber (see page 29), on the other hand, said that firms should only choose a market location when the market was at the point of lowest transport cost. This would occur when a product gained weight or volume in the manufacturing process.

The UK's brewing industry

The brewing industry uses malt (made from barley) mixed with water, yeast, hops and sugars to produce a number of different beers. At a global scale the industry is strongly market-oriented. In the UK only 2 per cent of production is exported, while imports account for 7 per cent of consumption.

In the early twentieth century, high distribution costs tied the brewing industry to local markets. Almost every small town had its own brewery. However, the introduction of large 38-tonne tanker trucks and the expansion of the motorway and dual-carriageway network have caused distribution costs to fall steeply. The improved road network means that a single brewery can now serve a much wider market area.

?

1 Why do you think that beer plays so small a part in international trade? Use the information in Table 3.3 in your answer.

2 Read through the section on the UK's brewing industry then study Figure 4.1

a Describe the site of Whitbread's brewery at Samlesbury.

b Using the evidence of Figure 4.1 explain two advantages of this site.

c Look at section 11.2 and comment on the Samlesbury site in relation to the urban–rural manufacturing shift.

Figure 4.1 Whitbread's Samlesbury brewery

Falling transport costs have allowed breweries to increase in size and reduce unit costs through scale economies. Thus, instead of breweries supplying localised areas, a large firm like Whitbread, which in 1999 supplied 16 per cent of the market, can serve the entire UK market from just two main plants (Fig. 4.2): Samlesbury in Lancashire serves the North, and Magor in Gwent serves the South. Both are large, modern plants with good motorway access.

Table 4.1 Markets and the location of the brewing industry in the UK

	Population (millions)	Employment in brewing (thousands)
Greater London	6.77	7.8
Rest of South-East	9.68	4.2
East Anglia	2.01	1.6
South-West	4.59	2.8
West Midlands	5.20	5.2
East Midlands	3.94	2.5
Yorks and Humberside	4.90	3.6
North-West	6.37	6.0
North	3.07	2.9
Wales	2.84	1.8
Scotland	5.11	4.8

Is the location of the brewing industry market-oriented? You can answer this question by completing the following exercises.

3 Plot employment in brewing (y-axis) against the population of each region (x-axis) as a scattergraph, using the data in Table 4.1.

4 Describe the relationship between the two variables.

5 Calculate the Spearman rank correlation coefficient (see Appendix Al) for the two variables. What does the coefficient tell you about the location of the brewing industry?

The soft drinks industry in south-west Ontario

Like brewing, the soft drinks industry is strongly market-oriented. Its main input is water, with syrup or concentrate accounting for the remainder. The product is heavy, weight-gaining and low-valued, and cannot bear high transport charges. Consequently, bottling plants are located close to their markets. A study of Coca-Cola bottling plants in south-west Ontario in Canada compared the actual distribution of plants (Fig. 4.3) with the optimal distribution (Fig. 4.4). (In this case the optimal pattern was the one which minimised total costs.) The study concluded that the plants were located in a cost-minimising manner, and confirmed that the industry was clearly market-oriented.

Figure 4.2 Whitbread breweries, closures and markets in the 1980s (*Source:* Watts, 1991)

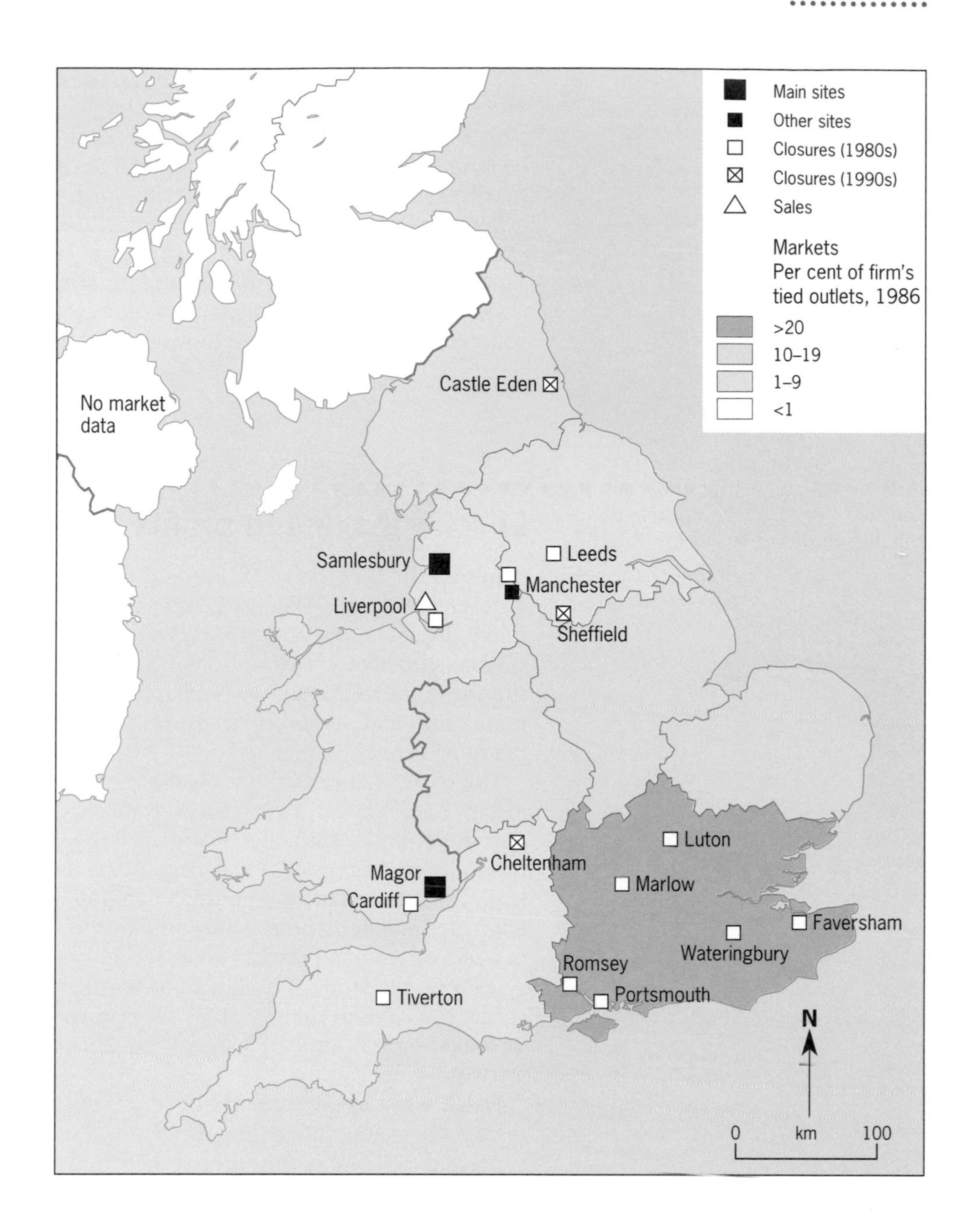

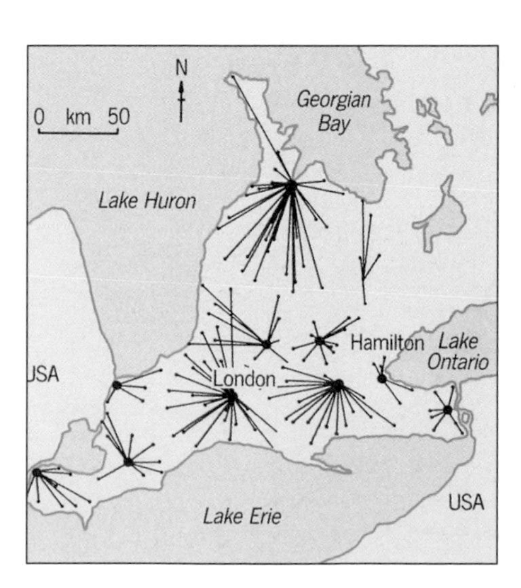

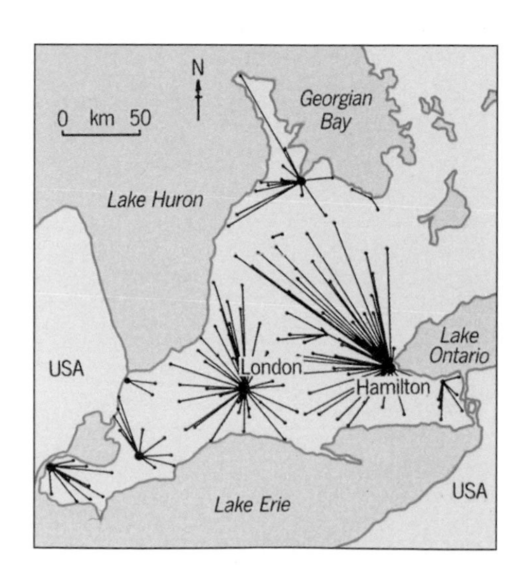

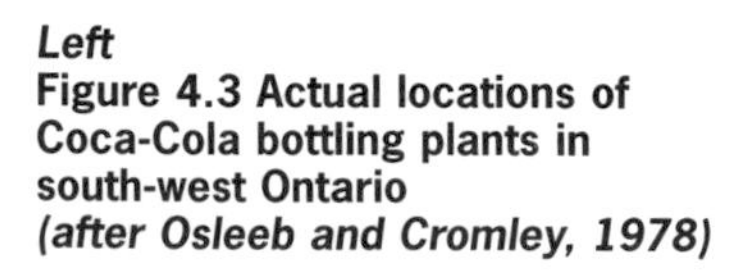

Left
Figure 4.3 Actual locations of Coca-Cola bottling plants in south-west Ontario
(after Osleeb and Cromley, 1978)

Right
Figure 4.4 Optimal locations of Coca-Cola bottling plants in south-west Ontario
(after Osleeb and Cromley, 1978)

Market locations and non-economic considerations
The location of Japanese vehicle assembly plants (or transplants) in Europe and North America suggests that nearness to the market is not always dictated by economic considerations. Although large companies are becoming more global, markets still differ in taste, especially for cars. Toyota, Nissan and others partly established production in Europe and North America in order to design their cars to match local tastes and fashions. In addition to production, all the big Japanese firms are currently setting up research centres in Europe and the USA. Soon, Japanese firms will build cars which are designed, as well as sourced, with components wholly from within Europe and the USA.

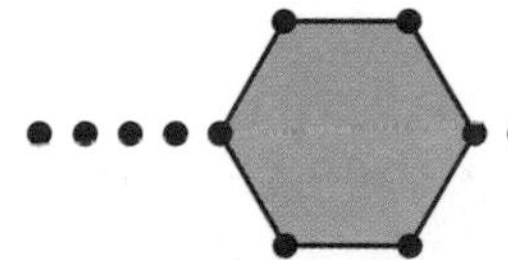

Lösch's maximum profits theory

Whereas the theories of Weber and Hoover were concerned only with spatial differences in costs, August Lösch said that the optimal location was one which maximised demand and profits. To predict the optimal location, Lösch made several simplifying assumptions. They are similar to those adopted by the theories that we looked at in Chapter 3. They include: a uniform distribution of population and resources; transport costs equal in all directions; and decisions made by economic man.

The basis of Lösch's theory is the modified demand curve (Fig. 4.5). As distance from a factory increases, transport costs (and therefore prices) rise, and the quantity sold falls. Eventually the price becomes too high to sell any production. This point (F) represents the limit of the market area served by a firm. Lösch adapted the modified demand curve into a cone, by rotating F about P. Total demand is now equal to the volume of the cone, of which P,F,Q is a section. Similar market areas served by competing firms would form a system of overlapping circles. In order to avoid overlap, and fill all the available space, the circular market areas are reduced to a hexagonal pattern (Fig. 4.6). There is a strong similarity here with Christaller's central place theory in settlement geography.

Lösch went on to argue that every industry had a unique network of hexagonal market areas. He also showed that, when each network is superimposed one on another and rotated around a common centre, a distinctive pattern emerges. Six 30-degree sectors have large numbers of plants, and six have relatively few. The former he referred to as city-rich sectors, the latter as city-poor sectors. This is Lösch's economic landscape (Fig. 4.6). Given the objective of maximising sales, the optimal location for a plant is at the market. In Lösch's economic landscape, this is in one of the city-rich sectors.

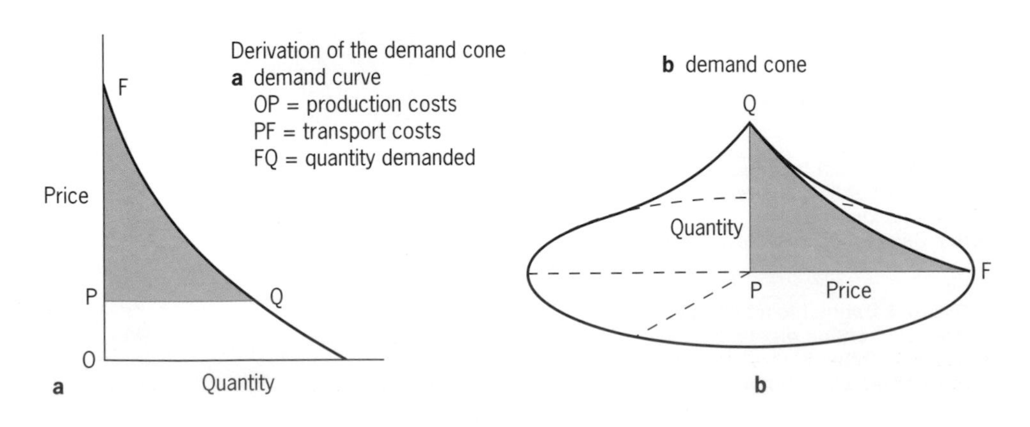

Figure 4.5 Lösch's demand cone

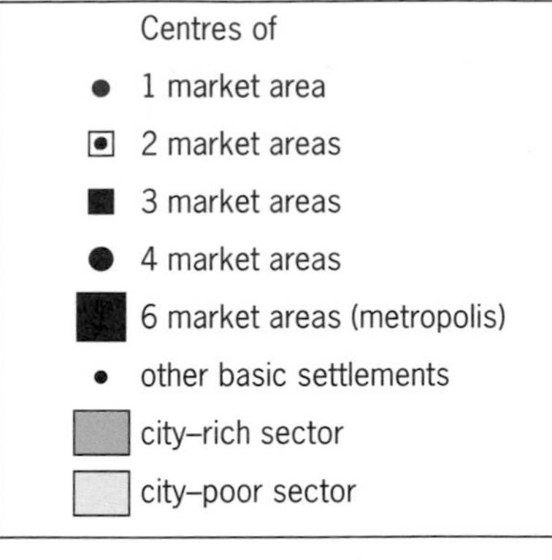

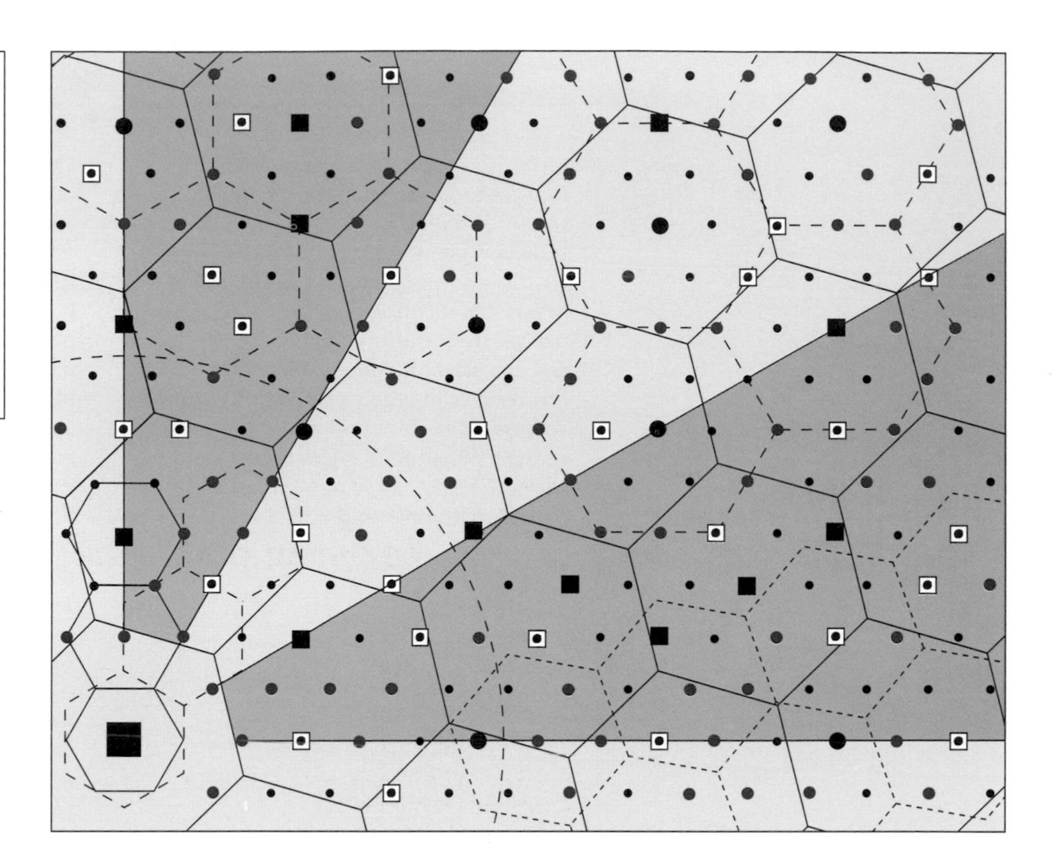

Figure 4.6 Löschian landscape *(after Saey, 1973)*

Calculating market potential

One of the simplest ideas for measuring access to markets is the **market potential** model (Fig. 4.7). Based on the gravity concept, it assumes that the attraction (A_i) at any one point is proportional to the size (S_j) of a region, and inversely proportional to the distance between it and that region (D_{ij}):

Equation 1

$$A_i = \left[\frac{A_j}{D_{ij}^k}\right]$$

Equation 2

$$A_i = \sum_{i=1}^{10} \left[\frac{S_j}{D_{ij}^k}\right]$$

where k is an exponent representing the frictional effect of distance. The sum of all influences at one point is termed the potential.

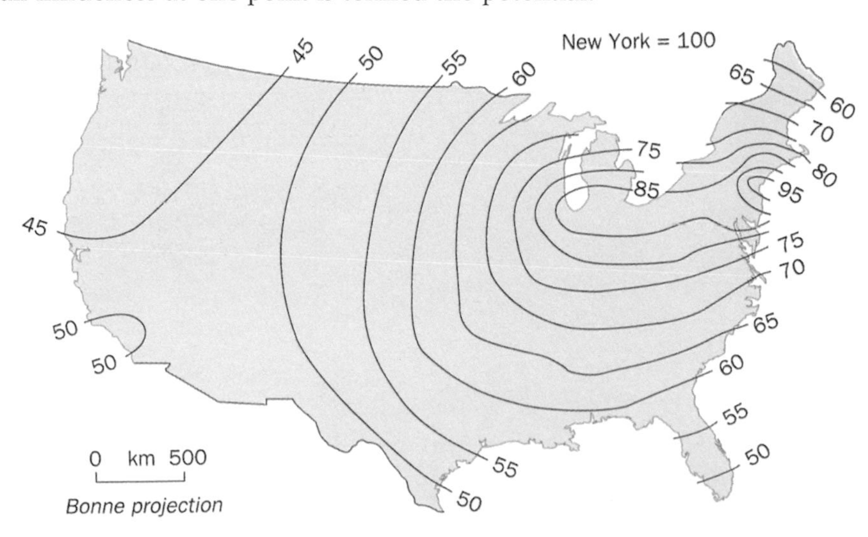

Figure 4.7 United States market potential map

?

6 Working as a class group, construct a market potential map for Britain, using the information on GDP in each of the standard planning regions (Table 4.2), and road distances between the centres of gravity in each region (Table 4.3). The procedure is outlined below.
a Each student should select one of the ten centres of gravity (Table 4.2) which represents each region and calculate its market potential using equation 2. Assume that $k = 1$. For Newcastle (A_1), the calculation would be:

$$(A_1) = (2)\frac{49{,}334}{154} + (3)\frac{43{,}479}{260} + (4)\frac{23{,}233}{414} \ldots + (10)\frac{53{,}659}{240} + (1)\frac{28{,}260}{0.5 \times 154}$$

The last part of the calculation takes account of Newcastle's own region, the North. (For a city's own region, distance is usually expressed as half the distance to its nearest centre (i.e. Leeds) or region.)
b On an outline map of Britain, mark the position of each centre of gravity. Collect the market potential scores from the rest of the class, and mark these alongside the appropriate centres of gravity on your map.
c Standardise the scores by dividing the highest score into 100, and weighting each by this value.
d Choose three or four isoline values which cover the range of potential scores, and draw these by interpolation (i.e. like fitting contours to spot heights) on your map.
e Write a paragraph to describe the distribution of the potential market for manufacturing industry in Britain.
f Where is the optimal location for a market-oriented industry in Britain?

Table 4.2 Gross domestic product (GDP) by region and centres of gravity

Region	Regional GDP (1996)	Centre of gravity
North	28,260	(1) Newcastle upon Tyne
Yorkshire and Humberside	49,334	(2) Leeds
East Midlands	43,479	(3) Nottingham
East Anglia	23,233	(4) Norwich
South-East	235,163	(5) London
South-West	50,539	(6) Taunton
West Midlands	53,659	(7) Birmingham
North-West	62,499	(8) Manchester
Wales	26,253	(9) Cardiff
Scotland	54,141	(10) Glasgow

Table 4.3 Shortest distance by road between centres of gravity (km)

	1	2	3	4	5	6	7	8	9	10
Newcastle	—	154	260	414	448	538	342	230	512	240
Leeds		—	115	280	310	410	216	70	378	344
Nottingham			—	197	205	299	94	115	262	450
Norwich				—	184	419	261	294	402	606
London					—	267	189	318	245	643
Taunton						—	203	333	136	658
Birmingham							—	139	173	467
Manchester								—	304	342
Cardiff									—	630

The automobile industry

At the global scale the automobile industry is market-oriented: most cars are sold near to where they are made. In 1911 Ford located its first assembly plant outside the USA, choosing to serve the European market from Manchester. The advantage of this move was simple: lower distribution costs.

Today Ford's operations are worldwide and its plants are located to serve specific markets. However, distribution costs are no longer very important. Instead, a market location has two advantages: it allows manufacturers to get round tariff barriers fixed by governments, and it gives them close contact with the demands and tastes of local markets. The latter is important, as the vehicles which meet the needs of Europeans are often different from those preferred by Americans or Japanese.

In the period since 1980 market locations have increased in attractiveness. This has been due to Japanese car makers undertaking huge overseas investment projects in North America and Europe. Both markets could have been supplied more cheaply with vehicles made in Japan. This underlines the crucial role of governments and consumer preference in the location of the modern automobile industry.

Global production

Automobile production is dominated by MEDCs, especially Japan, the USA and the countries of Western Europe (Fig. 4.8). Western Europe is the largest manufacturer (Fig. 4.9), followed by Japan and the USA (Table 4.4). However, Europe's leading car maker, Germany, produces only half of Japan's output.

Table 4.4 Car production and markets, 1997 (millions)

	Production	Home sales	Balance
Western Europe	13.733	13.417	+0.316
Asia–Pacific	13.131	8.367	+4.764
North America	8.152	9.333	–1.181

?

Study Table 4.4 and answer the following questions:

7 Which regions are (a) net exporters and (b) net importers of cars?

8 Describe the trade balance in cars in Western Europe.

9 What do the figures in Table 4.4 tell you about the global trade in cars?

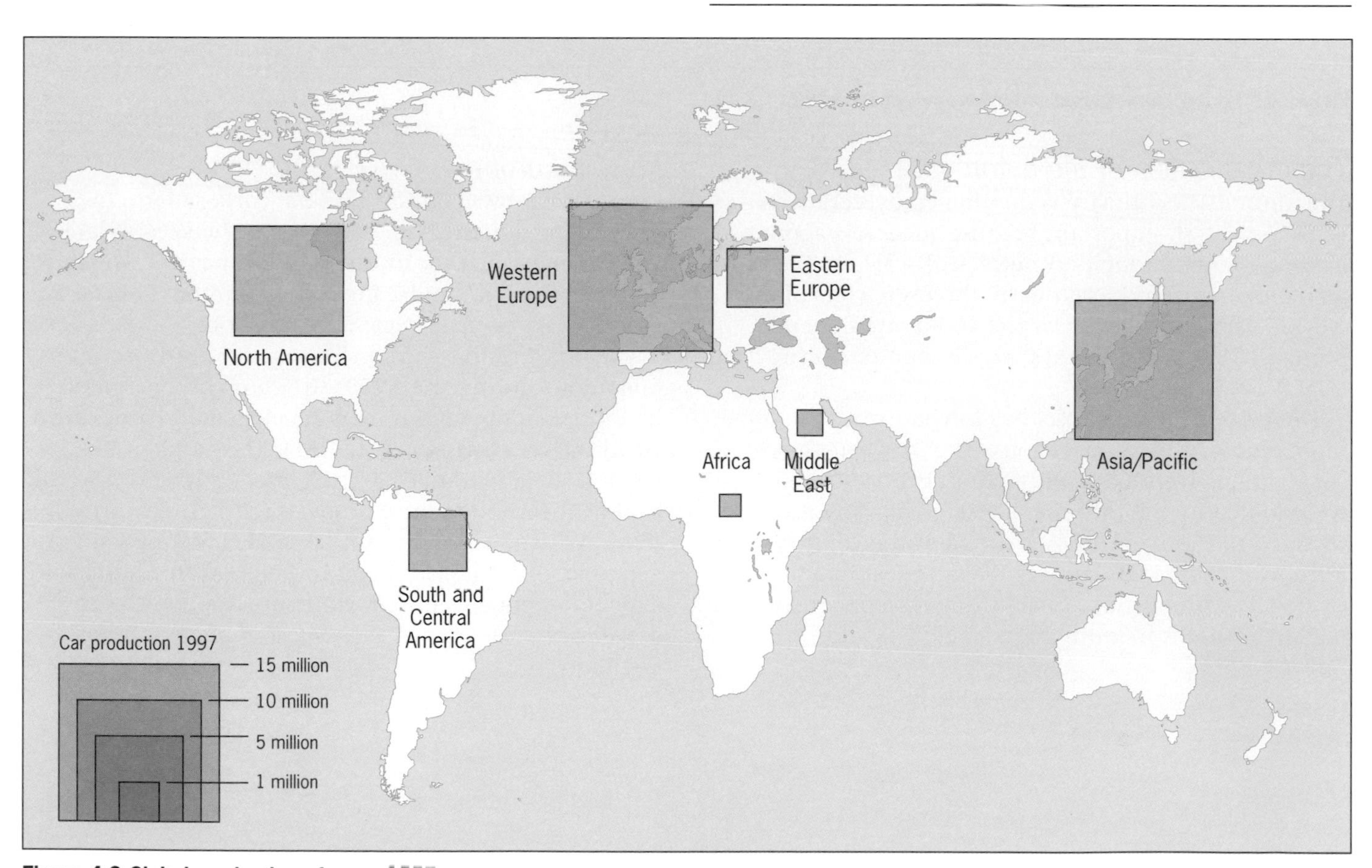

Figure 4.8 Global production of cars, 1997

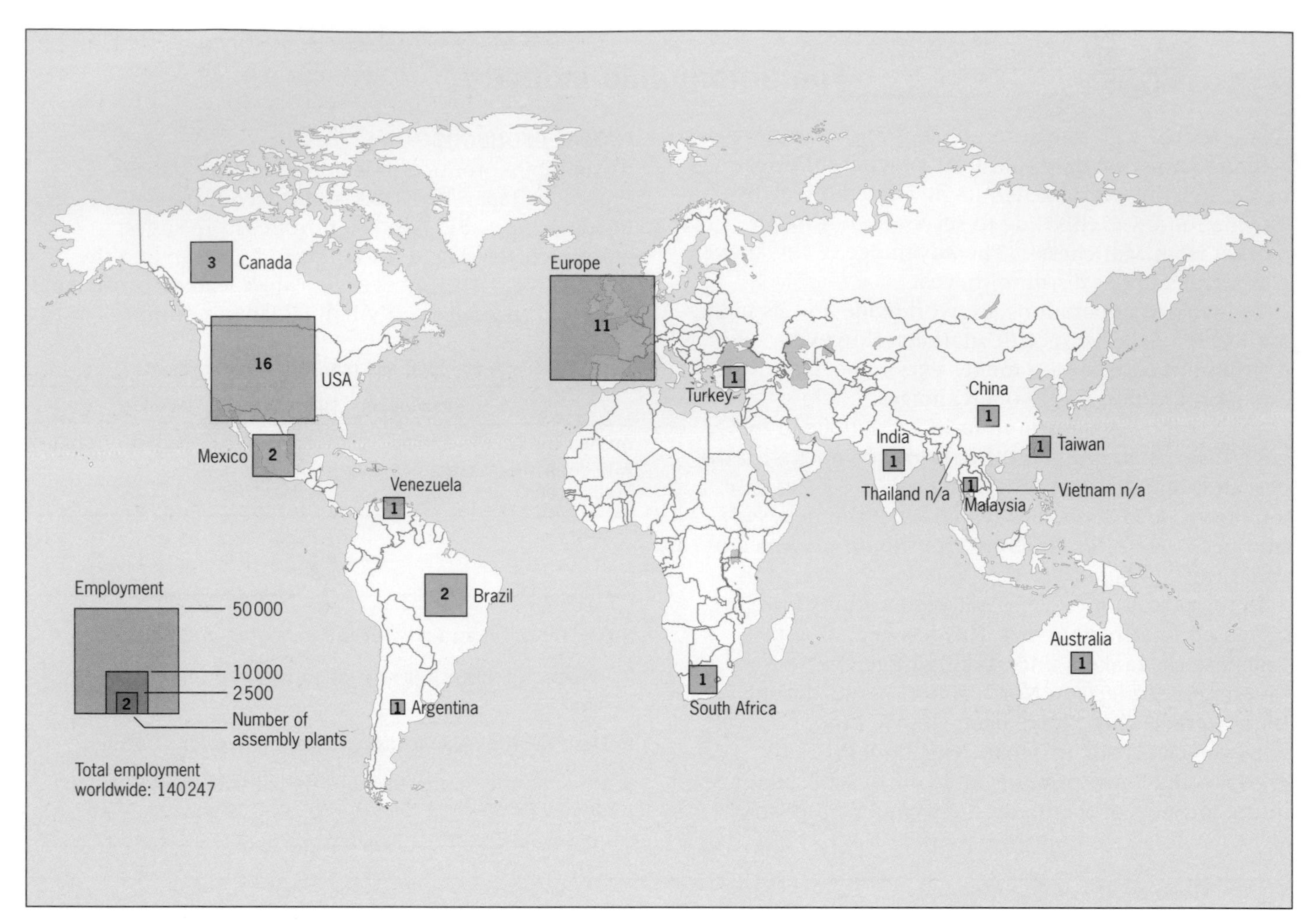

Figure 4.9 Ford Motors: globalisation of production, 1999

The organisation of the automobile industry

The automobile industry is dominated by very large firms. Size is all-important because success is about lowering costs through economies of scale. In 1998 just ten companies produced nearly three-quarters of the world's cars. Indeed, the largest company, General Motors (GM), is the world's largest manufacturing firm (Fig. 4.10).

The drive to reduce costs has forced firms to become more global in their operations. The first truly global corporations were Ford and GM. In the 1990s they were joined by the three Japanese giants: Toyota, Nissan and Honda. And by the end of the century European companies such as VW and Fiat had begun to globalise production, establishing assembly plants in North America, South America and eastern Europe.

Ford of Europe

Ford is a US-owned transnational corporation (TNC) with its headquarters in Detroit. It is the second largest vehicle manufacturer in the world. It has 16 assembly plants in the USA and 23 others in Europe, Central and South America, Asia, Oceania and Africa.

Western Europe is the main focus of Ford's overseas operations. In the late 1960s the company restructured its European operations, and developed an **integrated production strategy** (Fig. 4.11). Thus a plant like Genk in Belgium builds a single model (in this case the Mondeo) for the entire European market. The same strategy is applied to the sourcing of components. For example, the Bridgend plant in South Wales is the sole supplier of engines for all Escort models, which are assembled at Halewood (Merseyside) and Saarlouis in Germany.

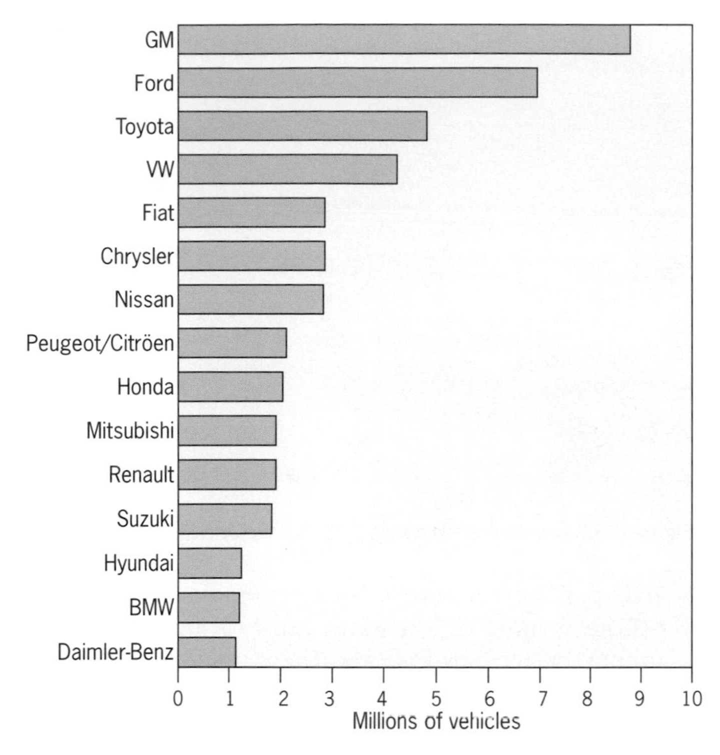

Figure 4.10 The world's leading motor vehicle manufacturers, 1998

The advantage of integrated production is lower unit costs through large-scale production. Ford has also favoured dual production (i.e. the same model built in two separate plants). This gives greater flexibility, allowing production shortfalls in one plant to be made up from elsewhere. GM has a similar integrated production strategy in Europe, based on assembly plants in Germany, Britain, Belgium and Spain.

Figure 4.11 shows the assembly plants and parts operations owned by Ford in Europe (Ford also sources parts from many other manufacturers not shown on Figure 4.11).

10 Despite Ford's integrated production strategy, its manufacturing operations in Europe show a degree of spatial clustering. Can you suggest an explanation for this?

11 Figure 4.11 shows that Ford's car components are sourced throughout Western Europe. What does this suggest to you about the importance of both transport costs and scale economies in the automobile industry?

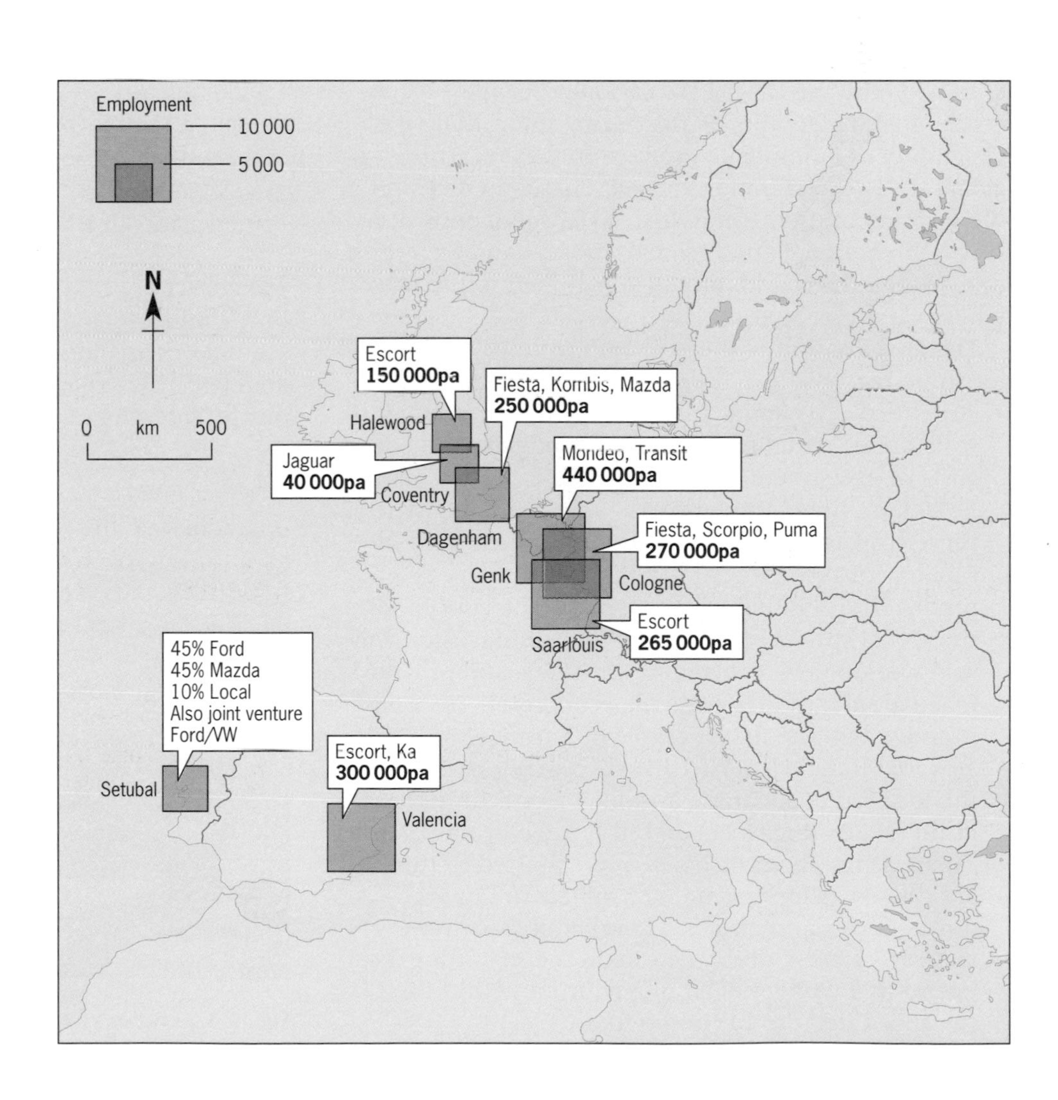

Figure 4.11 Ford's EU assembly plants

The importance of the automobile industry

Despite the relative decline of manufacturing industry in the post-industrial economies, the automotive industry remains a leading sector in many MEDCs. Globally, the industry employs 2.5 million directly in assembly plants and three times that number in parts manufacturers. In the UK, the automobile industry is worth £55 billion a year (or 5 per cent of GDP) and is a major employer, especially in the West Midlands. Its importance was underlined in 1999, when the UK government gave BMW financial support worth £110 million. The money was to have secured a £1 billion investment by BMW, to build a new generation of Rover cars at Longbridge in Birmingham. Without this backing BMW would probably have transferred production to Hungary. The attraction of Hungary was lower wages (less than one-sixth of those in the UK) and lower labour costs (e.g. no pensions, shorter paid holidays etc.).

In May 2000, however, BMW found that the Longbridge plant was no longer cost-effective. The Rover group was eventually purchased by the Phoenix Group for the nominal fee of £10, saving many jobs.

Globalisation of production

The trend towards globalisation – a world economy, dominated by the USA, East Asia and Europe – accelerated during the 1990s. During this decade the geography of automobile production was recast on a global scale. Globalisation caused a fundamental shift in the location of production. Instead of companies supplying national and regional markets and exporting cars around the world, production shifted to the places where cars were sold.

The globalisation of the automobile industry occurred for three reasons.

- To reduce costs and increase efficiency. Economies of scale are essential to volume car producers. The fixed costs of investment in plant, machinery, R&D and design are huge. Production is only profitable when these costs are spread over an output measured in millions of cars (experts say the minimum scale of production is 5 million cars a year). The concept of the global car – a standard model manufactured and sold worldwide – evolved in the 1990s and provided massive savings through scale economies. The Ford Mondeo was the first example of a truly world car.
- To avoid trade barriers. This was a major reason for the location of Japanese assembly plants in the USA in the 1980s (Fig. 4.12) and in Europe in the 1990s. Japanese cars built in Japan and exported to the EU are classed as foreign and are subject to import duties, tariffs and quotas. However, if Japanese companies build the same cars in the EU, using local labour and components sourced from within the EU, the cars qualify as EU in origin.

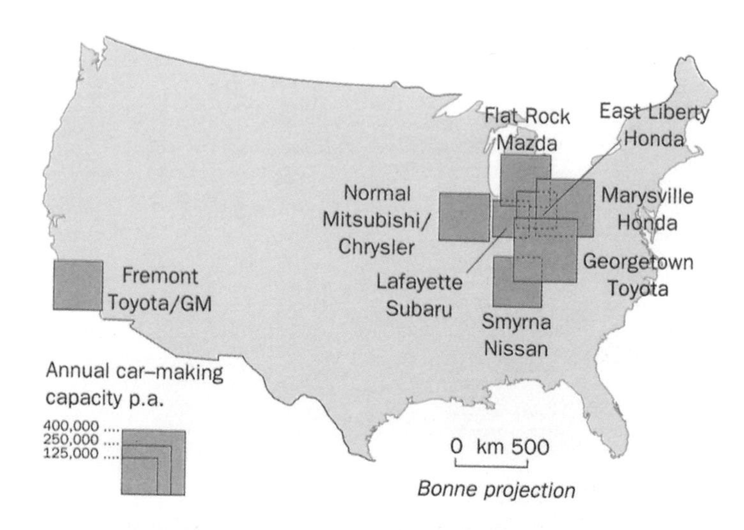

Figure 4.12 Location of Japanese car transplants in the USA

- To exploit new markets. Today's demand for cars in MEDCs is more or less static. Most car purchases simply replace existing cars. These 'mature' markets offer few prospects for growth. However, emerging markets in Central and South America, the Middle East and Asia give scope to expand production. This explains the flurry of investment by leading automobile manufacturers in countries such as Brazil, Argentina and Mexico during the 1990s. For example, car production in Brazil increased from 3.5 million to over 5 million between 1996 and 2000. By the end of the century European manufacturers such as VW, Fiat and Peugeot had all established transplants in Brazil.

The impact of globalisation in the UK

In the mid-1980s the UK automobile industry produced around one million cars a year. By 2000 output had doubled. This expansion was due largely to Japanese inward investment. Between 1984 and 1993 three

Figure 4.13 Honda's car transplant at Marysville, Ohio

major Japanese manufacturers – Nissan, Toyota and Honda – established transplants in the UK (Fig. 4.14). All were built on greenfield sites and all received significant financial support from the government of the day. The UK was seen as the ideal location from which to serve the EU market. It had a skilled workforce; relatively low wages; a friendly government, which encouraged foreign investment; an excellent transport and communications infrastructure; and generous grants to assist foreign investors, especially in peripheral regions.

New production methods

Lean production and just-in-time

The 1990s saw the adoption of Japanese **lean production** methods by most automobile manufacturers. These methods had given the Japanese industry a huge competitive advantage in the 1980s. Lean production involved closer cooperation between car assemblers and parts suppliers and the outsourcing

?

12 Study the photograph of the Nissan car assembly plant at Sunderland (Fig. 4.15).

a From the evidence of Figure 4.15 identity two features of the site.

b Explain how these features give advantages to the modern automobile assembly industry.

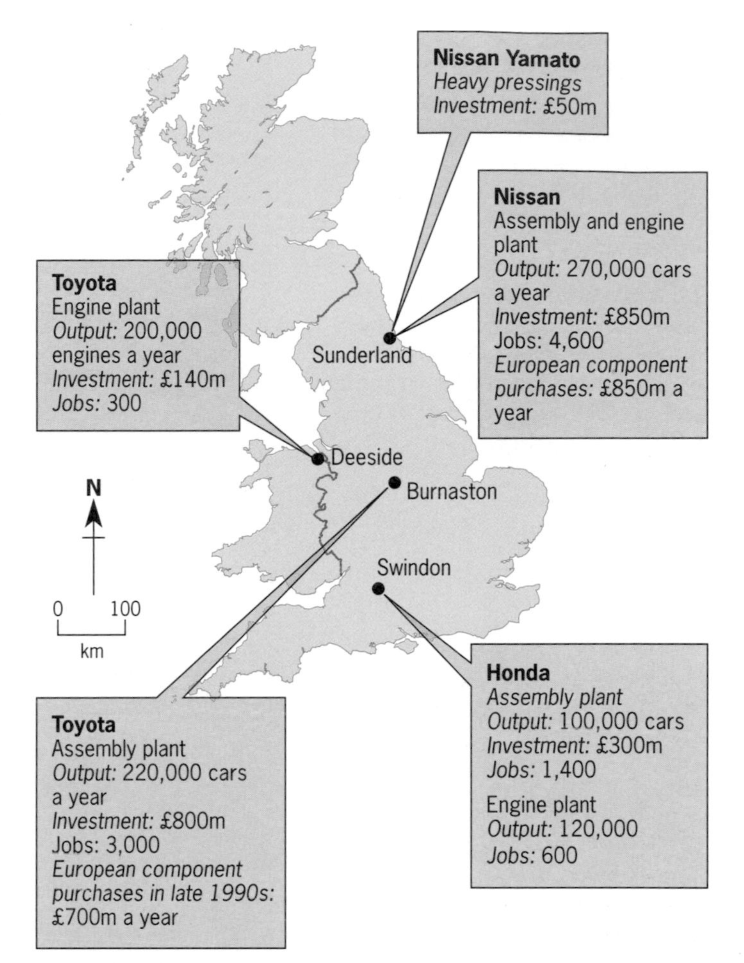

Figure 4.14 Japanese car makers' investment in the UK

Figure 4.15 Nissan's car transplant in Sunderland

?

13 Eighty per cent of the content of Nissan cars made at Sunderland have to be sourced in the UK. Study Figure 4.16 and explain how the North-East has benefited from the car components industry.

14 What is your attitude towards inward investment to Britain by Japanese car makers? Consider this issue by outlining your own values and beliefs on the subject. To help you arrive at a viewpoint, use the framework provided in Table 2.3. Discuss your view in debate with other students in your class.

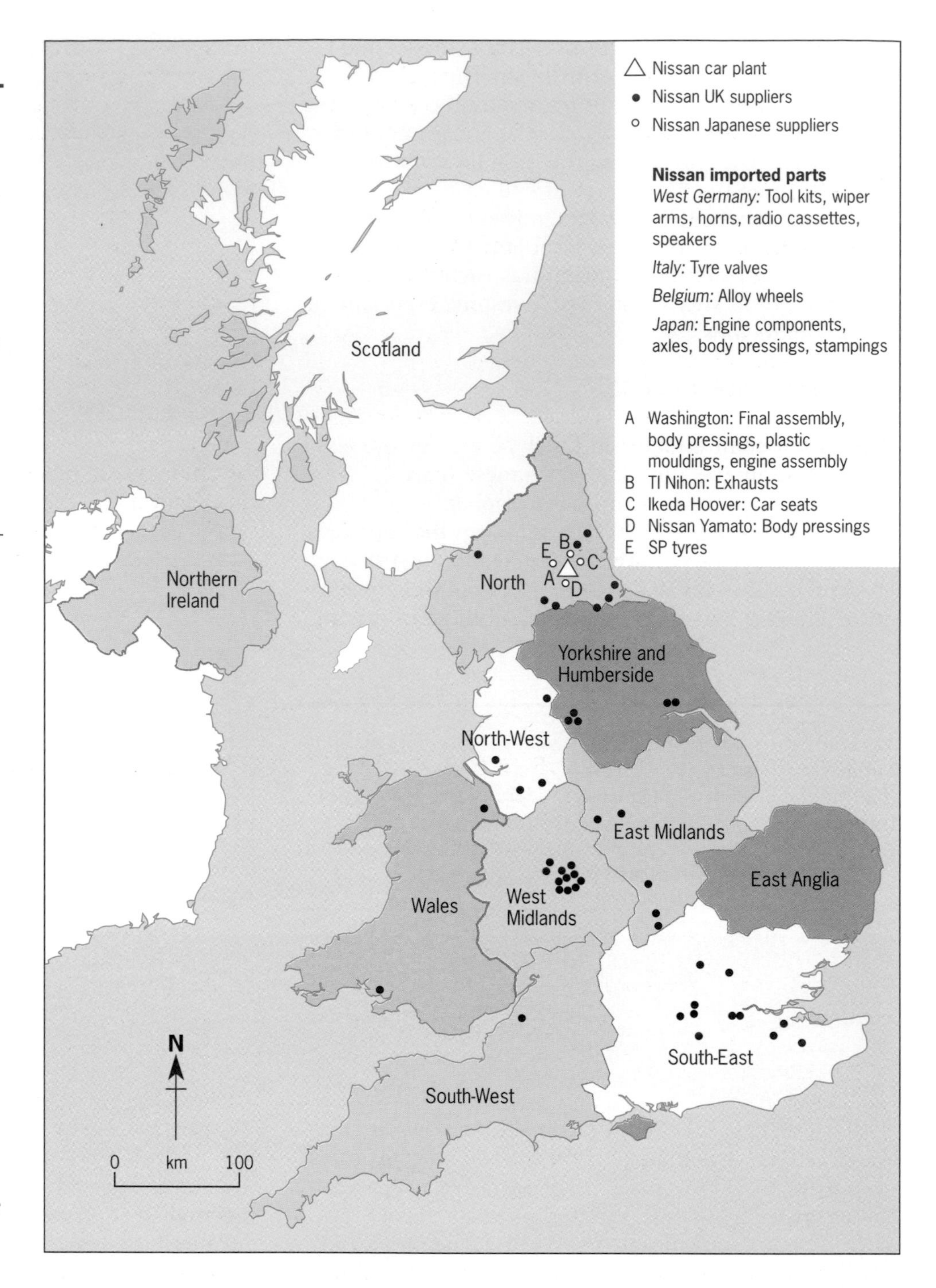

Figure 4.16 Nissan's principal supplier locations in the UK (*Source:* Peck, 1990)

of parts on a large scale. Assemblers held very small stocks of parts. Nissan's factory at Sunderland, for example, holds parts made by European suppliers for less than one day before they are used on assembly lines. Meanwhile, orders for parts are made on average just five days before they are needed.

By keeping only minimal stocks of components and ordering them as needed, assemblers can cut costs significantly. However, such a system depends on high-quality components with zero defects and their reliable delivery to assembly plants **just-in-time** (JIT). It also requires close collaboration and good communications between supplier and consumer. JIT is a feature of the motor vehicle industry in Tokyo and in the Toyota City region in Japan where it has produced a distinct geography of linked clusters of assemblers and suppliers.

While JIT has produced some clusters of parts suppliers around Nissan's assembly plant in North-East England (Fig. 4.16), it has not created large interlinked clusters. There are three possible reasons for this. First, parts suppliers in the UK existed before the establishment of Japanese transplants. Second, UK parts suppliers are more autonomous than their

Japanese counterparts. And third, the motorway network is less congested than in Japan, making geographical proximity between assemblers and parts suppliers less important.

Flexible production

Production became more flexible in the 1990s. Increasingly, manufacturers had to incorporate customisation into their mass production systems. At Nissan's Sunderland plant literally hundreds of different variants of the Primera and Micra models are produced to meet demand at short notice. Henry Ford's famous statement – 'you can have any colour you want so long as it's black' – summed up the age of mass production. The early twenty-first century will be the age of mass customisation.

Big is beautiful

Economies of scale are the driving force behind change in the global automobile industry. During the 1990s smaller companies such as Rover found it harder to compete. As a result, a large number of mergers and takeovers occurred. BMW acquired both Rover and Rolls Royce; Chrysler merged with Daimler-Benz; and Ford acquired Jaguar. Collaboration and joint projects were also popular. These trends are sure to continue in the future.

Overcapacity

Overcapacity was the major problem facing automobile manufacturers at the end of the 1990s. Most companies operated at only 70 or 80 per cent of their capacity. The establishment of new transplants in the 1980s and 1990s at market locations contributed to overcapacity. So too did the mature car market in MEDCs. Overcapacity even led to the closure of modern assembly plants, such as Renault's factory at Vilvoorde in Belgium in 1997. With the emergence of new manufacturers in Asia (e.g. South Korea and Malaysia), overcapacity is likely to remain a problem in the future.

The clustering of automotive parts suppliers in Belgium

Belgium accounts for 9 per cent of the EU car industry. Four assembly plants operated by Ford, Opel, VW and Volvo provide direct employment for 30,000 workers. In addition, a further 260 automotive parts suppliers provide 15,000 jobs. Belgium's advantages for car assemblers and parts suppliers include good access to EU markets, a skilled workforce and access to major seaports such as Antwerp and Rotterdam.

One feature of the Belgian car industry is its spatial concentration in Flanders. The four assembly plants are all within an hour's drive of Brussels. Clusters of parts suppliers and sub-contractors have grown in this area to provide a JIT network for the assembly plants. The government has promoted interlinked clusters of suppliers and sub-contractors (Fig. 4.17). Two special JIT industrial areas, at Antwerp and Limburg, aim to attract parts suppliers by offering grants and providing industrial premises and sites. The advantages of JIT networks include: lower costs to assemblers because of reduced stocks of parts; more space on the factory floor; higher-quality parts and punctual delivery by suppliers; and greater flexibility of production. If a shipment arrives late or is defective, the assembly line shuts down and the supplier risks losing the business.

Proximity between assembly plants and sub-contractors in Belgium is long established. In 1991 a Swedish parts supplier – Perstorp AB – built a new factory in Ghent, just 4 kilometres from Volvo's assembly plant. The location allowed Perstorp to provide JIT deliveries of consoles and sound insulation material. The government gave financial support to the venture with grants for plant and machinery (worth 9 per cent of total costs) and the Flanders Development Agency also provided assistance. In the late 1990s the US-owned car components firm Collins and Aikman acquired the Perstorp factory. Collins and Aikman specialises in carpeting and upholstery for the automotive industry. It operates JIT systems at its six factories in the USA as well as in Ghent. The clustering of the automotive industry in Flanders continues to gather pace. Between 1996 and 2000 major parts suppliers from Germany, France, the USA, Spain, Sweden and Switzerland built factories in the region.

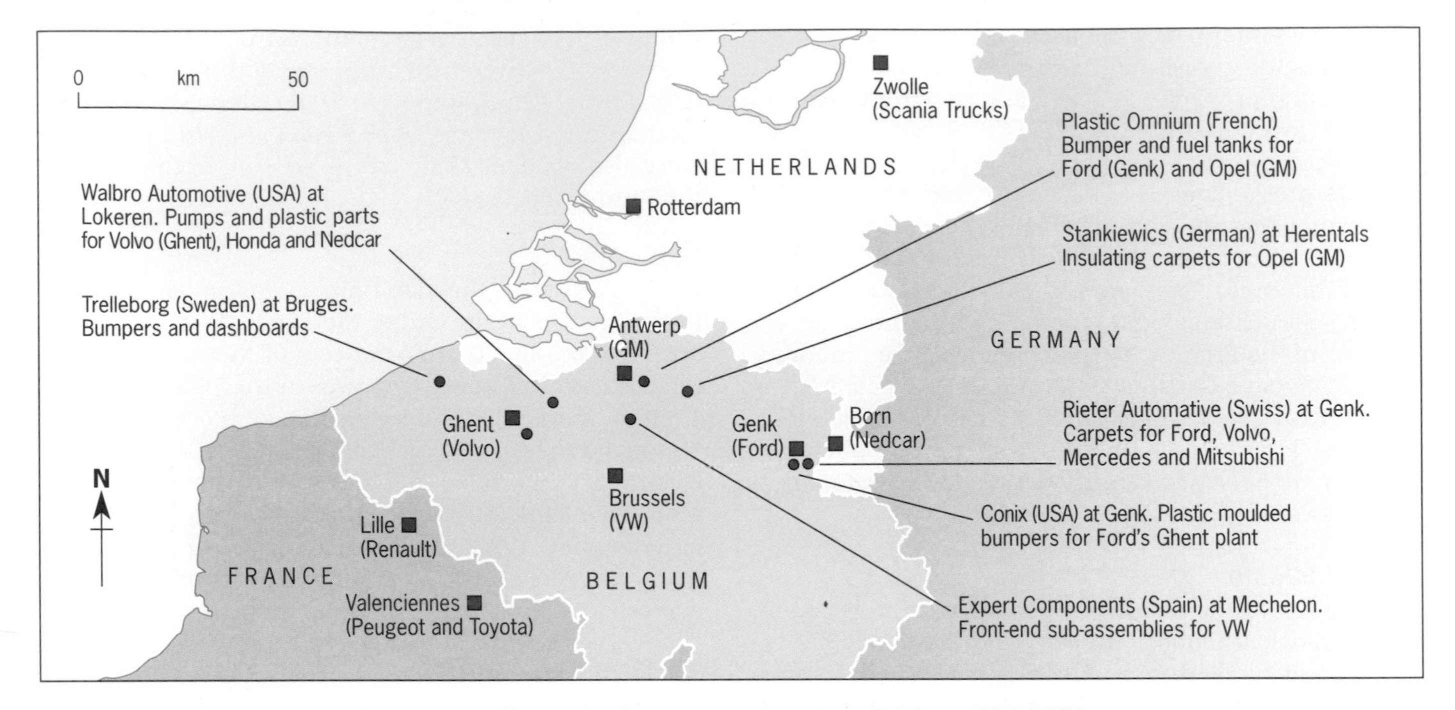

Figure 4.17 Location of automotive parts suppliers and sub-contractors in nothern Belgium, 1996–2000

?

15 Outline the advantages of JIT for (a) assembly plants and (b) parts suppliers and contractors.

16 Describe how JIT is changing the geography of the motor vehicle industry.

Summary

- Lösch's theory says that the optimal location for an industry is at the market where sales and profits are maximised.
- Industries which manufacture weight-gaining products, such as beer and soft drinks, should locate at the market to minimise transport costs.
- At a global scale the automobile industry is market-oriented.
- The automobile industry is dominated by very large, transnational organisations.
- Large firm size is important in the automobile industry in order to achieve scale economies and lower unit costs.
- Smaller automobile manufacturers, struggling to compete in the world market, have been forced into mergers and joint ventures with larger firms in order to survive.
- The automobile industry is increasingly organised on a global and continental scale.
- Inward investment by Japanese car manufacturers in the UK has been followed by the arrival of Japanese parts manufacturers.
- Market locations may be chosen for non-economic reasons: to cater for local tastes and preferences, and to avoid trade restrictions.
- Just-in-time (JIT) delivery systems favour the geographical concentration of assembly plants and parts manufacturers.

5 Labour and capital

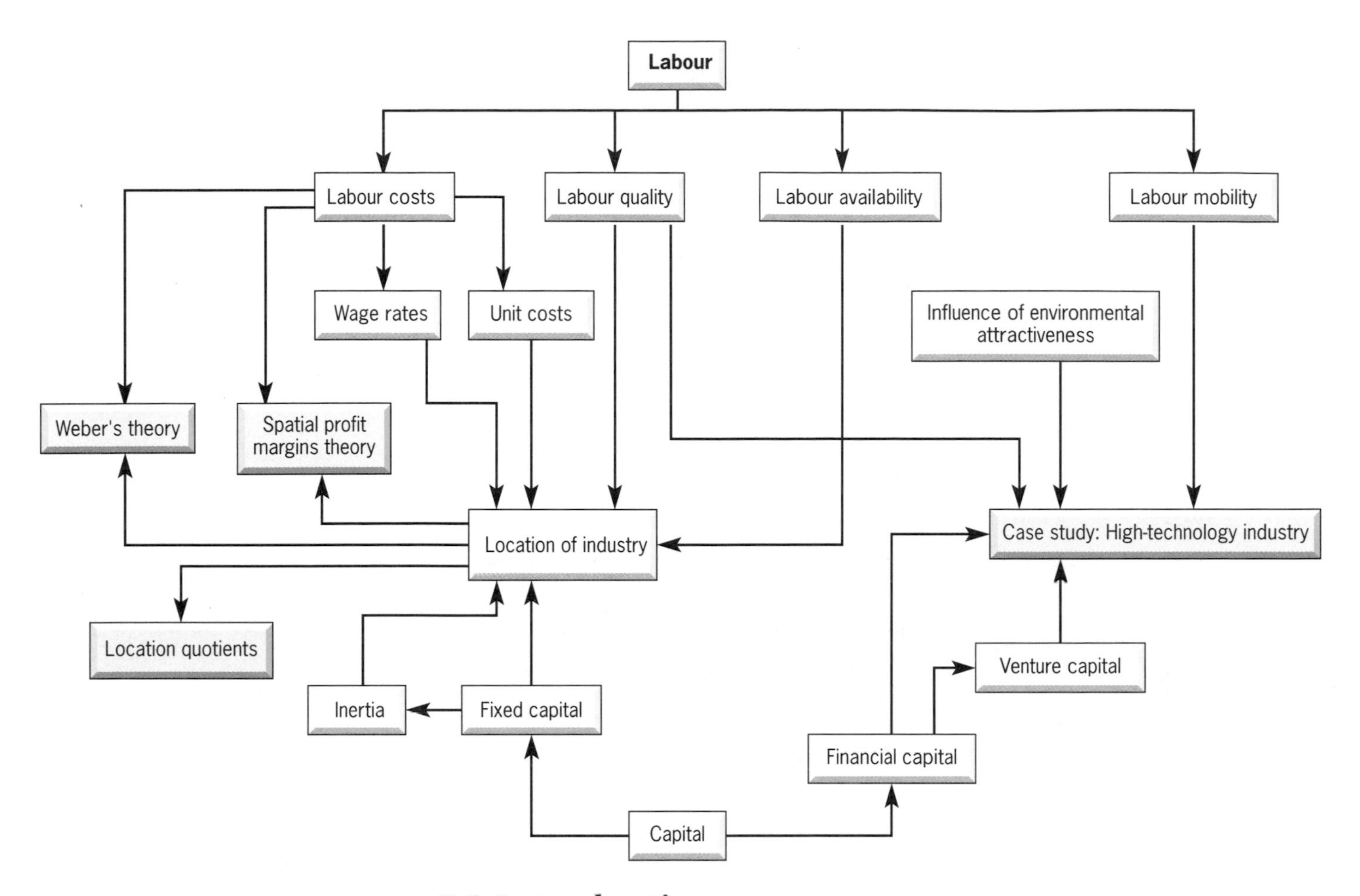

5.1 Introduction

Labour and capital are essential to manufacturing. Both vary geographically in cost and availability. In addition, labour shows spatial variation in its quality. In this chapter we shall look at the effect of labour and capital on the location of industry.

5.2 Labour costs

On average, labour accounts for nearly one-quarter of total costs in manufacturing industry. At the extremes, labour costs vary from around 8 per cent of total costs in the food industry, to 40 per cent in a research-based industry like defence.

The importance of labour costs is acknowledged in location theory. Weber showed how an optimal location could be one which minimised labour costs (see page 29). In the spatial profit margins theory (see page 32) a firm may locate outside the normal spatial profit margins, if low labour costs offset the higher production and transport costs found there.

The global scale

Economists view labour costs in two ways: as wage rates (Fig. 5.1), and as unit costs (Fig. 5.2). Wage rates vary enormously at the global scale. Wages are high in the economically developed world, and low in LEDCs. These differences have a direct influence on the locational decisions of large transnational

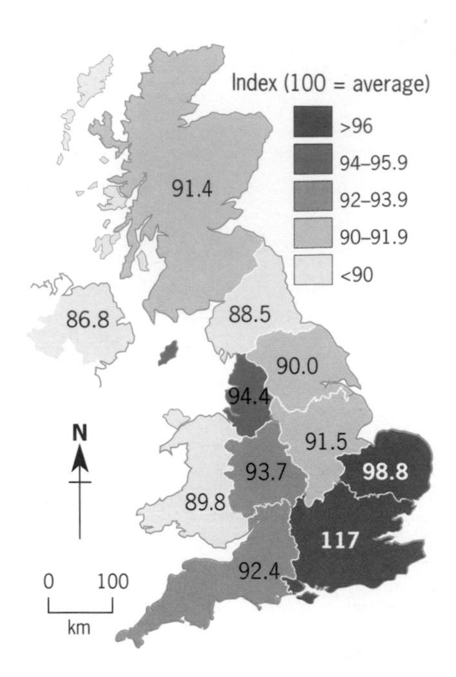

Figure 5.1 Regional average earnings in the UK

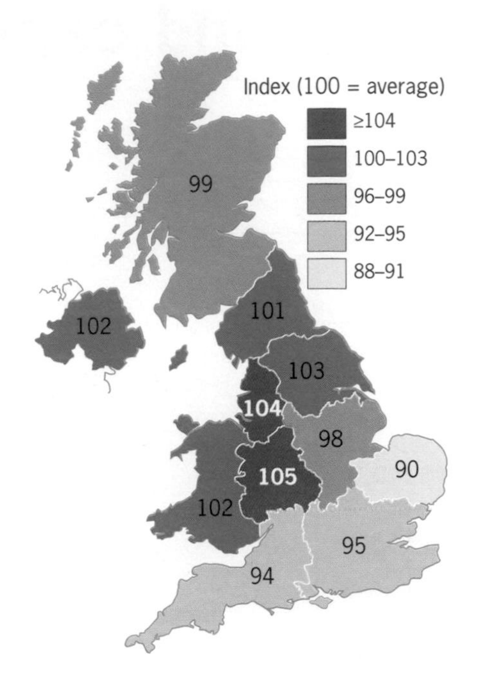

Figure 5.2 Regional unit labour costs in the UK

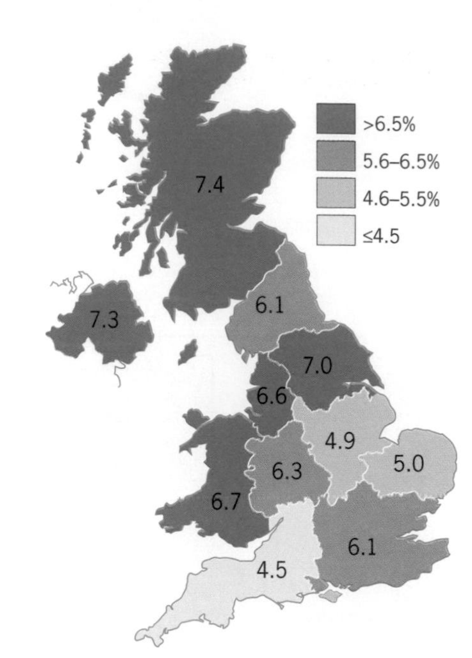

Figure 5.3 Regional unemployment in the UK, Spring 1998

companies (TNCs). In the past 30 years, many labour-intensive industries (especially consumer electronics making TVs, radios, cassettes, etc.), were relocated from the USA and Japan to low-wage countries in Asia. Initially, newly industrialising countries (NICs) like South Korea, Taiwan and Singapore were favoured. But as wages in these countries rose, TNCs moved to even lower wage economies such as Thailand and Indonesia. For example, wages in China in the late 1990s were only one-twentieth of those in Taiwan.

The continental scale: the EU

Large geographical differences in wage rates are also found at the continental scale. In the EU, if we take average wage levels in manufacturing as 100, then Denmark has the highest wages at 175, and Portugal the lowest at 25. It was this kind of difference that persuaded General Motors (GM) to locate a plant making ignition systems in Portugal in 1990. Apart from Portugal's low wages, GM was also attracted by the high productivity and good industrial relations of Portuguese workers.

The following exercises are based on Figures 5.1–5.3.

1 Describe the geographical pattern of unemployment, wage levels and unit labour costs in the UK.

2 Test the hypothesis that levels of unemployment in the UK are related to unit labour costs rather than wage levels. To do this,
a Draw two scattergraphs, plotting unemployment against unit labour costs and wage levels.
b Calculate the Spearman rank correlation coefficients (see Appendix A1) for the two sets of variables.
c State your conclusions in a couple of paragraphs.

The national scale: the UK

The smaller the scale, the smaller the geographical difference in wage rates. In the UK, taking the average national wage level as 100, regional wages range from 87 in Northern Ireland to 117 in the South-East (Fig. 5.1). However, the differences shown in Figure 5.1 take no account of the industrial structure of the regions. For example, in areas with many low-wage industries, wage levels are automatically depressed.

The relatively small regional differences in wages in the UK are explained by national pay bargaining. Thus, in multi-plant firms wage levels are usually the same in all plants, regardless of location. As a result variations in wage rates are often greater between firms than between regions.

In the UK regional differences in wage rates are too small to influence industrial location (Fig. 5.4). More important than wages are **unit labour costs**. These costs relate wage levels to output or productivity. After all, most firms are concerned less with what they have to pay their workers than with what they get in return. Thus, a region of low wages may discourage

Figure 5.4 Regional variations in labour costs in the UK

investment if its workforce has a poor record of productivity and high unit labour costs.

Labour costs are often higher in urban than in rural areas. This is due to greater labour turnover in urban areas where workers have more job opportunities. When turnover is high, firms have extra costs through recruitment, training and lower productivity levels.

5.3 Labour quality

Labour skills are usually more important to firms than labour costs. As automation has taken over many manual jobs, the demand for low-wage labour has decreased. In 1989, Toyota chose Derby instead of Humberside and South Wales for its first European assembly plant. Although unemployment in these regions was higher than in Derby, simple labour availability was not important to Toyota. What did impress Toyota was the engineering skills of the Derby workers. The company was also interested in the number of workers living within 45 minutes' commuting time of the plant, and the absence of competition for labour from other Japanese firms.

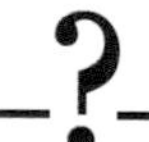

3 According to Figure 5.4, which regions are most attractive and which are least attractive for industry? Give a short explanation for this.

Highly skilled scientists, engineers and technicians are crucial to high-tech industries, which depend on R&D and product innovation. Most high-tech industries locate in areas where these skilled people are found. In the UK, this usually means the M4 and M11 corridors, and central Scotland. There is a parallel here with traditional nineteenth-century industries like shipbuilding and pottery. They became strongly localised in regions with specialist workforces. We call such a concentration of skills in particular regions the **sectoral spatial division of labour**. It exists at an international as well as a national scale. In the EU, for instance, France, Germany and the UK are favoured for engineering skills, rather than Italy, Greece and Spain.

Contrary to locational theory, the locations chosen by high-tech firms in the UK are in no sense least-cost. Indeed, they are among the most expensive in the country. However, the high costs of labour, rents and local taxes are more than offset by the quality of the workforce.

?

4 What relationship would you expect to find between unit labour costs and the frequency of industrial disputes? Present these data as a scattergraph.

5 Using the information in Figures 5.2 and 5.5 and the Spearman rank correlation coefficient, test the relationship between unit labour costs and the frequency of industrial disputes in the UK's standard planning regions. Comment on your result.

6 Study Table 5.1.
a Test the hypothesis (using the Spearman rank correlation) that trade union membership and employment in manufacturing are related.
b Suggest possible reasons why there might be a relationship between union membership and employment in manufacturing.

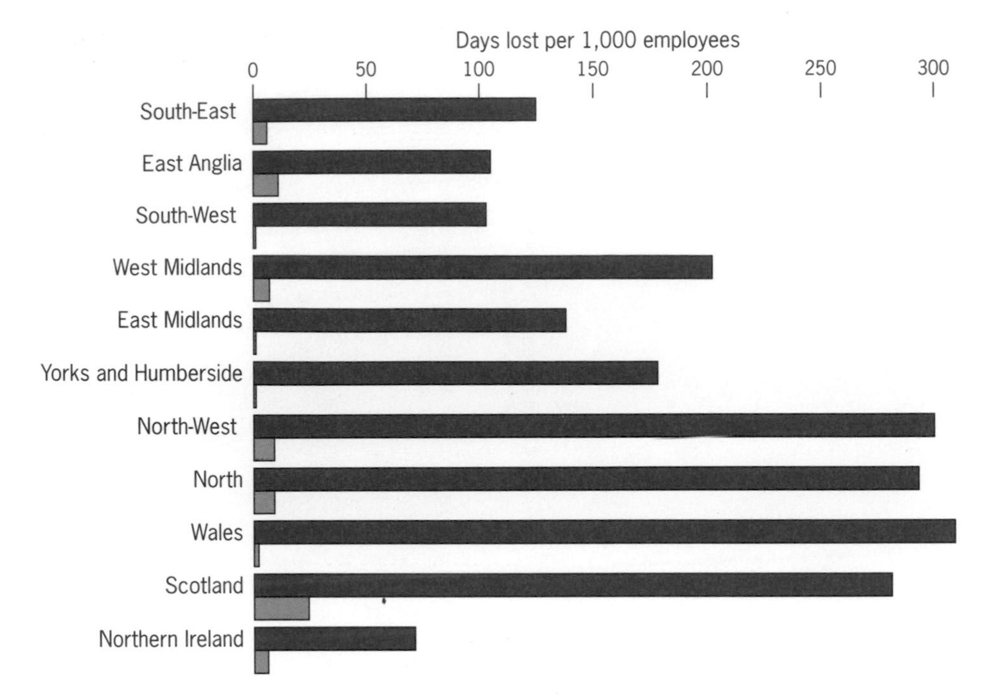

Figure 5.5 Regional pattern of industrial disputes in the UK, 1989 (red) and 1998 (blue). (*Source: Labour Market Trends and Regional Trends*)

Good labour relations are an important ingredient of labour quality. Until the 1980s the older industrial regions, heavily dependent on smokestack industries and a male workforce, were strongly unionised. This is evident in the regional figures for industrial disputes for 1989 (Fig. 5.5). In some regions the number of days lost to strikes exceeded 250 per 1000 employees. Yet by the late 1990s industrial relations had been transformed (Fig. 5.5). In many regions strikes were practically unknown. Among the factors which explain this transformation are: deindustrialisation in the 1980s and the decline of heavy manufacturing industries; the growth of female employment, part-time

Figure 5.6 Not all demonstrations are in support of strikes – many are protests against unemployment

Table 5.1 Trade union membership and employment in manufacturing in UK regions

	% Union membership
North	40
North-West	35
Yorkshire and Humberside	33
East Midlands	29
West Midlands	31
East Anglia	23
London	25
South-East	22
South-West	26
Wales	41
Scotland	35
Northern Ireland	40
	% Employment in manufacturing
North	30.26
North-West	26.48
Yorkshire and Humberside	26.91
East Midlands	30
West Midlands	30.67
East Anglia	23.04
London	11.24
South-East	15.66
South-West	20.31
Wales	29.3
Scotland	22.33
Northern Ireland	20.85

working and small businesses; legislation against the unions in the 1980s; single union agreements and industrial democracy; and the increasing importance of the service sector in the UK economy.

5.4 Labour availability

The availability of labour, as indicated by high rates of unemployment, has little attraction for most industries. Assisted areas in the UK have had large pools of unemployed labour for over fifty years, but have always struggled to attract new industry. There are, however, exceptions. Sparsely populated regions like central Wales or the Scottish Highlands rule out labour-intensive enterprises. Conversely, in densely populated regions of full employment (like the South-East in the late 1990s), some firms could only find the workforce they needed by relocating in regions of labour surplus.

5.5 Labour mobility

In most LEDCs there are regions of full employment and regions of unemployment close together. This leads us to conclude that the geographical mobility of labour is limited. Government policies aimed at assisting regions of high unemployment have always acknowledged this fact by taking jobs to workers.

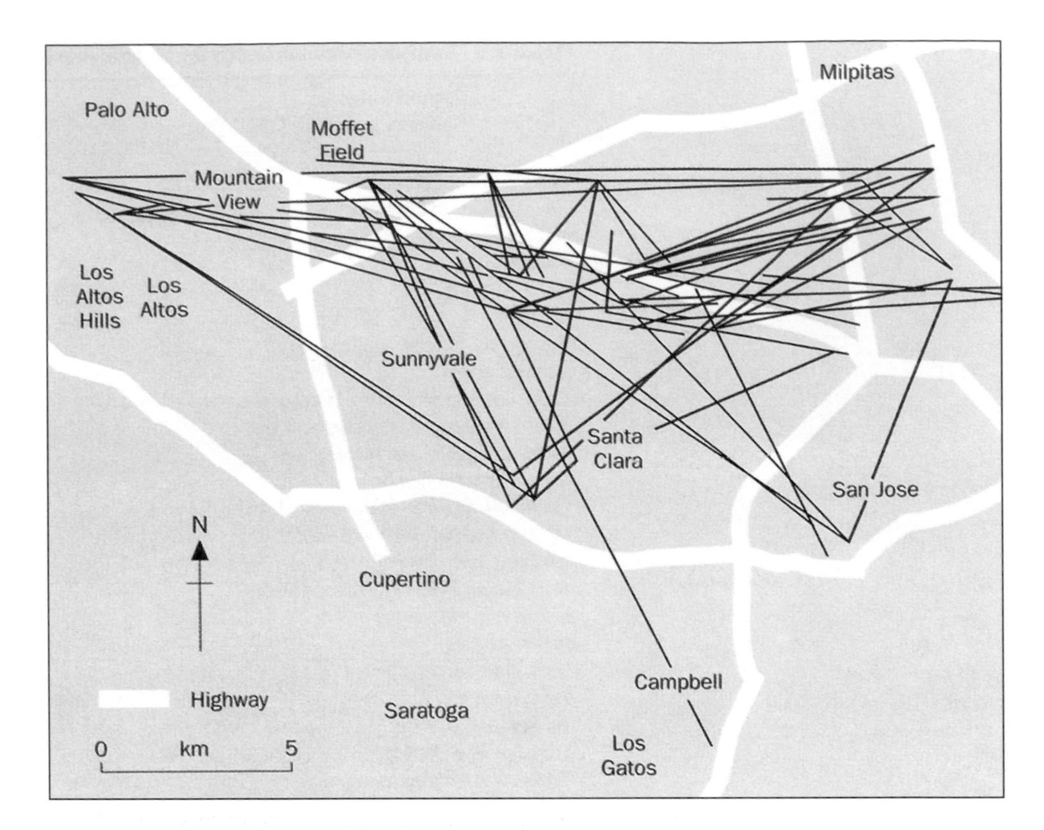

Figure 5.7 Inter-firm mobility of semiconductor engineers in Silicon Valley
(*Source:* Angel, 1989)

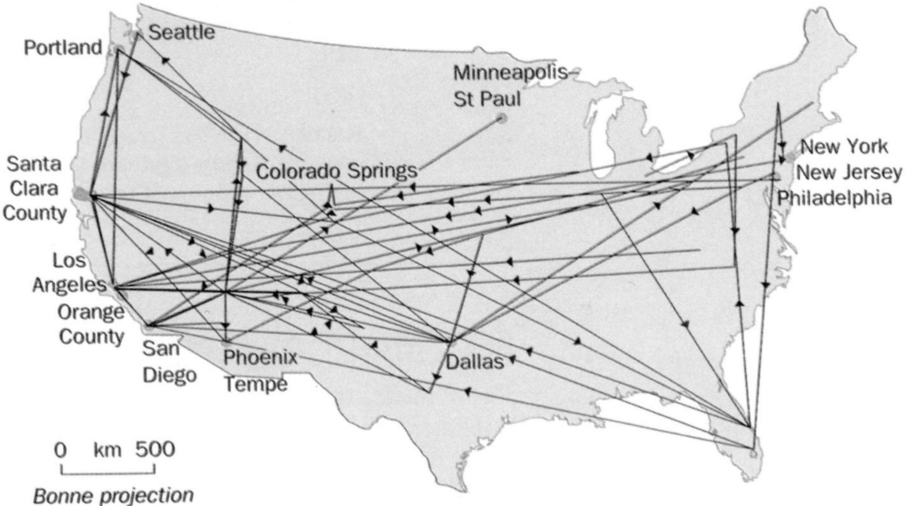

Figure 5.8 Inter-firm mobility of semiconductor engineers in the USA
(*Source:* Angel, 1989)

However, some groups of workers are more mobile than others. The more highly qualified and skilled a group of workers, the greater their mobility. This is illustrated by the US semiconductor industry, where electronics engineers have extremely high levels of geographical mobility. Although nearly half of all job changes occurred within Silicon Valley in California (Fig. 5.7), the pattern of moves extended across the USA (Fig. 5.8).

Given difficult economic conditions, even unskilled workers can show a remarkable degree of mobility. For example, huge numbers of migrants from southern Italy find work in northern Italian cities. In Germany, 8 per cent of the workforce are *Gastarbeiter*, mainly from Mediterranean countries. Thirty per cent of these are Turks, employed as cheap labour in unskilled occupations.

Mobility is also possible between occupations. For example, redundant steelworkers might be retrained as carworkers. But overall, the number of skilled workers moving between occupations is small. Once again this emphasises the relative immobility of labour.

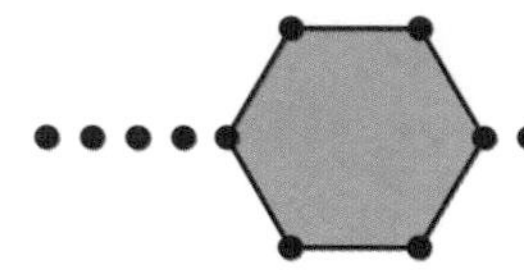

Location quotients

7 a Comment on the absolute and relative importance of manufacturing to the UK economy in 1991 and 1996.
b Calculate the location quotients (LQs) for manufacturing industry in 1991 and 1996 for each region in Table 5.2.
c Describe the spatial distribution of (i) manufacturing industry's relative importance (based on regional LQs) and (ii) absolute importance (based on regional GDP values) in the UK.

Location quotients (LQs) are a measure of the concentration of an economic activity in a region compared with the national average. For example, if 30 per cent of all employment in region i is in manufacturing, and nationally the average is 15 per cent, the location quotient for manufacturing in this region is 30/15 or 2. This means that manufacturing in region i is twice as important as the national average.

$$\text{Location quotient (LQ)} = \frac{X_i / X}{Y_i / Y}$$

X_i = employment in an economic activity in region i.
X = total employment in all industries in region i.
Y_i = national employment in economic activity i.
Y = total employment nationally in all industries.

A location quotient of 1 shows that an economic activity is represented in a region in exactly the same proportion as nationally; less than 1 suggests under-representation; and more than 1 indicates that a region has more than its fair share of a particular economic activity.

Table 5.2 Regional gross domestic product, 1991 and 1996 (£ billions)

	Primary activities		Manufacturing		Services	
	1991	1996	1991	1996	1991	1996
UK	108.2	141.00	15.90	17.04	374.48	482.96
North	6.59	8.55	1.08	0.92	15.55	18.79
North-West	13.93	16.55	0.80	0.77	34.67	45.12
Yorkshire and Humberside	10.21	13.28	1.75	1.57	26.55	34.48
East Midlands	9.30	13.05	2.03	1.6	22.02	28.83
West Midlands	12.26	16.46	1.35	1.42	27.99	35.93
East Anglia	3.83	5.35	1.02	1.27	12.86	16.61
South-East	27.77	34.66	2.46	2.62	151.91	197.88
Greater London	9.66	11.05	0.70	0.53	67.52	86.71
Rest of South-East	18.11	23.61	1.76	2.09	84.39	111.17
South-West	7.46	10.27	1.87	2.66	29.31	37.61
Wales	5.91	7.70	0.84	0.85	14.06	17.70
Scotland	8.63	12.09	2.08	2.57	31.48	39.48
Northern Ireland	2.27	3.03	0.63	0.80	8.7	16.56

5.6 Capital

The location of industry is influenced by two types of capital: financial and fixed.

Financial capital

Funds available for investment are called financial capital. Firms raise **financial capital** in a number of ways: through the issue of new shares; through loans from banks and other financial institutions; and from profits. A feature of these funds is their mobility. As large **multi-plant firms** come to dominate the global economy, TNCs move capital around the world. They invest in regions where profits are high, and disinvest where profits are low.

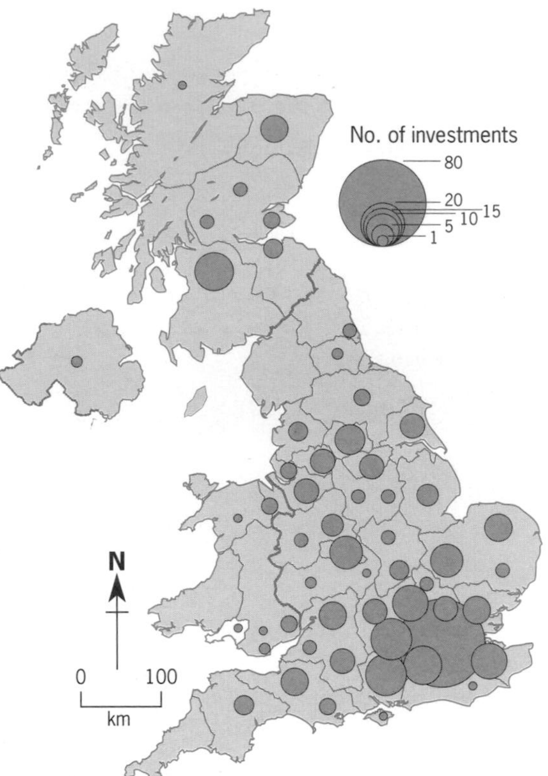

Figure 5.9 Number and distribution of venture capital investments in the UK (*Source:* Mason 1987)

Geographical differences in capital supply are greatest at the global scale. Shortages regularly occur in LEDCs and are a major obstacle to economic development. Even within MEDCs, some forms of capital are not equally available everywhere. Venture capital groups, providing loans to new businesses considered too risky by banks, are often highly localised. In the USA, private **venture capital** is concentrated in California arid Massachusetts, where it has been crucial in the growth of high-tech industries. Similar regional variations are found in the UK. The South-East, with one-third of companies, attracts around two-thirds of all venture capital investment (Fig. 5.9). What is responsible for these regional differences?

Regional differences in venture capital investment in the UK arise from the location of most venture capital groups in London. They favour investments in the South-East, arguing that there are fewer investment opportunities in peripheral regions. Support for this view is provided by the low rate at which new firms are formed in peripheral areas, and the concentration of innovative high-tech businesses in the South.

There are also practical reasons for venture capitalists and clients to locate close together. Spatial proximity allows frequent contact between them. This is important because venture capitalists usually take a close interest in the performance and running of the firms they are backing. There is also evidence that potential investment opportunities in peripheral regions are not correctly appreciated by decision-makers based in London (see Chapter 7).

Fixed capital

Fixed capital refers to investment in plant and machinery. Unlike financial capital it is immobile and a major reason for the survival of past industrial patterns. Just as the forces of the past are felt in present-day industrial patterns, so the greatest influence on the future location of industry is likely to be its present location.

We refer to this phenomenon as **industrial inertia**. It occurs when an industry remains concentrated in an area long after its initial advantages have disappeared. Firms are reluctant to write off fixed capital investments so long as they remain profitable. Thus a number of European coalfields – the Ruhr, the Saarland and South Belgium among them – retain important iron and steel industries. In terms of least cost, these locations can no longer compete with coastal sites. However, because of the immobility of fixed capital, it may prove more costly to close down an inland works and transfer production to the coast. Moreover, long-established industrial regions may offer other **acquired advantages** such as economies of agglomeration (see section 6.1), markets, skilled workforces and good transport links.

?

8 Draw an annotated diagram to show the relationships between financial capital, fixed capital and the location of industry.

High-technology industry

High-tech industry is mainly associated with the production of micro-electronics. However, it covers a wide spectrum of industries, from the production of space vehicles and medical instruments to biotechnology and pharmaceuticals.

High-tech industries have a number of distinguishing features. They make products which are highly sophisticated; they rely heavily on research and development (R&D); and they are continually developing new products and new technologies. Most high-tech industries spend at least 1 per cent of their turnover on R&D, and at least 5 per cent of their employees work in this area.

High-tech activities have grown rapidly since the 1980s. In the UK, the electrical and optical equipment industries, which include most high-tech activities, account for 10 per cent of all employment in manufacturing.

Agglomeration of high-tech activities

The main feature of the geography of high-tech activities is their tendency to cluster, at all scales, in particular areas. This preference for certain areas suggests that high-tech industries are not entirely footloose. Although traditional locational factors such as materials, energy supplies and transport costs have little influence on their location, high-tech industries are none the less affected by locational constraints.

In order to understand the locational patterns of high-tech industries, we need to appreciate that high-tech establishments fall into two types. First, there are establishments which contain headquarters, administration, R&D and design functions; and second, there are branch plants concerned only with assembly and routine production. These two groups have different labour requirements, and therefore different locational patterns.

The global scale

At the global scale, high-tech industries are concentrated in MEDCs, especially Japan and the USA, and in Western Europe. The need for highly qualified engineers, scientists and managers confines headquarters and R&D functions exclusively to the developed world. Where high-tech industries are found in LEDCs, they are always branch plants involved in routine assembly. Many Japanese and US companies have transferred these operations overseas to countries like China and Mexico, where the attraction is cheap labour.

The national scale

At the national scale, high-tech industries also have a tendency to form well-defined clusters (Fig. 5.10). Some of these, such as Silicon Valley in California, Route 128 in Massachusetts and Cambridge in England are known worldwide. Equally apparent is the difference in locational patterns between headquarters and R&D, and branch plants. In the UK, the former are concentrated in the South-East, the latter in assisted areas, especially in central Scotland ('Silicon Glen') and South Wales. IBM, for example, operates a large factory at Greenock near Glasgow (Fig. 5.11) assembling PCs, while its headquarters and research centre is in Hampshire (Fig. 5.12).

The attraction of the UK's assisted areas to foreign high-tech firms are government grants for plant and machinery. By comparison, the reasons behind the

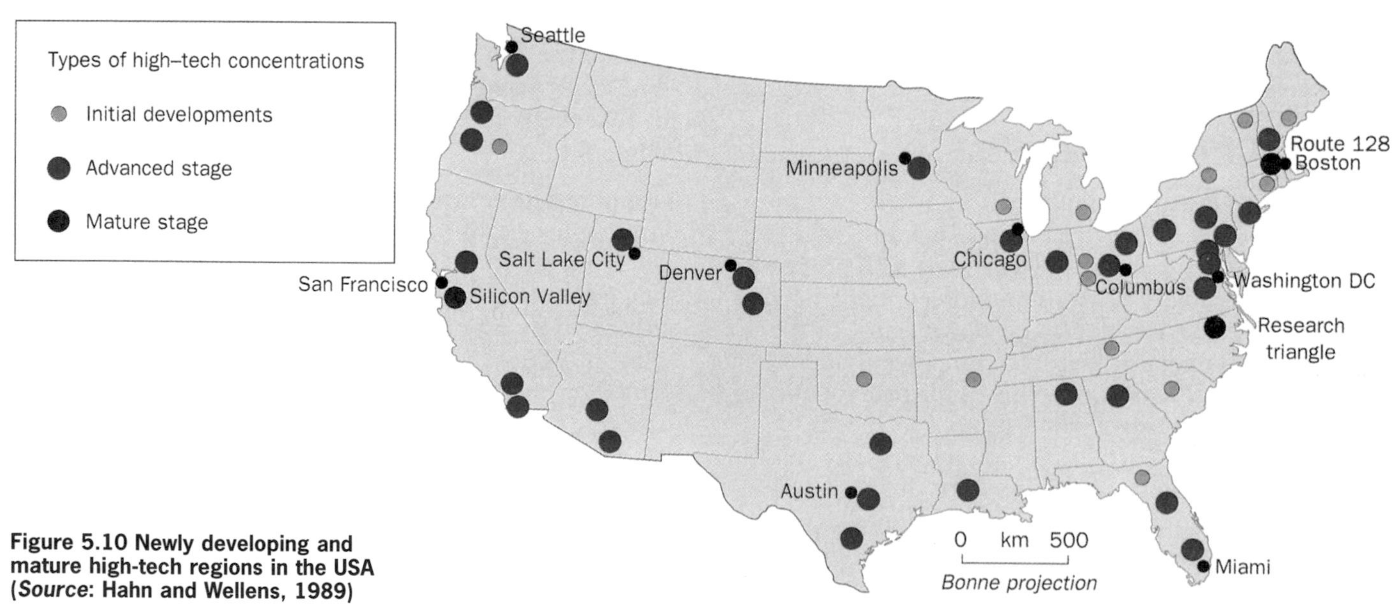

Figure 5.10 Newly developing and mature high-tech regions in the USA (*Source*: Hahn and Wellens, 1989)

Figure 5.12 IBM's HQ building in Hampshire

Figure 5.11 IBM Greenock: the company's European centre for PC manufacturing

location of headquarters and R&D functions are far more complex and varied. Several of them are discussed below.

Table 5.3 Locational factors in high-tech industry in Plymouth, Exeter and Bristol (*Source:* Gripaios, 1989)

	% firms listing each locational factor
Pleasant working environment	63
Access to good workforce	51
Available sites	50
Proximity to other high-tech firms	17
Supportive local authorities	17
University/polytechnic links	9

Labour and environmental quality

The raw material of high-technology is knowledge. This means that the success of any high-tech enterprise depends on its ability to recruit and keep highly skilled research scientists, engineers and technicians. In the USA, 40–60 per cent of high-tech employees are either graduates or qualified technicians. High-tech labour skills tend to be narrowly concentrated in specific geographical areas. Furthermore, employees are often very selective when it comes to where they are prepared to live and work. An environmentally attractive location (good climate, landscape, townscape housing, services, etc.) is essential to hiring the right kind of workforce. Companies must therefore go to where their workers want to live. In the USA this is often California or Colorado, and in the EU it is the less-industrialised 'sunbelt' regions such as the Côte d'Azur and Rhône–Alpes. As new high-tech firms locate in areas where high-tech labour skills are available, it reinforces existing centres of production and encourages clustering.

Links with suppliers and services

The importance of face-to-face meetings between manufacturers and their suppliers, consultants and bankers, is a major reason for clustering. In the USA, ideal business partners for high-tech producers are thought to be within a range of 30 and 50 kilometres.

Birth of high-tech firms

Working for established firms, scientists often develop new ideas for which they spot a commercial market. They leave their employers and start up their own businesses, usually in the same locality. These 'spin-offs' also lead to clustering.

Silicon Valley, near San Francisco, is an example of this process. William Shockley, one of the inventors of the transistor, moved to this area shortly after the war, and brought with him many top US scientists. Several of these people eventually broke away from Shockley and, with the financial backing of Fairchild Camera and Instruments, founded Fairchild Semiconductor. This corporation in turn gave rise to 50 new companies between 1959 and 1979, all of which located in Silicon Valley (Fig. 5.13).

A similar process has occurred in other high-tech clusters. However, to achieve self-sustained growth a cluster needs to be a minimum critical size. In the UK, only Silicon Glen in Scotland, the M4 corridor and Silicon Fen in Cambridgeshire are thought to be large enough to generate 'spin-off' growth (Figs 5.14 and 5.15).

Universities

University research activity accounts for high-tech growth in cities like Cambridge and Grenoble (France). The 'Cambridge Phenomenon' describes the growth of dozens of fast-growing high-tech companies in the city since the 1980s. Trinity College provided the initial stimulus, establishing the UK's first science park in Cambridge in 1972 (Fig. 5.16). The university also helped by encouraging links between its scientists and high-tech businesses.

By 2000 Silicon Fen – the high-tech cluster within a 30 km radius of Cambridge – comprised approximately 1,000 companies and generated $3 billion in revenue. Silicon Fen could not have happened without Cambridge University. Many companies began as Cambridge University spin-offs, started by graduates

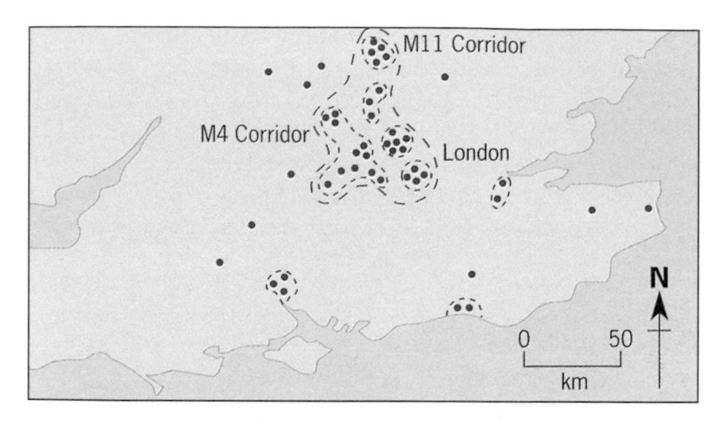

Figure 5.14 Electronics and instruments firms in South-East England (*Source:* Oakey and Cooper, 1989)

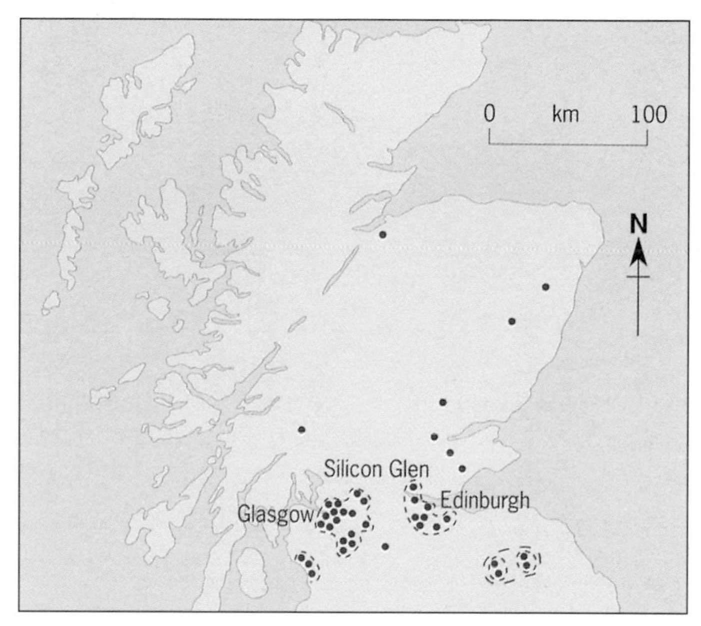

Figure 5.15 Electronics and instruments firms in Scotland (*Source:* Oakey and Cooper, 1989)

Figure 5.13 Silicon Valley, California

Figure 5.16 Cambridge Science Park

and academic staff. Most companies concentrate on research, design and development rather than production. By the late 1990s two new trends emerged: first, a shift towards telecoms and biotechnology; and second, the development of links between the university and large TNCs such as Glaxo (pharmaceuticals) and Microsoft (research).

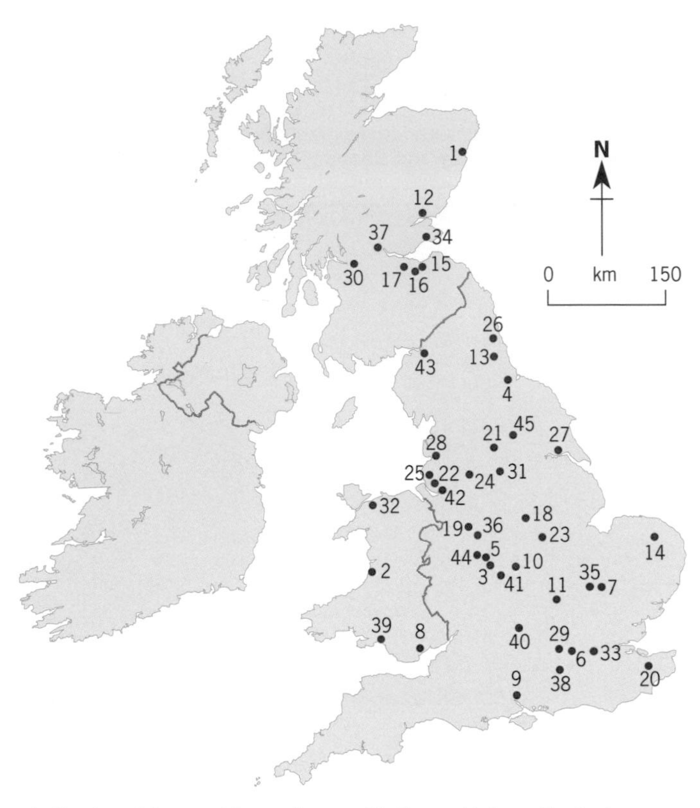

1 Aberdeen Science and Research Park
2 Aberystwyth Science Park
3 Aston Science Park
4 Belasis Hall Technology Park, Billingham
5 Birmingham Research Park
6 Brunel Science Park
7 Cambridge Science Park
8 Cardiff Business Technology Centre
9 Chilworth Research Centre
10 Coventry University Technology Park
11 Cranfield Technology Park
12 Dundee Technology Park
13 Durham University Science Park
14 East Anglia University Industrial Liaison Unit
15 Edinburgh Technopole/Bush Research Park
16 Elvington Research Park, Edinburgh
17 Herriot Watt Research Park
18 Highfields Science Park, Nottingham
19 Keele University Science Park
20 Kent University Applied Statistics Research Unit
21 Lister Hills Technology Development
22 Liverpool University R&D Advisory Unit
23 Loughborough Technology Centre
24 Manchester University/Manchester Science Park
25 Merseyside Innovation Centre
26 Newcastle Technopole
27 Newlands Science Park, Hull University
28 Preston Technology Management Centre
29 Reading University Innovation Centre
30 Scottish Enterprise Technology Park
31 Sheffield Science and Technology Park
32 Snowdonia Technopole Eryri Ltd, Bangor
33 South Bank Technopark, London
34 St Andrews Technology Centre
35 St John's Innovation Park, Cambridge
36 Staffordshire Technology Park
37 Stirling University Innovation Park
38 Surrey University/Surrey Research Park
39 Swansea Innovation Centre
40 The Oxford Science Park
41 Warwick University Science Park
42 Wavertree Technology Park, Liverpool
43 Westlakes Science and Technology Park, Cumbria
44 Wolverhampton Science Park
45 York Science Park

Figure 5.17 Science research parks in Britain

However, the success of Silicon Fen is not just down to Cambridge University. The growth of London as a global financial centre since the early 1980s attracted foreign banks and venture capitalists. They provided hundreds of small high-tech companies with finance that previously was unavailable.

Science parks

Science or 'incubator' parks, which are often joint ventures between universities and local authorities, have helped to promote high-tech development (Fig. 5.17). They were first developed in the USA to encourage academic scientists to exploit the industrial applications of their research. For instance, in Baltimore (USA) the city provided land, buildings and roads, while the university was in charge of laboratories, equipment and recruiting staff.

Government research institutions

The location of research and federal institutions in the USA, particularly aerospace and weapons, has influenced locational patterns in high-tech industries. Being close to federal institutions, such as the Department of Defense and the Goddard Space Center in Washington DC, increases the chances of high-tech firms securing contracts. Biotechnology, one of the new generation of high-tech industries, owes its growth in Washington DC to the presence of the National Institute of Health with its National Library of Medicine. The development of the UK's first high-tech cluster – the M4 corridor – owes much to nearby government research establishments such as nuclear weapons at Aldermaston, aircraft at Farnborough and atomic energy at Harwell.

Capital

As well as government grants and research programmes, many high-tech firms rely on private venture capital. The availability of such capital was central to the growth of California and Massachusetts as the two leading areas of high-tech manufacturing in the USA.

Transport

Two of the UK's three high-tech agglomerations are in South-East England. An advantage of this region is its well-developed motorway system and access to international airports. This is particularly important for an industry which serves international markets. Thus, in the South-East, high-tech firms often choose locations within a short drive of Heathrow. Among the high-tech TNCs located along the M4 corridor close to Heathrow are IBM, Digital, Racal and Apple. Rank-Xerox recently moved its European headquarters from central London to Marlow because it was near the M40 and only a short drive from Heathrow.

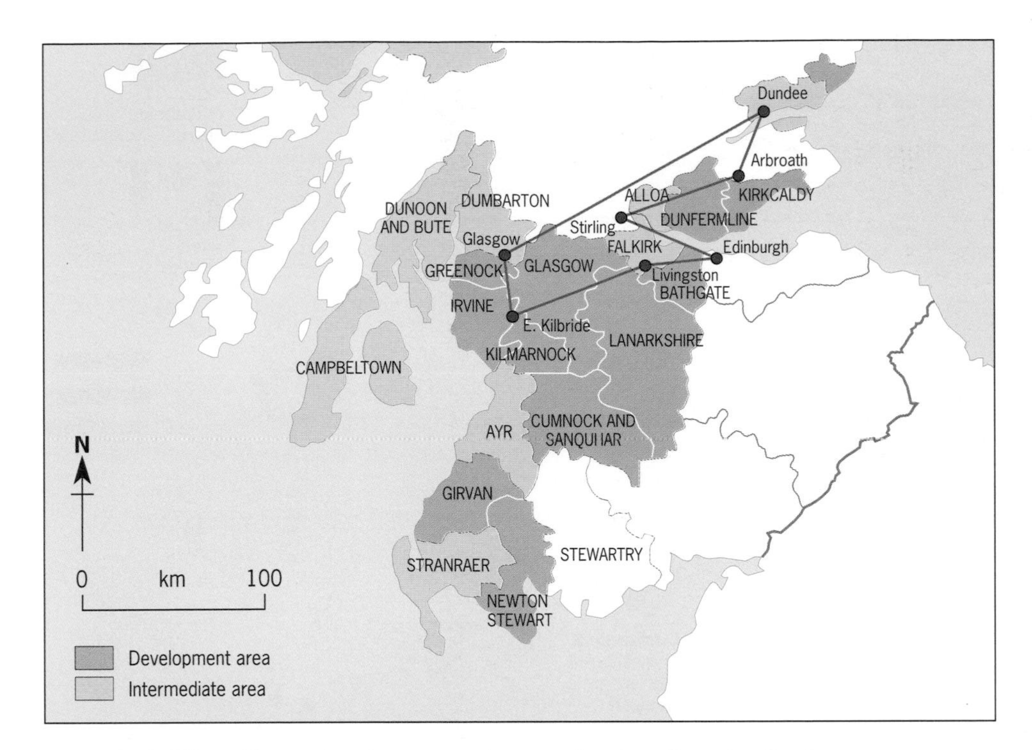

Figure 5.18 Silicon Glen and assisted areas in central and southern Scotland.

Figure 5.19 The Alba Centre – a high-tech research centre sponsored by Scottish enterprise.

Silicon Glen: high-tech industry in a peripheral region

Silicon Glen is the name given to the cluster of electronics industries in central Scotland (Fig. 5.18). Unlike the high-tech cluster around Cambridge, Silicon Glen is primarily a centre of production. Products range from semiconductors to consumer electronic goods such as TVs, CD players, mobile phones and computers. Today the region has international importance as a centre of electronics manufacture. It accounts for 15 per cent of Europe's output of semiconductors; 28 per cent of its PCs; and 80 per cent of its workstations. In 1999 total employment in the electronics industry stood at 76,000.

The growth of electronics in Silicon Glen is largely the result of inward investment since 1980 by, EU and East Asian TNCs (Motorola, IBM, Compaq, JVC, Mitsubishi, etc.). Nearly 60 per cent of employment in electronics in Silicon Glen is in foreign firms. The advantages of Silicon Glen for foreign companies include:

- a powerful regional development agency – *Locate in Scotland* – which markets the region and encourages inward investment. In 1998-9 the agency attracted 78 inward investment projects, which created 11,000 jobs;
- its location within the EU, which gives access to a potential market of 290 million people;
- an existing cluster of electronics companies in Scotland with a nexus of suppliers and buyers of components;
- a workforce that is skilled, flexible and relatively cheap;
- the availability of regional aid (grants) in assisted areas in central Scotland (Fig. 5.18);
- the proximity of several Scottish universities including over 3,500 students enrolled on electronics and software courses;
- easy access to outstanding mountain and coastal scenery and excellent recreational opportunities;
- the availability of venture capital for small innovative companies through *Scottish Development Finance*.

Although Silicon Glen remains dependent on electronics production from foreign-owned branch plants, successful attempts are being made to diversify into areas of research, design and development. *Scottish Enterprise*, a government economic development body, set up the Alba Centre in 1998. This initiative aims to make Silicon Glen a world centre for micro-electronics design. Located in Livingston in a purpose-built science park, the Alba Centre is a partnership between four Scottish universities and the private sector.

The regional scale

Urban areas

In towns and cities high-tech firms prefer locations in the suburbs. This preference is seen in the Washington–Baltimore corridor (Fig. 5.20). The reasons for this are:

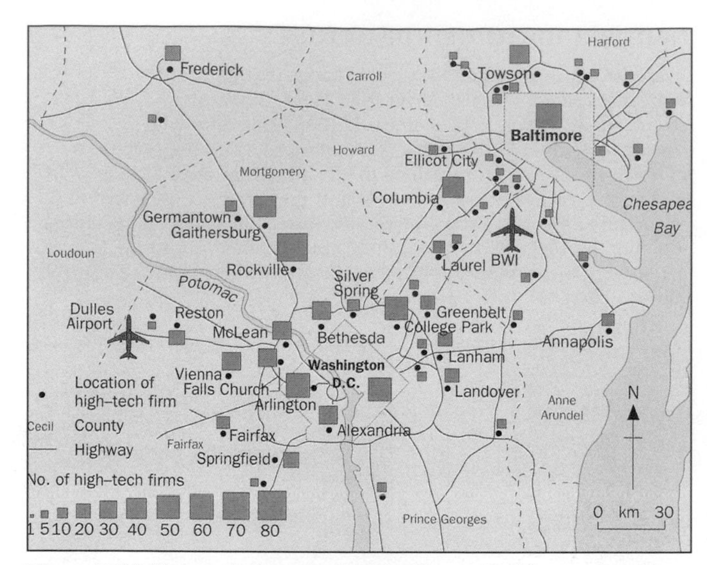

Figure 5.20 High-tech firms in the Washington-Baltimore Corridor, 1987 (*Source*: Hahn and Wellens, 1989)

1 Many of the more recent firms were established in the suburbs because that was where the founders and most of the employees lived.

2 Suburban locations give easier access to the three main airports of the region and to information centres such as research institutions and regional service companies. They are also easier to reach for their customers, which is especially important in an industry where frequent face-to-face meetings are essential.

3 There are many new industrial premises and high-tech parks which offer excellent accommodation at low rentals in the suburbs.

4 Better-quality housing and schools are found in the suburbs. The latter are especially important for young middle-class families.

Remote rural areas

Although most high-tech firms are found in clusters (Table 5.4) a sizeable number are located in semi-rural, rural and peripheral areas. A study of biotechnology in the UK found that 70 per cent of firms were in rural non-industrial locations.

A peripheral location is possible for firms which are highly footloose, and which rely on knowledge rather than material inputs. Firms operating in remote locations tend to make sophisticated, high-value, low-weight products. Blue-collar production workers are recruited and trained locally, and additional skilled workers are brought in from outside. Skilled workers are easily recruited to environmentally attractive locations, far from the congested core areas.

The personal preferences of the founder explain the location of many high-tech firms in remote but

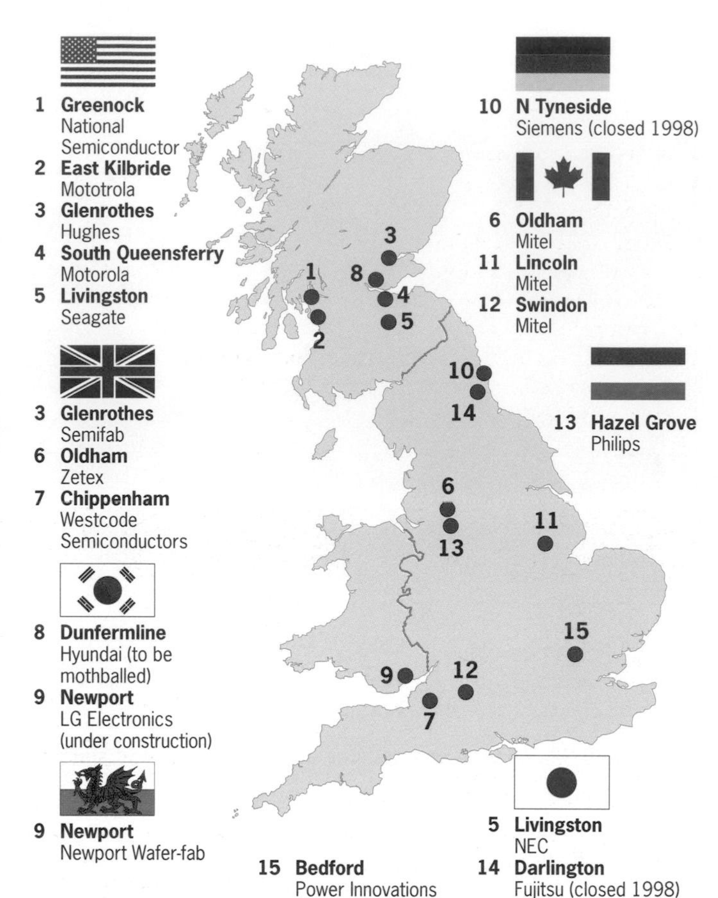

Figure 5.21 Main semiconductor plants in Britain (with nationality of parent company)

environmentally attractive areas like the Scottish Highlands. Most biotechnology firms in peripheral locations did not 'spin off' from large local firms. Instead they were deliberately sited in remote rural areas by founders who were prepared to trade off a proportion of profits against the **psychic income** (see section 7.1) of a location of high amenity. Gaeltec, a small electronics firm with 30 employees which makes transducers for the medical industry at Dunvegan on Skye, is a good example. Its managing director and founder chose to locate the firm 'in a beautiful part of the world, free from concentrated populations and the hassles of commuting'.

Table 5.4 Locational characteristics of electronics, instruments and biotechnology firms

	Clusters		Free-standing	
	No.	**%**	**No.**	**%**
Electronics and instruments				
South-East England	36	78.3	10	21.7
Scotland	31	73.8	11	26.2
San Francisco Bay	31	72.1	12	27.9
Biotechnology	13	30.2	30	69.8

?

9 Look at the figures in Table 5.4.
a Which types of high-tech activities are most likely to be found in clusters?
b What reasons explain this clustering?
c How does the location of biotechnology differ from that of electronics and instruments? Try to explain this difference.
10 With reference to Silicon Fen and Silicon Glen:
a Explain why high-tech industries form clusters;
b Examine the importance of the factors that promote the clustering of high-tech industry.

Summary

- For most manufacturing industries, labour costs are the largest element of total costs.
- Labour costs can be measured either as wage rates, or as unit costs, which link wage levels to output and productivity.
- Differences in wage levels are large at the global scale but are often small at the national scale.
- Wage rates have the greatest influence on industrial location at the global scale.
- The availability of labour as such (as shown by high rates of unemployment) in MEDCs is not a major influence on industrial location.
- Labour has limited geographical mobility, though the more skilled and highly educated a workforce, the greater its mobility.
- The regional concentration of manufacturing industry can be measured by the location quotient.
- There are two types of capital that influence industrial location: financial and fixed.
- Fixed capital is mainly responsible for the geographical inertia of industrial concentrations.
- High-tech industries, although comparatively footloose, show a strong tendency to cluster in geographical space.
- The location of high-tech industry is strongly influenced by the availability of labour skills and venture capital.

6 External economies and government

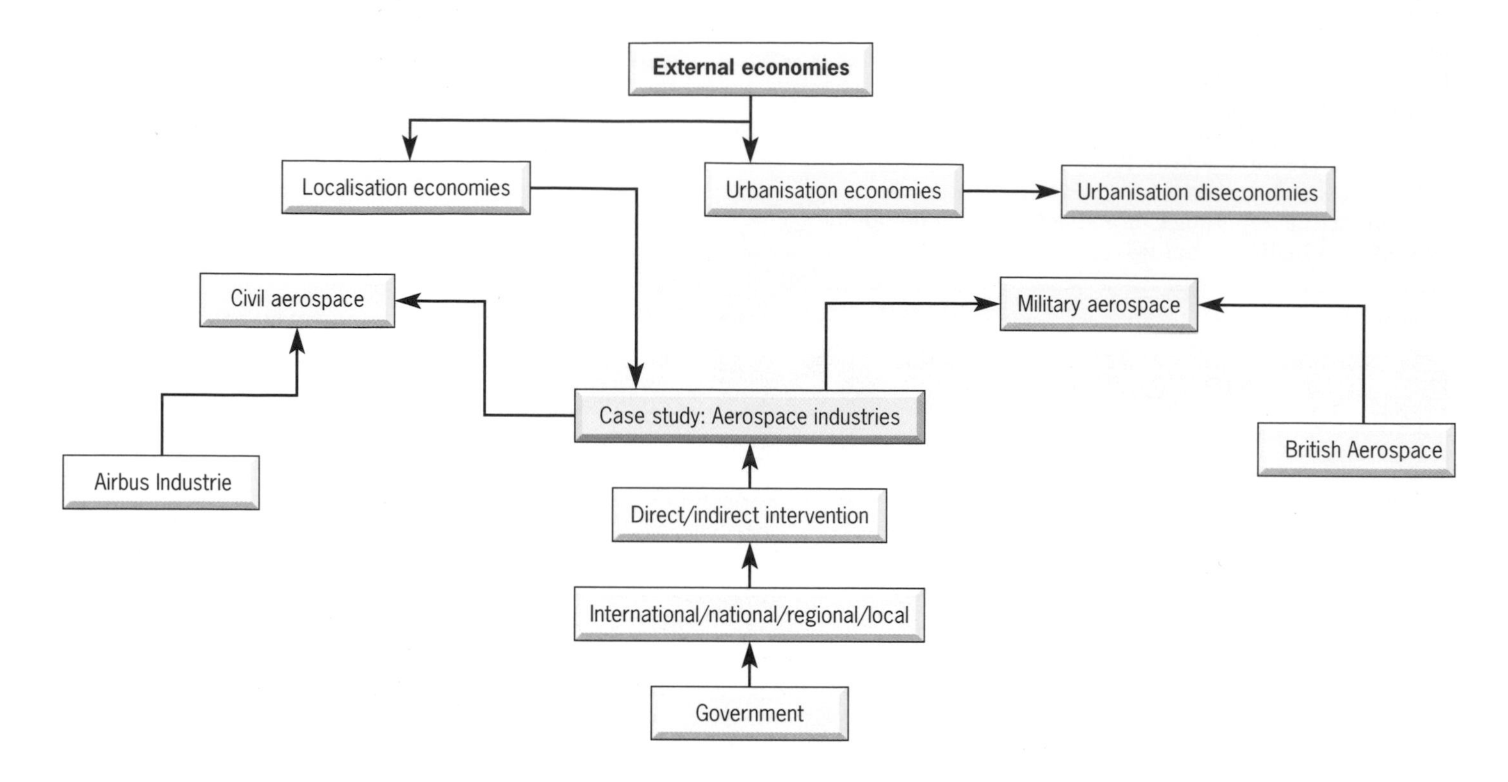

6.1 External economies of scale

In previous chapters we have seen that the distribution of manufacturing is highly uneven. This unevenness is found at all scales. Typically, most industries tend to cluster in certain preferred locations. Sometimes we can explain these patterns by reference to the location of materials, energy, markets or skilled labour. However, the most important factor causing industrial concentration is external economies of scale. Its significance is recognised in its alternative name: **agglomeration economies**.

External economies describe the advantages to a firm of locating in an existing urban-industrial centre. Unlike internal economies of scale (see section 3.5 on plant size) external economies arise from outside the firm itself. Geographers divide external economies into two components: **localisation economies** and **urbanisation economies**.

ABCDE = firms

Figure 6.1 Weber's analysis of the effect of external economies on industrial location

Localisation economies

Localisation economies occur when firms linked in the production chain by purchases of materials and finished goods locate close together. The main advantages of spatial proximity include reduced transport costs between suppliers and customers, faster delivery times (which may allow just-in-time delivery) and better communication allowing greater personal contact between firms.

Linkages of this type are particularly strong in the petrochemical industry (Fig. 6.2), and give rise to huge industrial complexes like those at Europoort (Netherlands) (Fig. 6.3) and on Teesside in North-East England. The basic materials for petrochemical manufacture are supplied by oil refineries through **backward linkages. Forward linkages** exist between petrochemicals and the industries it supplies with ethylene and propylene for the manufacture of plastics, paints, synthetic fibres and many other products.

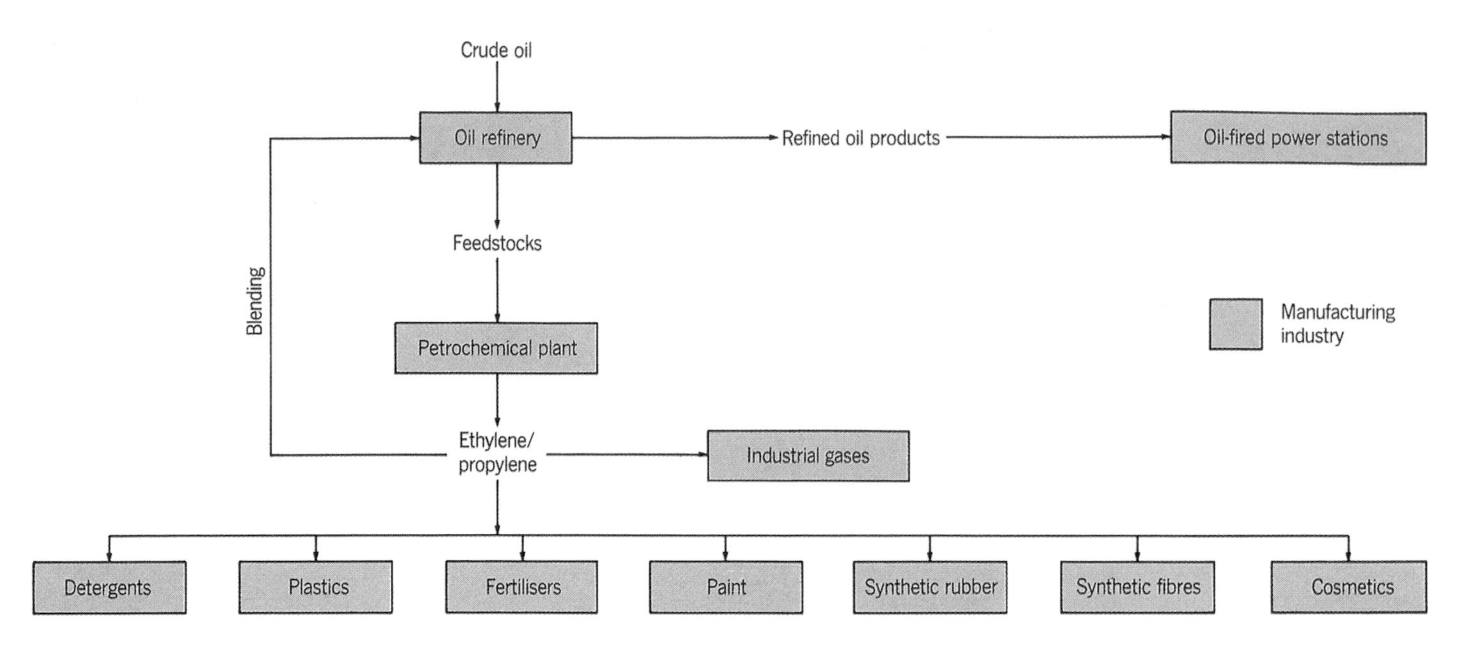

Figure 6.2 Some linkages in oil refining–petrochemical complexes

The aircraft and parts industry of southern California also forms an industrial complex brought about by localisation economies. Its focus is the major aircraft assembly plants of Boeing, Lockheed–Martin and General Dynamics (Fig.6.4). As the industry has grown, a vast network of linkages between assemblers, electronics, instrument, missile and space vehicle manufacturers has developed in the region. Such is the importance of localisation economies that an aircraft manufacturer locating outside southern California would be at a considerable disadvantage.

Urbanisation economies

Urbanisation economies are cost savings resulting from an urban location. Usually the larger the town or city, the greater the savings. Linkages between manufacturing and services is one aspect of urbanisation economies. Manufacturing industries depend on a range of producer services such as auditing, banking, advertising and industrial cleaning. Large companies can provide some of these services for themselves, but most find it cheaper to contract out the work to other firms.

Weber's theory considered a situation where external (or agglomeration) economies influenced the location of a firm. This is shown in Figure 6.1. Here it is assumed that if three firms locate together, they can save £20 per unit of production. However, it will only be profitable to do so if they incur extra transport costs of less than £20.

1 Refresh your memory of Weber's theory by reading pages 29–30, and study Figure 6.1.
a What does point P represent in each of the five triangles in Figure 6.1?
b What name is given to the £20 isoline around P? Explain its significance.
c Which firms will locate together to take advantage of external economies, and where will they locate? Explain your answer.

Figure 6.3 The oil refining and petrochemical complex at Europoort

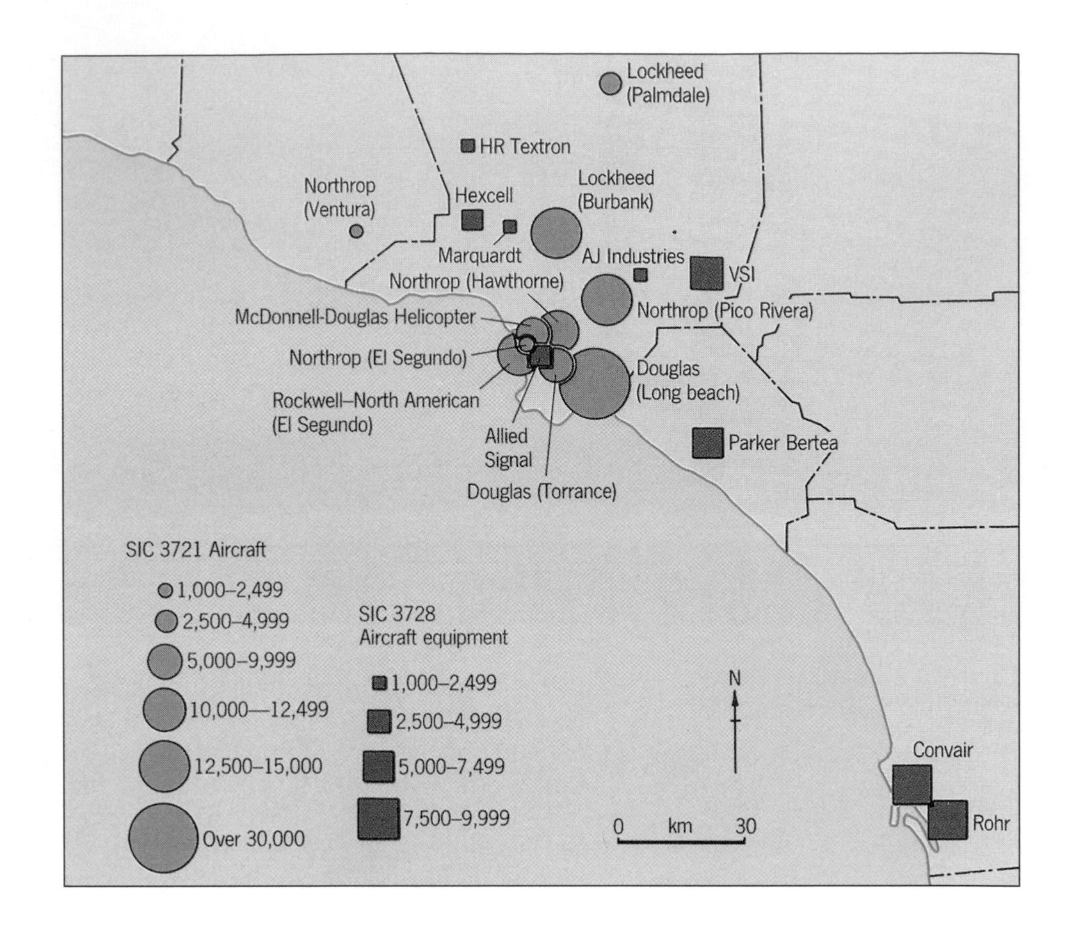

Figure 6.4 Principal aircraft and aircraft parts manufacturers in southern California (*Source:* Scott and Mattingly, 1989)

Some manufacturing firms choose city-centre locations in order to keep in close contact with services in the central business district (CBD). Examples include women's clothing, which is linked to fashion design and retailing, and printing which has strong links with many office functions.

Savings also arise from the economic and social infrastructure of urban areas. The **economic infrastructure** includes transport networks; water, gas and electricity grids; and factory space. The **social infrastructure** comprises housing, schools, hospitals and other services for workers and their families. Today, individual firms contribute (through local taxes) only a small proportion of the costs of providing and maintaining this infrastructure. In nineteenth-century Britain the position was often quite different. It was common for manufacturers to build roads, provide houses for their workers and organise their own energy supplies. Sometimes, as in Consett in County Durham and Saltaire (Fig. 6.5) in West Yorkshire, they built entire settlements to serve the needs of industry.

6.2 Urbanisation diseconomies

As cities increase in size. urbanisation economies may be outweighed by **urbanisation diseconomies**. There is no fixed threshold when economies turn to diseconomies, and the process is not inevitable. None the less, in very large cities costs often begin to rise. Land for new factories and factory extensions may be in short supply, forcing up land prices and rents. Cars in the central areas may cause congestion and increase transport costs. And competition between firms for labour may push up wage rates. Faced with these and other problems, many manufacturing industries have responded by transferring production to smaller towns and rural areas. This urban–rural shift has been widespread in MEDCs in the last 30 years. You will find a detailed account of this trend in section 11.1.

2 Make notes to show the differences between localisation economies and urbanisation economies.

3 Draw an annotated diagram to show the relationship between urbanisation economies and urbanisation diseconomies.

Figure 6.5 Saltaire – a nineteenth-century planned industrial village

6.3 Government

Table 6.1 Government and the location of industry in the UK

Scale	Direct	Indirect	
		Positive	**Negative**
International (EU)		Structural funds: ERDF, ESF, joint international ventures (e.g. aerospace)	
National	Nationalised industries	Regional policies. Ad hoc grants to attract FDI.	
Regional/local		Promotional measures. Rural development agencies in Wales and Scotland. Regional development agencies in Wales, Scotland and Northern Ireland	Local authority structure plans and land-use zoning.

The politics of government intervention

The nature and extent of government involvement in industrial location depends on the political system in any country. In **command economies** like China and Cuba, state control is absolute. In Western Europe, locational decisions are largely determined by profit and loss, though the state does intervene for social and political as well as economic reasons. The amount of intervention depends on the ideology of the government. Thus whenever the Labour Party is in power in the UK, the influence of the state often increases. In the past, industries have been nationalised, and more resources allocated to regional policies (see section 9.5). During a period of Conservative rule a reduction in government involvement is more likely. Regional policy may be given lower priority; nationalised industries may be privatised; and urgent problems, like inner-city decay, may be tackled through private rather than public agencies (see section 11.4).

?

4a Draw a table with the headings 'command economy', 'mixed economy' and 'free-market economy'. Under each heading write words to describe the type of government intervention you think happens there.
b Add notes to your table as you read the next Case Study.

Government policies and scale

Government intervention operates at different scales and can be direct or indirect (Table 6.1). In a free-market economy the emphasis is on persuasion rather than coercion. Intervention is usually greatest at local levels, and becomes less with increasing scale. At county and district levels in the UK, the location of new economic activities is subject to approval by local planning authorities. They have to prepare detailed structure plans and strategies for land-use change.

Development agencies created by government have been set up to revitalise decaying urban areas; promote the revival of remote rural areas in Scotland and Wales; and attract inward investment to the UK regions. The hope is that modest public funding of these agencies would encourage private investment on a much larger scale.

At the national scale, successive UK governments have followed regional policies over the past 70 years (see section 9.5). Their aim was to reduce regional economic differences, especially in unemployment. For the most part this was done by offering industry financial incentives to locate in less prosperous regions. However, between 1947 and 1981, controls were also imposed to turn investment away from the more prosperous and towards the less prosperous parts of the country.

In the last 30 or 40 years policies at the international scale have become more important. Within the EU there is co-operation between member states in matters of regional policy through the European Regional Development Fund (see section 9.4). The enlargement of the EU since 1981, and in particular the entry of Greece, Portugal and Spain, gave the EU's regional policies greater prominence in the 1990s. Common policies also exist for the coal and steel industries. Meanwhile, co-operation between EU governments in industries like aerospace and defence, which require huge capital investment, has become essential. As we shall see in the next Case Study, it is this political factor which increasingly shapes the geography of the aerospace industry in Western Europe.

Aerospace industries

The role of government

Aerospace is a leading industry in many MEDCs (Fig. 6.6). In the UK it employs nearly 100,000 people and exports 60 per cent of its production. The industry makes both civil and military equipment, and is dominated by national rather than transnational firms. For instance, BAE Systems (formerly British Aerospace), Europe's largest aerospace company, has no manufacturing capability outside the UK. Similarly, Boeing's 200,000 employees are concentrated exclusively in the USA and Canada.

The main reason for the absence of aerospace TNCs is strong defence links between the industry and governments. France and the UK spend almost 4 per cent of their GDPs on defence. This means that the state is usually the major customer for military equipment such as fighter aircraft, helicopters and guided missiles. Aerospace is also at the leading edge

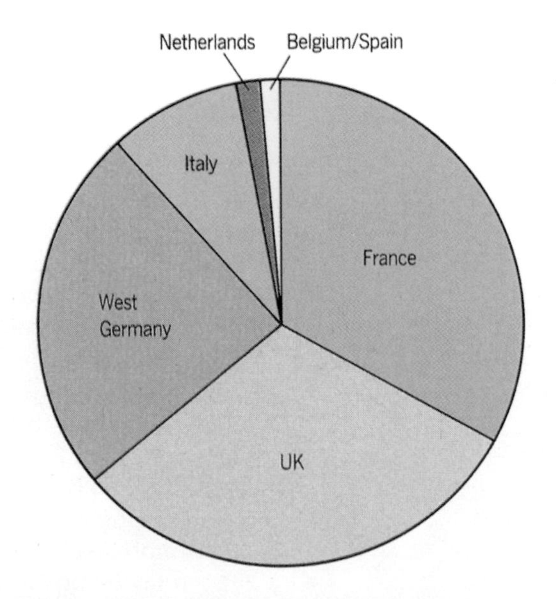

Figure 6.6 Aerospace production (by value) in the EU

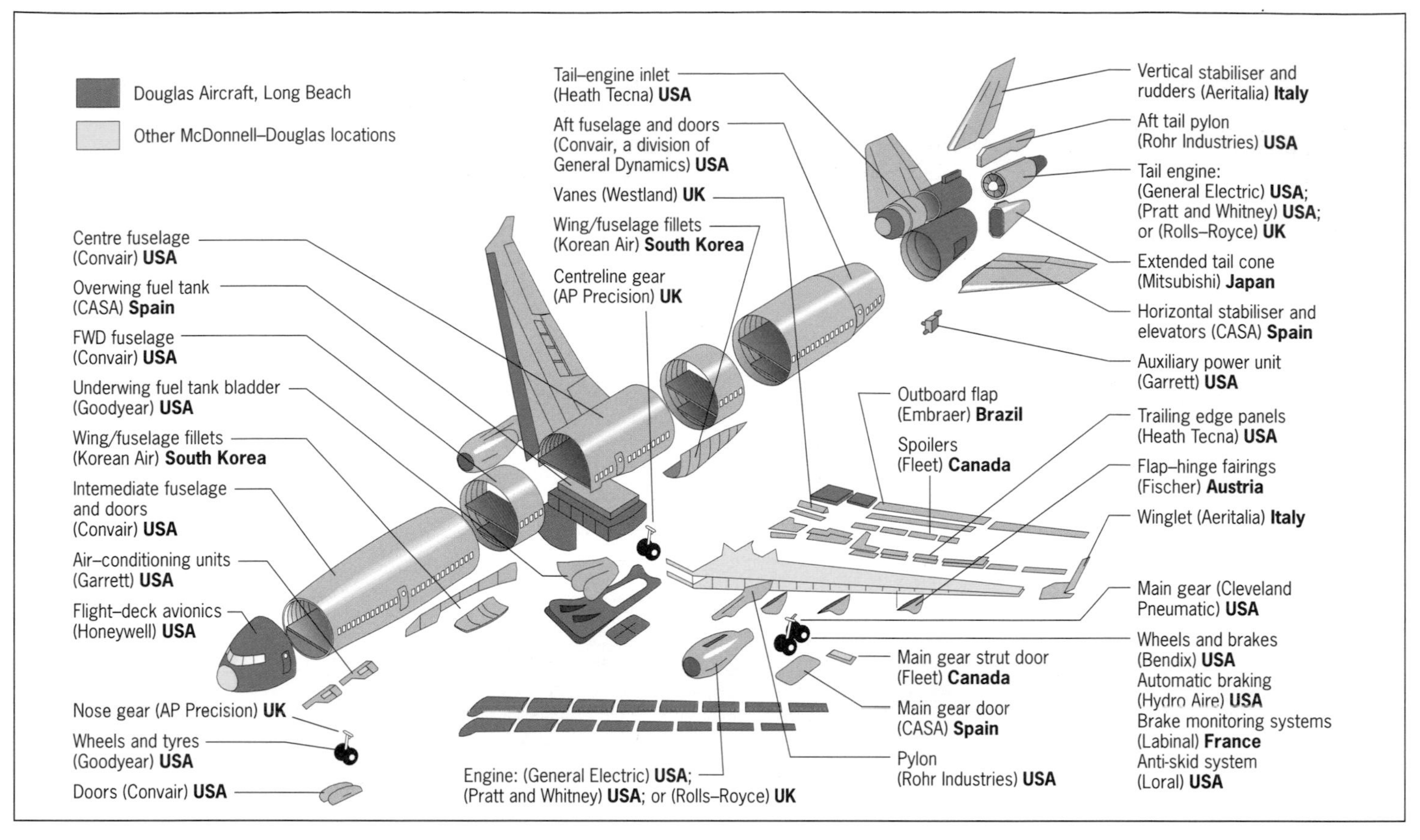

Figure 6.7 Global engineering: main MD-11 subcontractors (*Source:* McDonnell-Douglas and *The Economist*, 3 Sept. 1988)

of new technology. Much of its research is financed by governments who see it as vital to state security. In the UK, the Ministry of Defence spends half of all government expenditure set aside for R&D. Government involvement is also explained by the high risk and huge costs of many aerospace projects. Thus, at all scales, the geography of the aerospace industry tends to be influenced more by political than by economic considerations.

Civil aerospace

The civil aerospace industry manufactures airframes, engines and aero-equipment for commercial airplanes and helicopters. Two firms dominate the global aerospace industry: Boeing (USA) with a two-thirds market share, and Airbus (EU) with 30 per cent.

Booming production

Orders for new aircraft have grown rapidly since the late 1980s. This growth is largely due to the rising demand for air travel and the need for increased capacity. Also significant is the need for new aircraft to replace the ageing fleets of many of the world's airlines. Older aircraft like the 727 and DC9 are less safe, noisier and more polluting. Furthermore, congestion on busy air routes and at key airports in Europe and North America has created a demand for a new generation of larger, wide-bodied aircraft such as the Airbus A330 and the Boeing 777-300.

High costs and high risks

A distinctive feature of the aerospace industry is the enormous investment needed to develop new products. For example, a medium-sized airliner like the Airbus 300 costs over US$2 billion to develop, with a new engine costing a further US$1.5 billion. It may take 15 years to recover such a huge outlay. Given the complexity of aerospace products, manufacturers often have to decide years in advance what the market will want. Thus, commercial aerospace is a high-risk industry where companies bet their whole future on each new product.

The response of aerospace manufacturers

Manufacturers have responded to the high risks and costs by adopting three strategies. First, risks are shared by collaboration between firms in joint projects. Second, costs are reduced by global sourcing of components and assemblies (Fig. 6.7). And third, costs are also held down by making 'families' of airliners, with communality of parts. All three strategies are to varying degrees evident in the EU's Airbus Industrie.

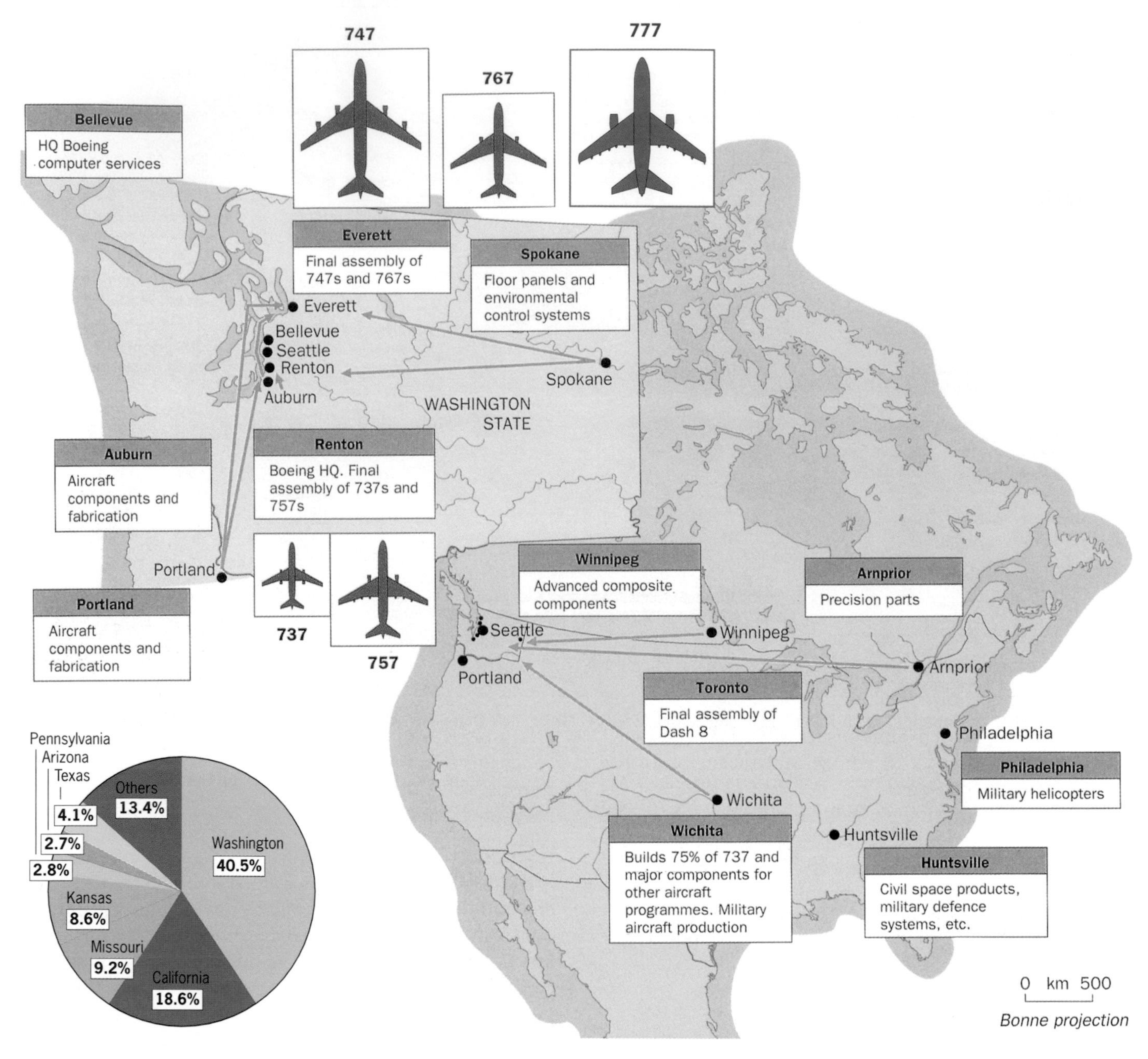

Figure 6.8 Boeing commercial airliners: geography of production

---

?

5 With reference to Figure 6.8 describe how the geography of production of the Boeing company is spatially concentrated at different scales.

6 Suggest how the geography of production of the Boeing company is likely to have been influenced by external and internal economies of scale.

Airbus Industrie

Airbus Industrie, with a 30 per cent market share, is the second largest manufacturer of civil airplanes in the world. It is a multinational consortium comprising four companies: Aerospatiale Matra (France); DaimlerChrysler Aerospace Airbus (Germany); BAE Systems (UK); and CASA (Spain). Airbus makes a complete 'family' of nine jet airliners including short-, medium-, and long-haul airplanes. The largest (the A340-300) carries approximately 400 passengers; the smallest (the A319) around 120. Work is allocated to each of the four partners according to their stake in the consortium (Fig. 6.9). Airbus Industrie's dispersed

Figure 6.9 The Airbus Industrie consortium

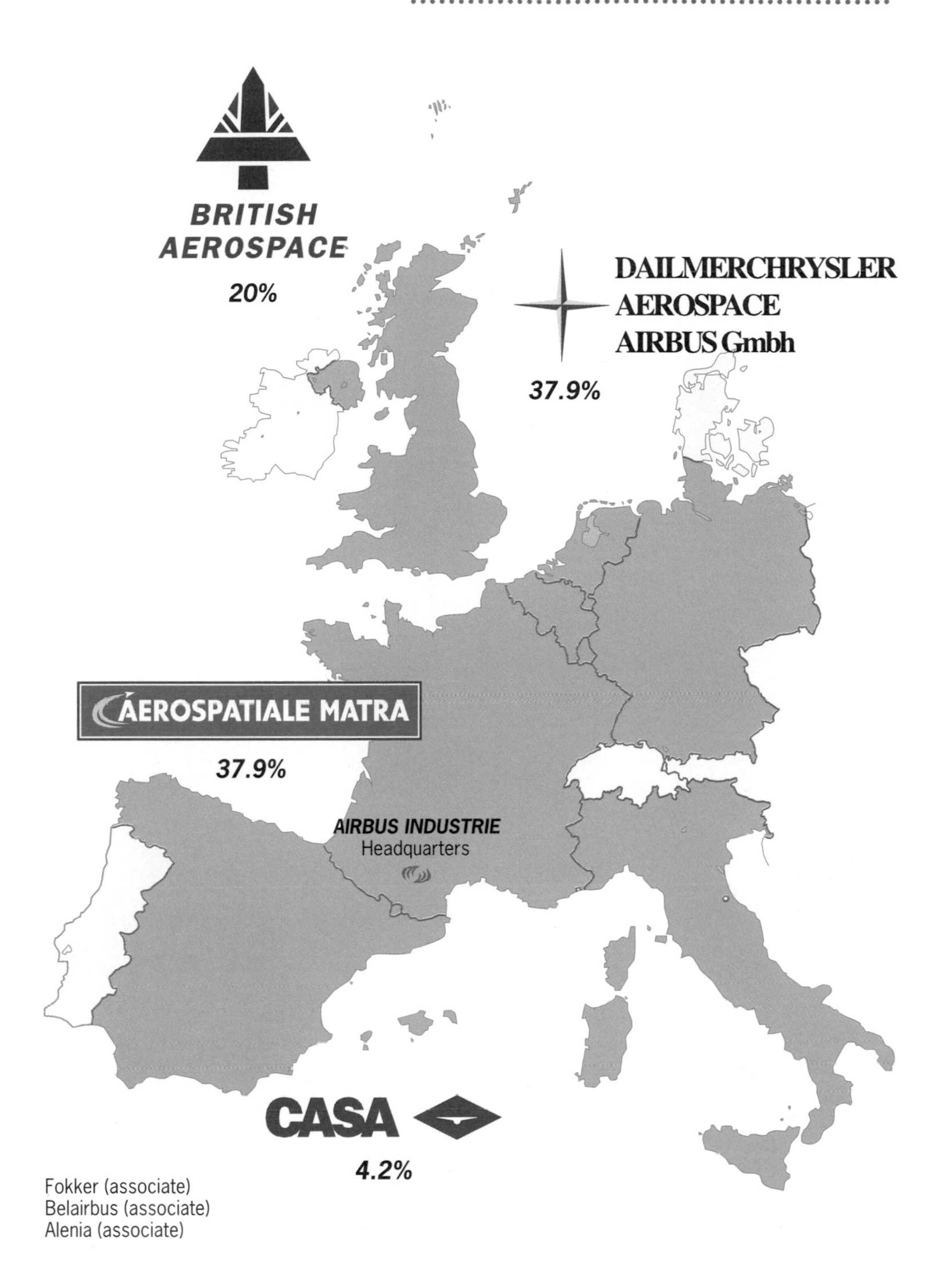

geography of production (Fig. 6.10) reflects the huge costs of designing and building modern airplanes and the importance of economies of scale. Europe can only compete with the USA in the aerospace industry if countries collaborate and undertake joint ventures such as Airbus.

Aerospace is an assembly industry, sourcing parts and sub-assemblies from a large number of suppliers. The organisation of Airbus reduces the costs of sourcing to a minimum. Whenever possible it relies on a single supplier for each assembly. This allows suppliers to optimise their output, achieve economies of scale and lower costs. However, instead of sourcing thousands of small parts, entire sections of airplanes are built in locations scattered across Europe (Fig. 6.10). As a result only 4 per cent of work-hours needed to build an airplane are spent on the final assembly lines at Toulouse in France and at Hamburg in Germany.

Transport of assemblies to Toulouse and Hamburg is by Beluga special transport airplanes (Fig. 6.11). Air transport is the only practical way of transporting bulky and awkwardly shaped assemblies such as wings, fuselages and tail planes. Each completed A300 and A310 requires a total of eight flights by the Belugas covering around 13,000 kilometres.

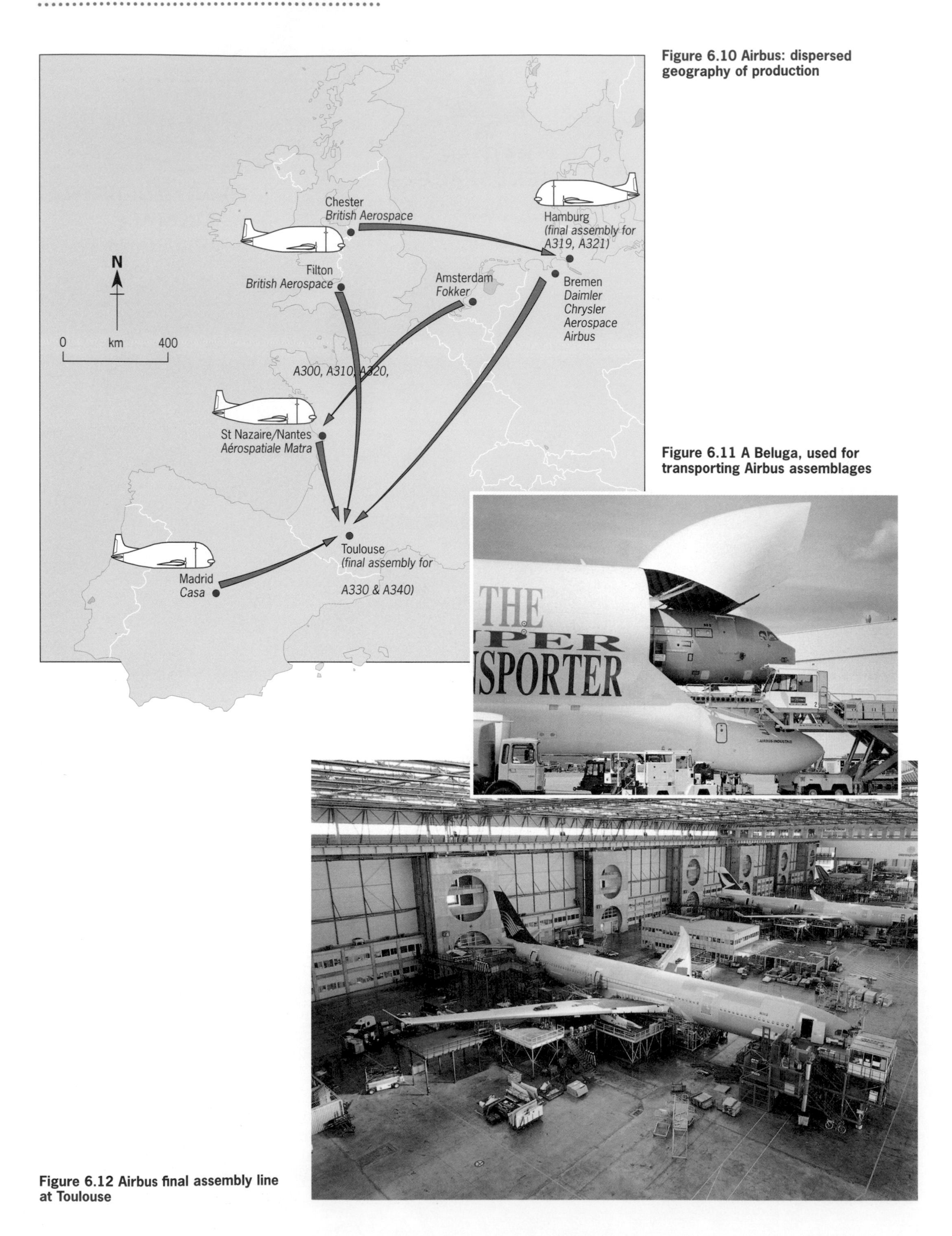

Figure 6.10 Airbus: dispersed geography of production

Figure 6.11 A Beluga, used for transporting Airbus assemblages

Figure 6.12 Airbus final assembly line at Toulouse

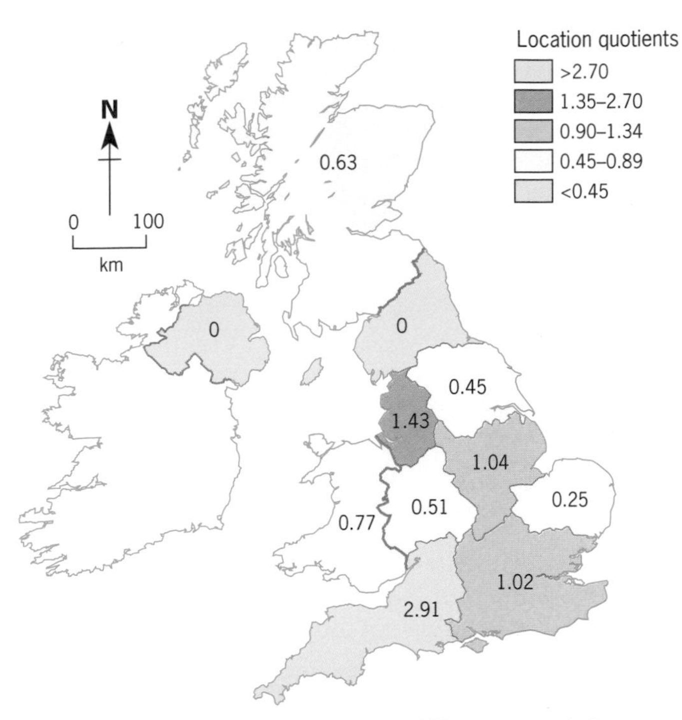

Figure 6.13 Regional distribution of the UK aerospace industry

In a further attempt to reduce costs, Airbus encourages communality of parts among its airplanes. For instance, the A330 and A340 use the same wings and many common systems and equipment.

Airbus is heavily subsidised by the four governments represented by the consortium. Its main rival, Boeing, argues that this gives Airbus an unfair advantage. But European manufacturers reply that massive defence contracts from the US government to Boeing and Lockheed are equally a kind of subsidy.

Military aerospace

Aerospace is a key component in defence industries. Indeed, nearly half the output of the UK's aerospace industry goes to the Ministry of Defence.

Because defence is such a major area of government spending in the UK it has an important effect on the geography of employment. Generally, those regions where aerospace and defence industries are concentrated have achieved greater prosperity. The South-East, South-West and North-West in particular benefited from UK military spending in the 1980s (Fig. 6.13).

BAE Systems is a massive conglomerate which includes Marconi electronic systems, the Royal Ordnance factories, and the manufacture of commercial airplanes and assemblies. However, over half BAE Systems' total turnover consists of military equipment, especially military aircraft, weapons and electronic systems.

Military aircraft are made in collaboration with other aerospace manufacturers. Again this is made necessary by the huge costs of developing military aircraft. Currently, the UK, Germany, Italy and Spain are collaborating in building the Eurofighter Typhoon, the world's most advanced combat aircraft.

7 Compare the aerospace industry with the automobile industry (see the Case Study on page 53) by completing a table using the following headings: type of industry (i.e. processing, fabrication, assembly); nature of firms; capital investment; economies of scale; government influence; spatial organisation.

Analyse the relationship between regional defence spending by the government and the importance of the regional aerospace industry in the UK by:

8 Plotting the data in Table 6.2 as a scattergraph. Calculating the Spearman rank correlation coefficient.

9 Writing a paragraph to explain the result.

Table 6.2 Regional defence spending and employment in aerospace

	MOD spending (%)	Employment in aerospace (% of national total)
South-East	49	28.1
East Anglia	3	0.5
South-West	12	16.8
West Midlands	4	8.1
East Midlands	4	12.4
Yorkshire and Humberside	2	4.3
North-West	14	9.5
North	3	0.5
Wales	2	3.8
Scotland	7	6.5

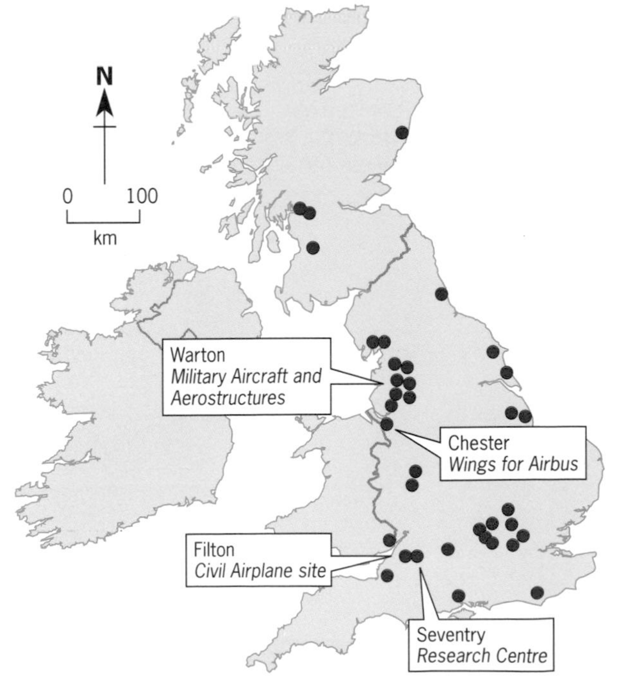

Figure 6.14 Distribution of BAE Systems' plants in the UK

6.4 Fordist and flexible industries

In terms of the organisation of production, geographers recognise two types of manufacturing industries: Fordist industries and flexible industries. Their main characteristics (Castree, 1992) are listed below. Before the 1970s manufacturing in MEDCs was Fordist (so-called after the assembly-line product techniques of Henry Ford's car plants in the 1930s). Recently many Fordist industries have declined and there has been rapid growth in industries using flexible production.

Characteristics of Fordist industries

- Mass production for mass consumption.
- Long production runs.
- Large workforce.
- Assembly-line production with workers given single tasks for the sake of productive efficiency (i.e. rigid division of labour).
- Standardised products.
- All or most of the components needed to assemble the final product made within the factory (i.e. vertical integration).

Characteristics of flexible industries

- **Flexible products:** rather than producing a few standardised products in bulk, a whole range of customised products are made.
- **Flexible production:** to meet even the smallest alteration in market demand, product types and quantities can be altered at short notice.
- **Flexible suppliers:** in order to make different products at short notice, flexible industries rely on several different parts suppliers on whom they can draw according to circumstances (i.e. horizontal organisation).
- **Flexible machines:** flexible production is partly achieved through using intelligent computerised machines which can make several products.
- **Flexible labour:** unlike the Fordist division of factory labour, flexible industries use a smaller workforce by promoting 'multi-tasking' (i.e. one person able to do several jobs).

?

10 Study the main characteristics of Fordist and flexible manufacturing techniques. Decide into which category you would place the following industries, and justify your choice: aerospace, iron and steel (Case Study, page 41), motor vehicles (Case Study, page 53), motor vehicle components and high-tech (Case Study, page 69).

Summary

- External economies describe the advantages to a firm of locating in an existing urban–industrial centre. They are known alternatively as agglomeration economies.
- Weber said that external economies could divert a firm from the least-cost transport location if the savings exceeded the extra costs of transport in locating there.
- Localisation economies refer specifically to the spatial clustering of firms and the resultant inter-firm linkages. These linkages help to reduce transport costs and improve direct communication between firms.
- Urbanisation economies are the savings made by manufacturing firms through location in urban areas.
- When cities exceed a critical size, urbanisation economies become diseconomies and unit costs start to rise. Rising costs result from traffic congestion, higher wages, high rents, high land prices, etc.
- Government policies influence the location of industry at varying scales, from the continental to the local.
- Strong links exist between governments and the aerospace industry.
- The aerospace industry is one of high risks and high costs. Manufacturers respond by undertaking collaborative projects and sourcing components and assemblies worldwide.
- The dispersed geography of production of Airbus is largely controlled by the huge capital cost of designing, developing and manufacturing airplanes.
- In MEDCs, defence spending has a major influence on the geography of manufacturing employment.

7 Decision-making

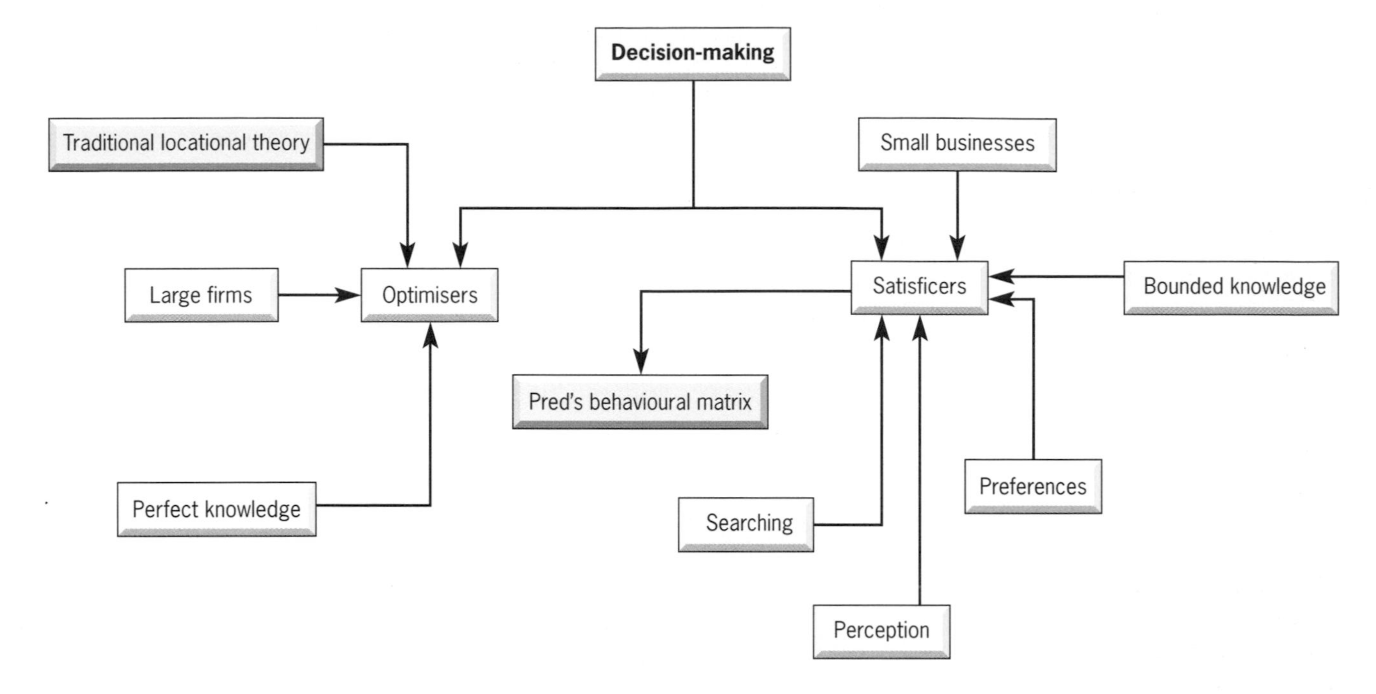

7.1 Introduction

In previous chapters we assumed that the location of industry was controlled largely by economic factors. Decision-makers selected locations which either maximised profits or minimised costs. In this chapter we shall question this assumption by taking a closer look at the way decision-makers behave in reality.

7.2 Optimisers or satisficers

The so-called behavioural approach, which considers the goals that individuals and companies have, grew out of dissatisfaction with traditional locational theories. As we saw in Chapter 3, these theories make a number of unrealistic assumptions about decision-making. In particular they assume that a decision-maker is an **optimiser**; that she or he has **perfect knowledge**; and that such a person seeks to maximise profits or minimise costs. We refer to this mythical being as **'economic man'**.

In reality, people usually aim for satisfactory, rather than optimal decisions. This satisficing behaviour should not surprise us. Decision-makers rarely have all the information they require to make the best choice. In other words, they have **bounded knowledge** rather than perfect knowledge. There are three reasons for this. First, complete information is rarely available. Second, people's perception of reality is inaccurate and gives them an imperfect image of the world. And third, searching for information is both time-consuming and costly, so that most searches are abandoned at an early stage.

A further aspect of satisficing behaviour is the importance it gives to non-material goals. Thus, a small industrial enterprise might choose to trade off lower profits from locating in a remote rural area for the extra **psychic income** provided by its environmental attractiveness. Unlike the optimisers of traditional theory, **satisficers** are influenced as much by personal preferences and values as by profit.

1 Read through section 7.2 and make a list of factors that might influence decision-makers. Start with 'need to maximise profits'. Add to your list as you read the next section.

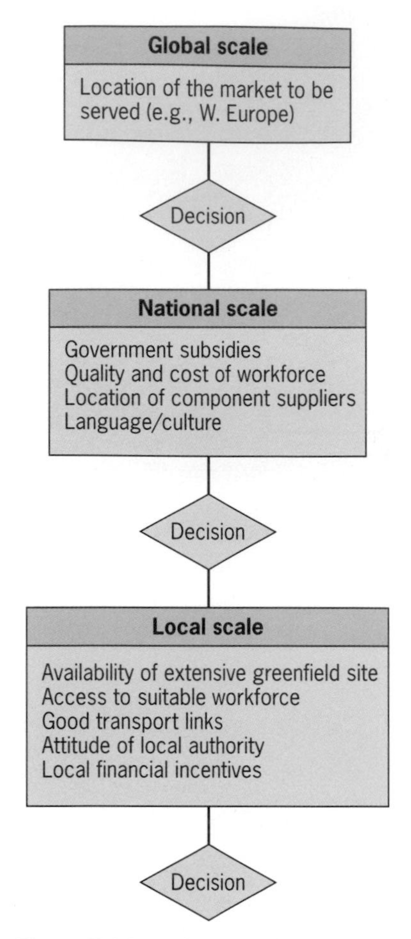

Figure 7.1 Locational decision-making: overseas investment by a transnational vehicle manufacturer

Clearly, the decisions taken by individuals and firms diverge from the ideal of economic man. However, it can be assumed that large, multi-plant firms will generally aim to locate a new plant where profits can be maximised. Locational decisions will be taken by boardroom managers and executives based at company headquarters, who will not have to work in the plant in question. In these circumstances it is unlikely that non-economic criteria such as the quality of the residential environment and recreational opportunities will have much significance.

It is also likely that multi-plant firms will undertake extensive searches into alternative locations at national and even international scales (Fig. 7.1). Toyota, for instance, investigated 28 sites in the UK before choosing Burnaston near Derby for its new assembly plant. In 1978, Peugeot drew up a shortlist of ten sites in France for a new gearbox plant (Fig. 7.2). After detailed investigations of each one, it finally chose Valenciennes in Nord–Pas-de-Calais. In comparison, small firms usually confine their search to a pre-determined region where it is already located. Their priority may be to minimise disruption caused by such moves to production and to the workforce.

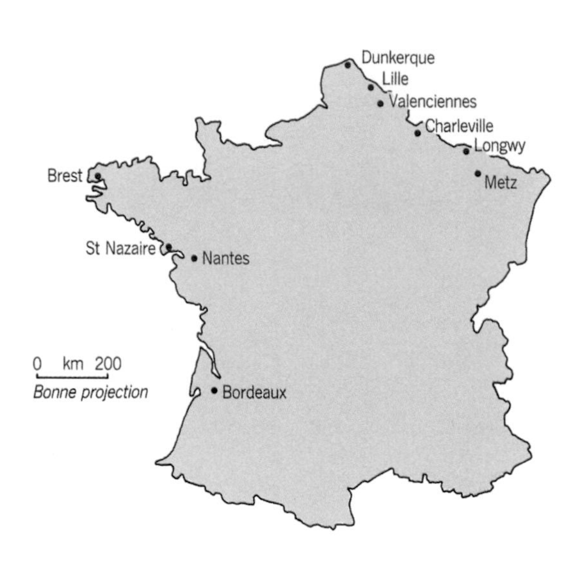

Figure 7.2 Sites investigated by Peugeot for the location of a new gearbox plant (1978)

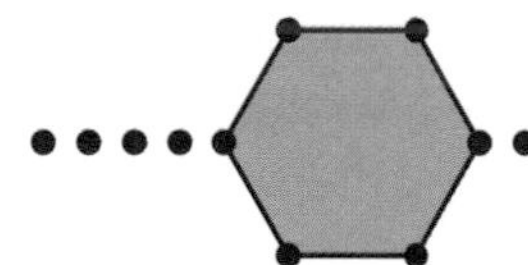

Pred's behavioural matrix

Pred's **behavioural matrix** (Fig. 7.3) summarises the behavioural approach to decision-making. In theory, every decision-maker (e.g. a business executive seeking a new location for a factory) can be allocated a place in the matrix, based on the quality and quantity of information they have, and their ability to use it. (In practice it is impossible to measure either variable precisely.)

According to Pred, a decision-maker positioned near the bottom right-hand corner, with plenty of information and ability, would probably make a sound locational decision close to the optimum. On the other hand, someone placed in the top left-hand corner, with little information and ability, would probably make a poor decision, a long way from the optimum. However, the whole point of the matrix is that nothing is absolutely certain: a favourable position in the matrix does not guarantee a good locational decision, any more than a poor position guarantees failure. Because of this element of uncertainty Pred's matrix is known as a **probabilistic theory**. Some of the theories we looked at in earlier chapters (Weber, Hoover) always have a fixed outcome. They are called **deterministic theories**.

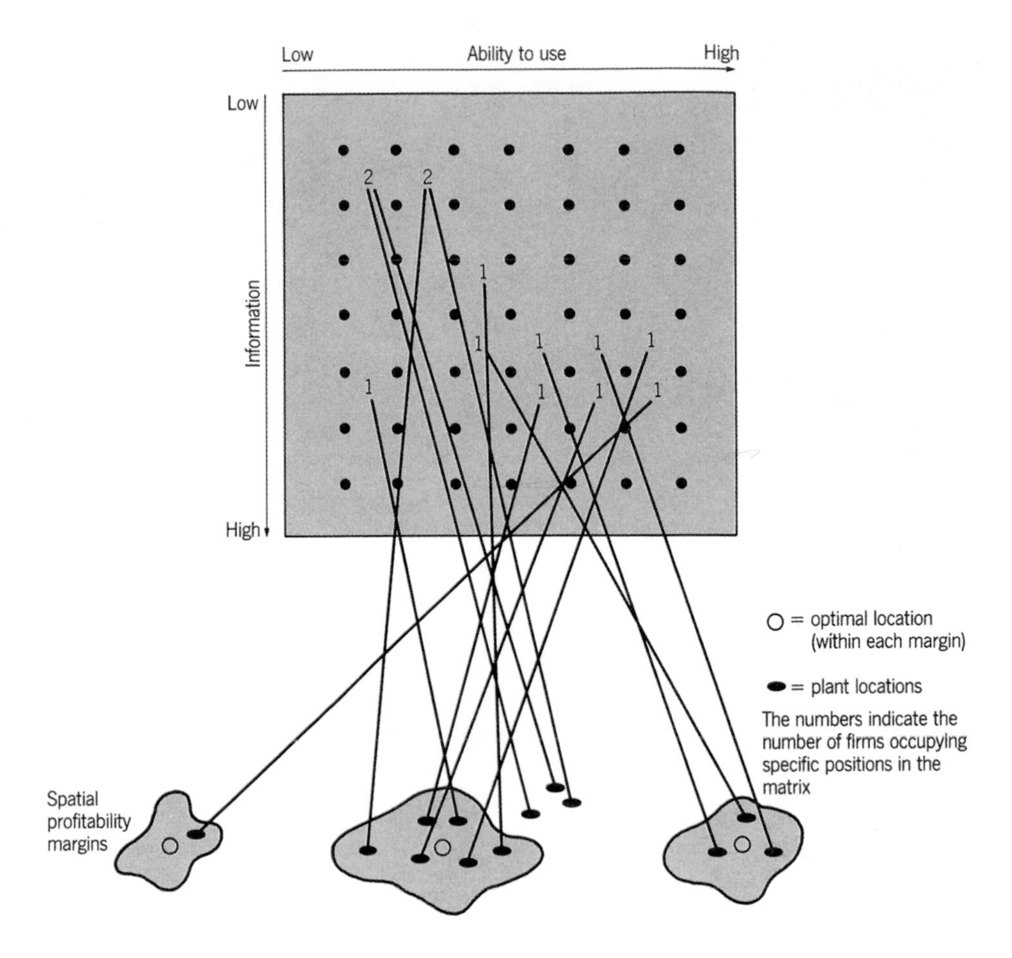

Figure 7.3 Pred's behavioural matrix and industrial location decisions (*Source:* Pred, 1967)

7.3 Personal preference

The personal preferences of decision-makers can have an important influence on industrial location. These preferences have most influence on small businesses and family-run firms, especially in the early stages of their development.

The home town effect

The influence on location of the home town of entrepreneurs is well known. Several large manufacturing centres owe their importance to this chance event. Examples include Josiah Wedgwood and the pottery industry at Stoke-on-Trent in the eighteenth century; Henry Ford, who set up the Ford Motor Company in his home town of Detroit in 1903; and William Morris's pioneering of the car industry in Oxford in the early twentieth century.

Early this century the town of Billund in central Denmark was dismissed as 'a god-forsaken railway stopping point where nothing could possibly survive'. Today, LEGO, the Danish toy manufacturer, has its headquarters and main factory there (Fig. 7.4). The firm has a total workforce of 6000, and operates branch plants in Switzerland, Germany, the USA, Brazil and South Korea. The importance of Billund is due solely to the historical accident that the company's founder, Ole Kirk Christiansen, was born and brought up there. Since its foundation in 1934, LEGO has remained a family firm. This form of organisation has given particular scope for personal factors to enter the

Figure 7.4 LEGO's Kornmarken factory at Billund, Denmark

decision-making process. For example, the decision to establish a branch factory in Switzerland was made because LEGO's technical director at Billund, who was Swiss, wanted to return home!

Thus entrepreneurs setting up small businesses have a freedom to act on their personal preferences in a way which is denied to large, multi-plant firms responsible to shareholders. In the UK, the personal motivation of founders in favour of environmentally attractive locations has fostered the growth of electronics, biotechnology and software firms in remote regions like the Highlands of Scotland.

For large firms, a location in areas of high environmental quality will only be chosen when it makes economic sense. This may happen when the aim is to attract highly qualified workers who are in short supply. Thus, many high-tech firms in the USA have moved to locations like southern California, Arizona and Colorado, where the quality of life and climate make it possible to recruit and retain key employees.

2 What images do you have of northern and southern Britain? Write down ten words that you associate with each region. Compare them with those of other students in your class to produce a collective image of each region. Discuss the accuracy of these images. What channels of information have contributed to your image of North and South?

3 Small footloose businesses have considerable freedom of location. Imagine that you are the managing director of such a firm which is planning to build a new factory.

a Place the following locations in rank order of preference (assume that they are all equally profitable).

Bath	Nottingham
Bradford	Oxford
Bristol	Salford
Cambridge	Stoke-on-Trent
Hull	Southampton
Leicester	York

b Compare your results with other students and produce an overall league table. Discuss the rankings and explain why some locations are perceived more favourably than others.

Mental images of place

Through their perception of geographical information, decision-makers build up a mental image of potential locations. Some places like Cornwall and the Lake District are associated with positive images. Others, such as the industrial North and South Wales, are perceived less favourably. Stereotype images of industrial Britain – militant workers, mines, smokestacks, grime, terrace houses and cobbled streets – are very persistent. These images go back a long way, to the nineteenth century and the Industrial Revolution. A classic example is Charles Dickens's vivid description of Coketown (Preston) in *Hard Times*:

> It was a town of red brick, or of brick that would have been red if the smoke and ashes had allowed it; but, as matters stood it was a town of unnatural red and black like the painted face of a savage. It was a town of machinery and tall mill chimneys, out of which interminable serpents of smoke trailed themselves for ever and ever and never got uncoiled. It had a black canal in it, and a river that ran purple with ill-smelling dye, and vast piles of buildings full of windows where there was a rattling and a trembling all day long... It contained several large streets all like one another, and many small streets still more like one another.

More recently these negative images have been reinforced by the cinema, and popular television soap operas and comedy series. Such images, although containing an element of truth, are grossly distorted. None the less, they are important: our decisions are based on how we think the world is, rather than how it is in reality.

?

4 Make a study of advertising which promotes places in the UK as locations for industry and services. You will need at least 20 different adverts. Search for them in national newspapers, colour supplements and magazines like *The Economist* in your school library.

5 Locate the places advertised on an outline map of the UK. Comment on their distribution.

6 Classify the places into new towns, old industrial areas, remote rural areas and others. Do any patterns emerge? Are some types of places more likely to advertise than others? If so, can you explain why?

7 Look at a recent copy of *Regional Trends*, and obtain information on a number of social and economic features for the places which are advertising. For example, in the section 'district statistics' there are data on unemployment, employment structure, housing tenure, gross value added in manufacturing, etc. Study these data, compare them with the national averages and comment on any patterns and trends.

8 Analyse the relative importance of the qualities claimed for each place (e.g. skilled labour supply) featured in the adverts.

9 Try to compare some of the claims made in the adverts (e.g. central location, good communications, etc.) with reality. To what extent do the adverts distort reality in order to promote a positive image of a place? Is this in your view justified?

Figure 7.5 Promoting places – advertising slogans and booster images

Local authorities and development agencies are aware of the power of mental images to influence inward investments. Since the 1970s many have launched vigorous advertising campaigns designed to create positive counter-images (Fig. 7.5). In this 'battle for the mind' places have been promoted through booster images, stressing qualities such as accessibility, the skills and loyalty of the local workforce, the availability of factory premises, government financial assistance and the attractiveness of the physical environment.

Summary

- Individual decision-making is likely to be satisficing, not optimising.
- Satisficers have bounded knowledge and often seek non-material goals.
- Large, multi-plant firms, employing professional managers and responsible to shareholders, are more likely to adopt optimising strategies than small businesses.
- Location decisions are influenced by the perception and mental images of places.
- Unfavourable perception of places may be countered by the use of booster images through advertising.

8 The organisation of firms

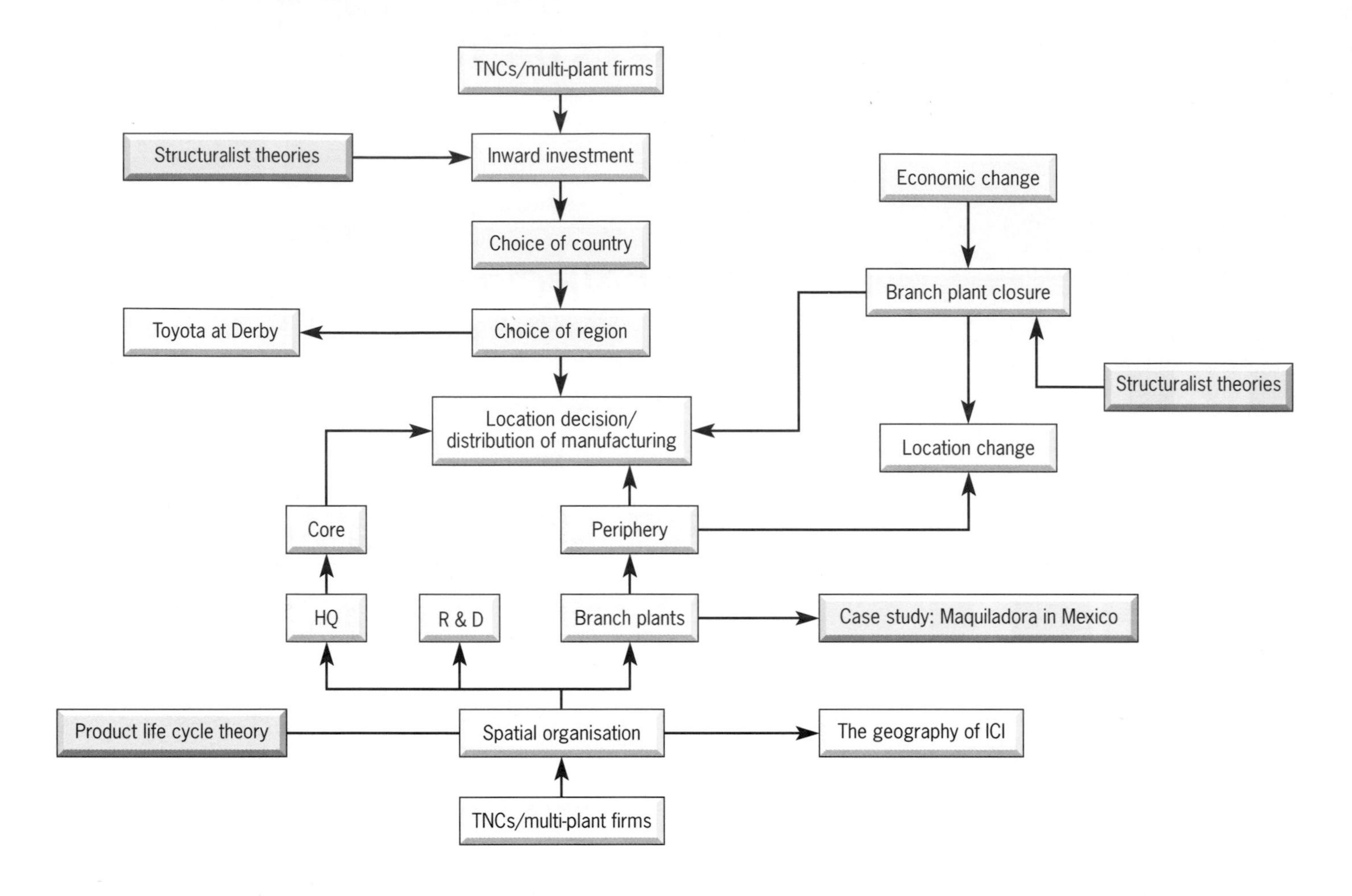

8.1 Introduction

The geography of modern manufacturing is only partly explained by the theory and locational factors we have met in previous chapters. Recently there has been a growing realisation that the organisation of firms (especially large ones) is a major influence on location. Large firms are often multi-plant, **multi-locational** and transnational. Their spatial organisation is complex, with different functions (headquarters, design, research and development (R&D), and production) frequently occupying separate sites. Indeed, very large firms operating worldwide have a geography of their own.

8.2 Transnational corporations

Scale and organisation

Transnational corporations (TNCs) are very large firms like Philips, IBM and Nissan, which market their products globally and have manufacturing operations in several countries. The major advantage of large size is economies of scale. As a result firms can reduce costs, finance new investment and compete in world markets. General Motors, the US car giant, is the world's largest industrial corporation. Its annual turnover exceeds the GDP of all except 20 countries in the world. One-third of its output is from plants outside the

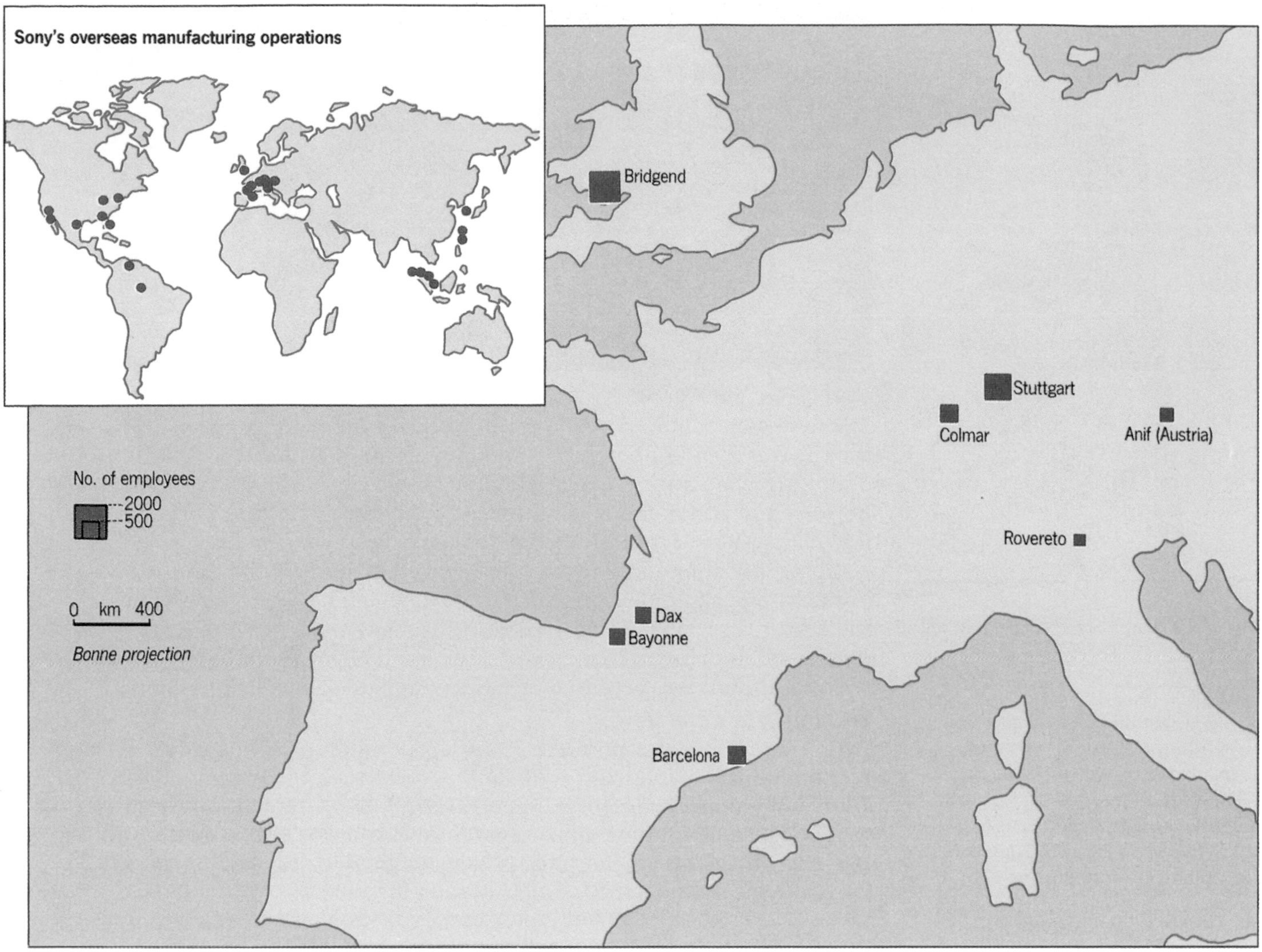

Sony's overseas manufacturing operations
Bridgend
Stuttgart
Colmar
Anif (Austria)
No. of employees
2000
500
Rovereto
0 km 400
Bonne projection
Dax
Bayonne
Barcelona

Figure 8.1 Sony's European manufacturing operations

34.1%
All other exports
33.3%
TNC intra-firm exports
32.6%
TNC exports to unrelated firms

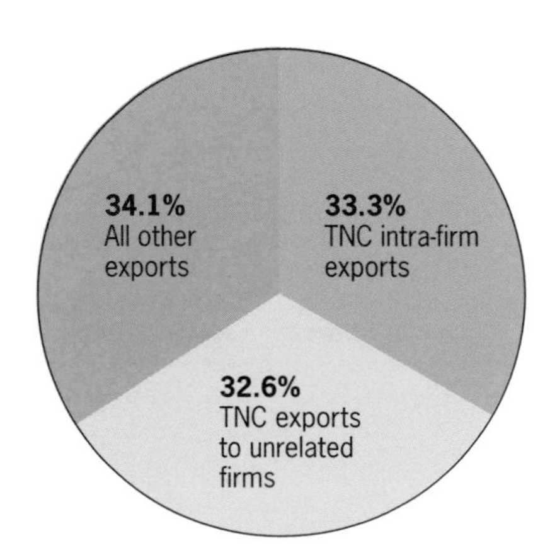

Figure 8.2 TNCs and world exports: value of world exports of goods and services

USA. Even so, few TNCs are truly global; most retain the bulk of their workforce, production and R&D in their home countries.

The largest firms serve global markets. To do this efficiently they often adopt a decentralised organisation, with control devolved to regional headquarters. Thus, Ford's operations are divided into five geographical groupings (North America, Europe, Asia–Pacific, Latin America and Middle East). Each one is largely independent of company headquarters in Detroit and is responsible for its own R&D, product development and investment plans.

The globalisation of manufacturing

In the last 30 years, globalisation has led to a relative shift of manufacturing from North America, Europe and Japan to LEDCs in Asia and Latin America (Fig. 8.4). This trend is set to continue: by 2005 it is likely that almost one in three jobs in manufacturing industry will be in LEDCs. The huge growth of inward investment by Taiwanese firms in East and South-East Asia between 1986 and 1997 is a recent example of this global shift (Fig. 8.8). This growth has been fuelled by the liberalisation of the Chinese economy in the 1990s, pro-market reforms in Vietnam and low labour costs throughout the region.

US firms were the first to globalise their operations (Ford set up its first overseas plant at Manchester in 1911). After 1945 many European firms also started to invest heavily overseas. By the late 1970s the relocation of East Asian firms in American and European markets had begun. The Japanese

Figure 8.3 Sony's branch plant at Bridgend, South Wales

electronics and automotive industries led the way. In the late 80s and 90s the Japanese were followed by the South Koreans and the Taiwanese. In Europe, the UK was the most popular location for Japanese **inward investment**. Most of this investment went to peripheral regions such as South Wales, North-East England and Scotland (Fig. 8.5). Indeed by 2000, 30 per cent of all jobs in manufacturing in Scotland were in foreign-owned companies.

Globalisation has not, however, affected all manufacturing industries. For example, a heavy, bulky item such as steel, which has high transport costs, is still a mainly regionally-traded product. In some industries proximity to market is essential. Despite high labour costs the fashion clothing industry remains important in MEDCs because it needs to react quickly to changes in market demand.

The advantages of global production

There are two main reasons why firms opt for global production: first, because by locating overseas they can get round trade barriers such as tariffs, quotas and voluntary agreements which protect home markets; and second, because production overseas (especially in LEDCs) often lowers costs. Thus, TNCs may gain access to cheaper labour and materials, and operate in an environment with fewer and less stringent pollution controls. Contrary to locational theory, there is little evidence that large firms locate close to overseas markets to reduce their transport costs.

The importance of TNCs

TNCs are probably the most important single influence on the global economy. World production and trade are bound up with their investment and disinvestment decisions. Today, much of the world's trade is intra-firm (see the section on Ford of Europe in the Case Study on p.53), taking place within TNCs. In future, this type of trade is likely to grow, as countries that belong to the World Trade Organisation (WTO) agree to lift tariffs and other barriers to trade.

1a Analyse the regional distribution of successful inward investment manufacturing projects in the UK (Fig. 8.5) by calculating the ratio between the number of persons employed in each region (Table 8.1) and the number of successful projects.
b Describe and suggest possible explanations for the spatial distribution of this ratio.

2a Describe the global pattern of Japanese overseas investment in Figure 8.6.
b How is the concentration of investment in the USA and Europe likely to differ from that in South-East Asia?
c Why do you think that there has been little Japanese interest in South America and Africa?

3 Table 8.2 shows that very large firms are important in the motor vehicle industry. Refer back to the Case Study on page 53 and try to explain why this is so.

Table 8.1 Number of persons in employment 1998 (millions)

Region	Millions
North-East	1.070
North-West	2.989
Yorkshire and Humberside	2.211
West Midlands	2.410
East Midlands	1.966
East	2.536
South-East	7.033
South-West	2.300
Wales	1.216
Scotland	2.278

Table 8.2 Top 10 TNCs by foreign assets, 1995 (*Source: The Economist*, 22 Nov. 1997)

Company	Industry	Foreign assets as % of total	Foreign sales as % of total	Foreign employment as % of total
Royal Dutch/Shell	Energy	67.8	73.3	77.9
Ford	Automotive	29.0	30.6	29.8
General Electric	Electronics	30.4	24.4	32.4
Exxon	Energy	73.1	79.6	53.7
General Motors	Automotive	24.9	29.2	33.9
Volkswagen	Automotive	84.8	60.8	44.4
IBM	Computers	51.9	62.7	50.1
Toyota	Automotive	30.5	45.1	23.0
Nestlé	Food	86.9	98.2	97.0
Bayer	Chemicals	89.8	63.3	54.6

Figure 8.4 Changing regional shares of global manufacturing, 1970–2005

Figure 8.5 Direct inward investment, 1997–98: number of project successes in manufacturing

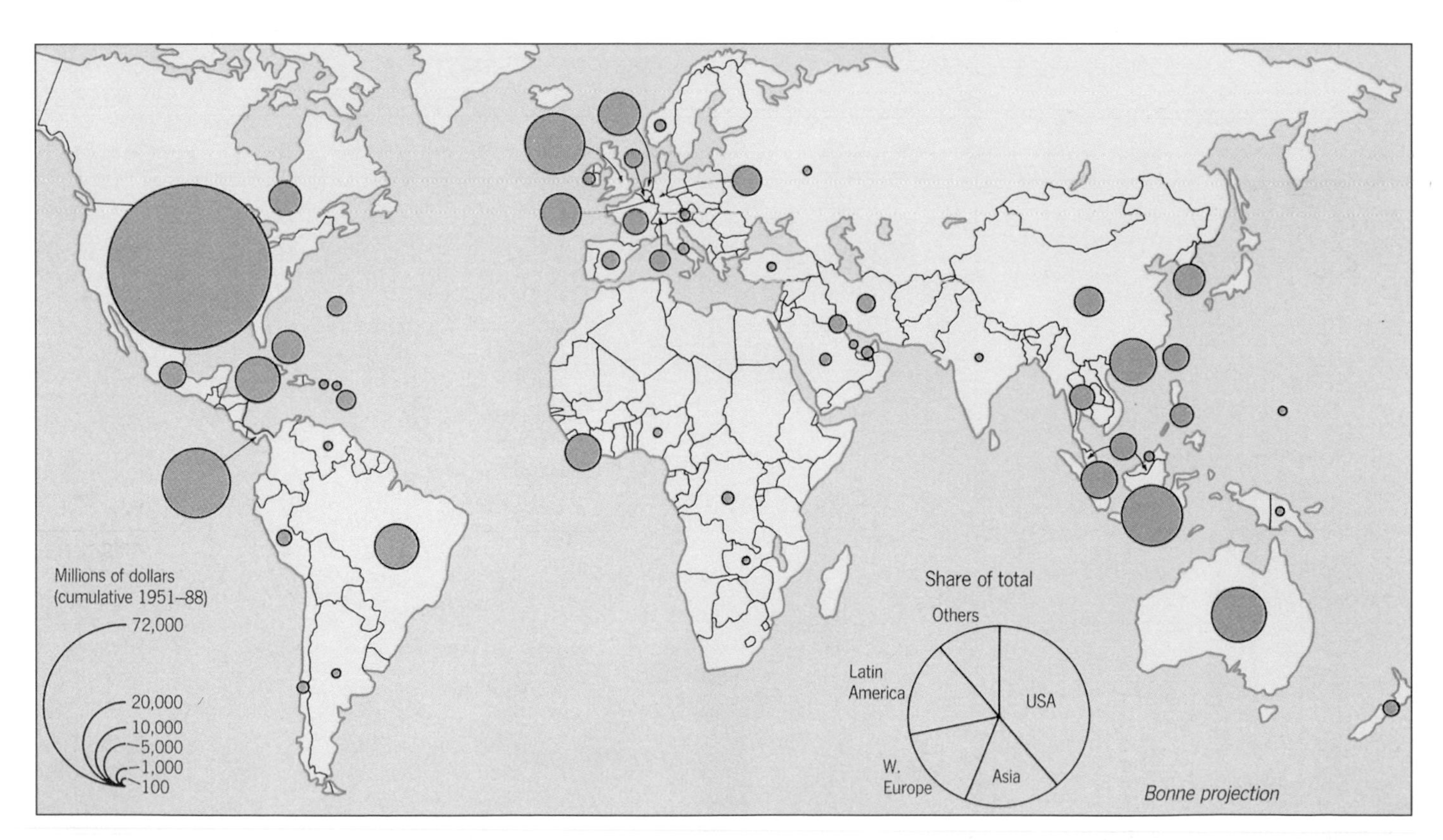

Figure 8.6 The global distribution of Japanese overseas direct investment, 1988 (*Source:* Dicken, 1990b)

Figure 8.7 Honda's car plant at Swindon: Honda was one of 3 major Japanese car makers to establish assembly plants in the UK between 1986 and 1993.

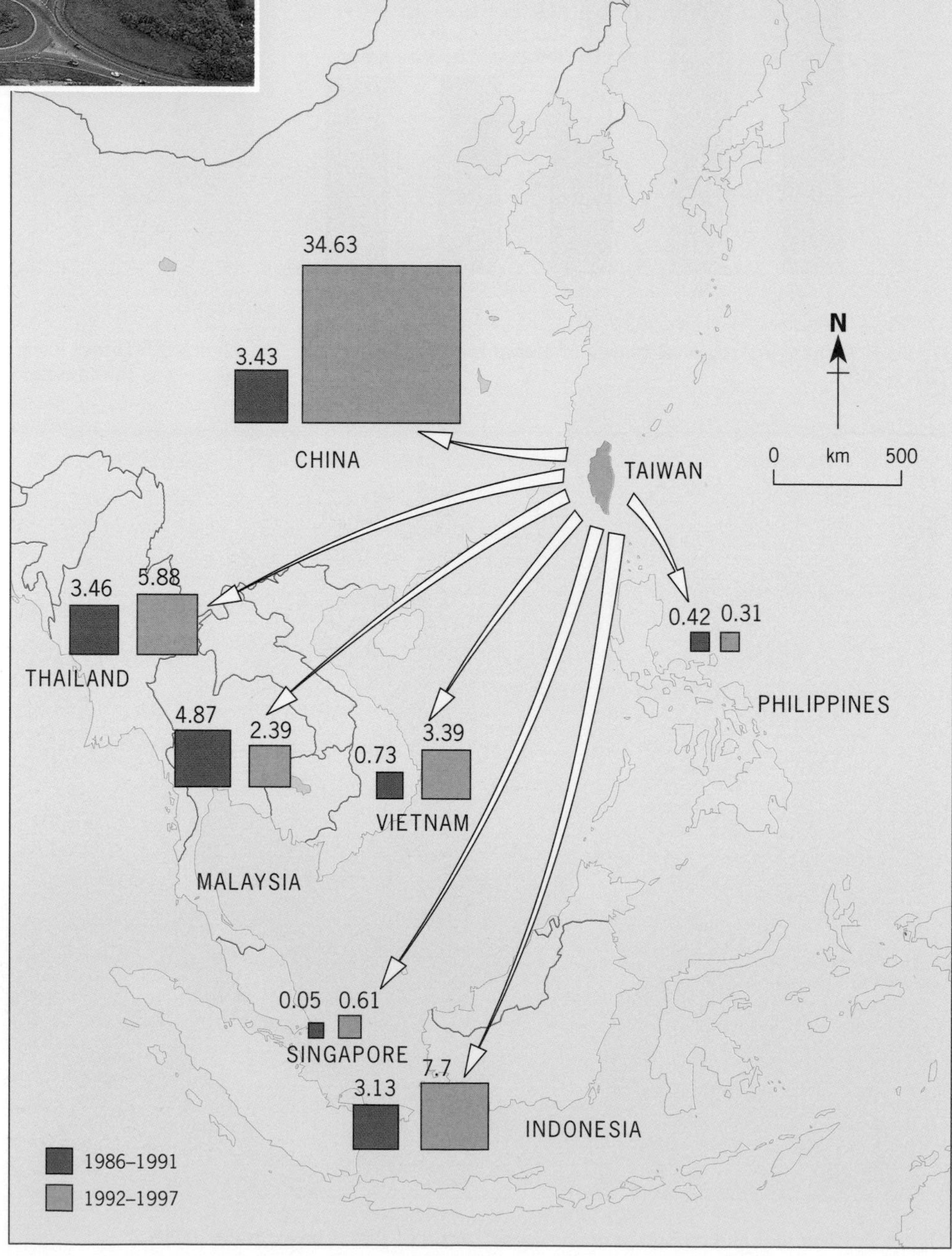

Figure 8.8 Inward investment in East and South-East Asia by Taiwanese small and medium-sized enterprises (billions $), 1986–97

?

4a Describe the changing pattern of investment in East and South-East Asia by Taiwanese small and medium-sized enterprises between 1986 and 1997.
b Explain why many Taiwanese firms relocating in East and South-East Asia are labour-intensive.

8.3 Large firms and locational decision-making

Unlike small firms, large TNCs have real choice when choosing a location for a new plant. Governments are able to influence the locational decisions of TNCs indirectly by imposing tariffs and quotas or by making voluntary agreements. The purpose of these measures is to protect domestic markets. Faced with these obstacles, TNCs have a choice: either they can continue to serve foreign markets through international trade, or they can opt for direct (overseas) investment. In the 1980s and 1990s Japanese, South Korean and Taiwanese TNCs increasingly chose the second option.

Direct investment results in locational decisions at two scales: the choice of country, and the choice of a region within the country.

Locational choice between countries

The choice of a particular country for investment often hinges on political factors. Governments are usually keen to encourage inward investment. This is not surprising: inward investment not only creates jobs (often in large numbers), it also boosts exports which help to balance a country's trade. Competition between countries to attract investment is often intense. Most governments offer a package of financial inducements to TNCs, including capital grants and tax concessions. TNCs have the economic power to trade off one country against another in order to get the best deal.

Yet TNCs do not have it all their own way. Governments usually link inward investment to performance requirements. This often means local content agreements, where a company agrees to source a given proportion of its materials from local suppliers. Without this, a product may be unacceptable in a market like the EU. Other agreements may lay down a minimum level of exports, and the establishment of R&D facilities as well as production. Investment in R&D has the advantage of providing a wider range of skilled jobs, and suggests a long-term commitment by a company. In times of falling profits this reduces the risk of closure.

Locational choice within countries

Within any country the final location will only be determined after detailed analysis of costs at several sites. Unlike smaller firms, TNCs have the resources to undertake relatively thorough searches of potential sites. Four main factors guide site selection: communications, the availability of labour, the cost of land and buildings, and the level of government subsidy (Fig. 7.1). The sites chosen by major Japanese car manufacturers in Britain – Nissan at Sunderland, Toyota at Derby and Honda at Swindon – show the influence of some or all of these factors.

German firms in Britain regard the South-East as the most favourable location. This is in spite of its relatively high local taxes, rents and wages. The perceived advantages of the South-East are its motorway network giving good access to ports and airports, and London's status as a capital city and financial centre.

However, even large firms can be influenced in their locational choice by non-economic factors. In 1980, ICI chose to locate a new polymer plant at Wilhelmshaven in Germany, rather than at Dunkerque in northern France. Although Wilhelmshaven offered lower energy costs and access to Germany's successful car industry, ICI attached more importance to the helpful attitude of politicians at both city and regional levels. In other words, it was a behavioural factor (see section 7.2) which tipped the balance in Wilhelmshaven's favour. The sites chosen by two major Japanese car manufacturers in the UK – Nissan at Sunderland and Honda at Swindon – show the influence of these factors.

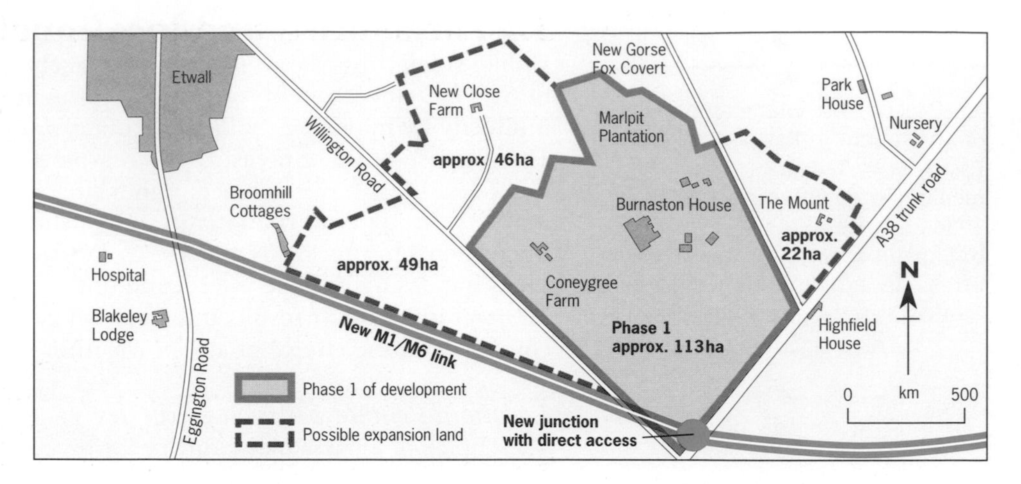

Figure 8.9 The site of Toyota's car assembly plant at Burnaston (*Source*: Divelly, 1990)

The location of Toyota's car assembly plants in Europe

Derby

In 1989 Toyota announced its decision to build its first European car assembly and manufacturing plant at Burnaston near Derby. With an investment of £840 million, it represented the biggest Japanese investment in Europe at the time. The plant, built to serve the European market, was producing 220,000 cars a year by the turn of the century.

Having chosen the UK, Toyota's first requirement was a large greenfield site with a minimum of 100 ha (Fig. 8.9). Toyota also wanted a location outside the South-East, with its high land prices and wage levels. In all, 28 possible sites were investigated. Eventually they were reduced to three: Burnaston, Immingham (North Lincolnshire) and Llanwern (South Wales).

In the end, three considerations weighed in Burnaston's favour: its skilled workforce (and lack of any local competition for these workers from other Japanese firms); its situation close to the car components industry in the West Midlands; and the co-operation and positive attitude of the local authority (Derbyshire County Council).

Valenciennes

In 1997 Toyota took the decision to expand its operations by building its second European assembly plant at Valenciennes in northern France. The plant, due to open in 2001, will assemble a new small car for the European market. The combined output of Toyota's Derby and Valenciennes plant will raise the company's production to 400,000 cars a year by 2003. The Valenciennes project will create 2000 new jobs directly in the assembly plant, and a further 2,500 among parts suppliers.

Toyota opted for the Valenciennes site against competition from 75 other locations in Europe. The advantages of Valenciennes included:

- a £30 million subsidy from the French government;
- the existing concentration of car assemblers (Fiat, Renault, PSA) and automotive suppliers in the region, with a large pool of skilled labour;
- France's membership of the single currency, making it easier and cheaper to sell cars on the Continent;
- unemployment levels of 20 per cent in North-eastern France (the legacy of deindustrialisation and the decline of staple industries such as textile and steel); it was thought that investment in an economically and socially deprived region would enhance Toyota's image among the French car-buying public;
- a well established infrastructure and a favourable geographical situation close to parts suppliers and the European market e.g. access to the Channel Tunnel for sourcing parts made in the UK.

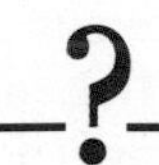

5 Draw an annotated diagram to show the general and specific factors that influenced Toyota in its choice of location at Derby and Valenciennes.

8.4 The spatial organisation of multi-plant firms

Large multi-plant firms often comprise three organisational elements: a headquarters, a research and development centre, and branch plants (Fig. 8.11). Control and policy-making are found at the headquarters, and routine production at branch plants. Most of the workforce based at the headquarters and R&D locations is highly qualified and highly skilled. Branch plants, on the other hand, employ mainly semi-skilled or unskilled labour.

These organisational divisions often have a distinctive geographical expression. Thus headquarters and R&D functions are typically located in prosperous **core** regions, which give access to large pools of skilled workers, producer services and good communications. Eighty per cent of the UK's largest firms are headquartered in the South-East. Branch plants, by comparison, are more footloose. Even so, peripheral regions with plentiful labour supplies and government grants and loans have proved especially attractive.

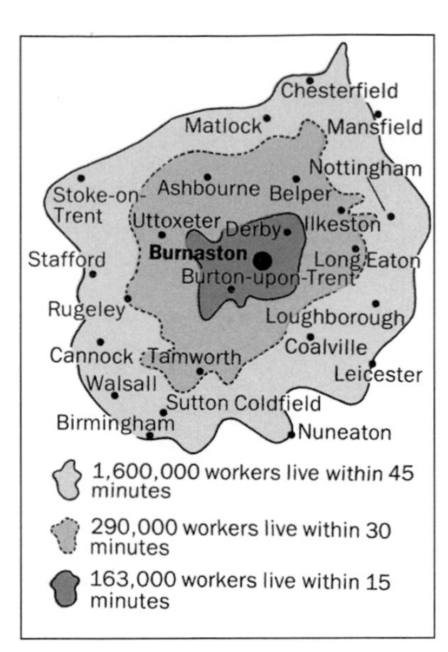

Figure 8.10 Access to labour supply: Toyota's car assembly plant at Burnaston

The geography of ICI

Imperial Chemical Industries (ICI) is one of the world's largest chemical firms. It manufactures a wide range of products, from fertilisers and explosives to plastics and pharmaceuticals. It has nearly 80 R&D establishments, and operates 200 factories in 59 different countries. The distribution of these activities results either from ICI's own direct investment policies or from takeovers.

At the global scale a large proportion of employment in TNCs is found in the home country. Thus ICI, as a British TNC, has 40 per cent of its workforce based in the UK. Subsequent growth has spread outwards from this core area (Fig. 8.12). ICI's traditional markets were the UK and the 'British empire' countries.

Figure 8.11 Multi-plant manufacturing firms: locational characteristics of different operations (*Source:* Humphrys, 1988)

	Function	Personnel	Location
Headquarters	Control Policy–making Strategic decisions	Salaried Managerial Skilled White collar	Office block High–order central place Usually CBD Sometimes urban fringe park
Research and Development	Control Product development Product improvement Equipment improvement Techniques and methods to improve productivity	Salaried Highly qualified Professional and technical staff	Urban fringe of large high–order central place or small town in metropolitan hinterland
Processing	Production Conversion of raw materials to more useful form	High proportion of wage–earners Operatives Blue collar Technically skilled Unskilled	Port/coast Domestic raw material source
Fabrication	Production Conversion of processed products into final form/subassembly		Peripheral region Area with government aid for industry available
Integration	Production Assembly of fabricated products into finished products		Small town/rural subregion in central region and/or close to motorway/main–line railway

Decisions
Information
Materials

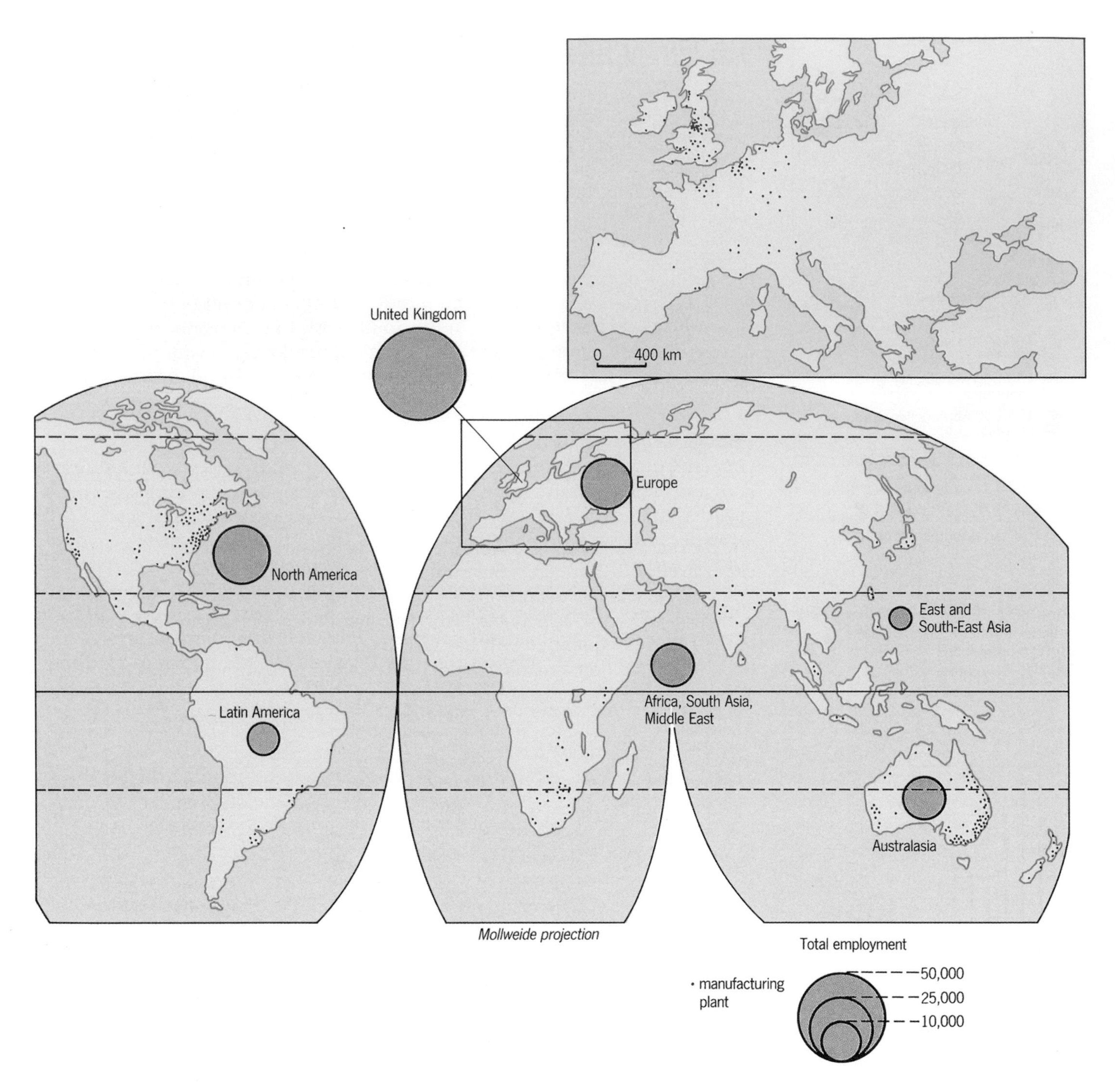

Figure 8.12 ICI worldwide: manufacturing plants and employment

Over the past 30 years ICI has grown in Western Europe, North America and is currently expanding in the growing Asian economies on the Pacific rim.

R&D is central to a firm like ICI which specialises in advanced technology. Currently, ICI spends 5 per cent of its annual turnover on R&D, a figure comparable to many high-tech industries. However, ICI's R&D functions have an uneven spatial distribution. They are mainly located in economically developed regions like Europe and North America, where the firm's demand for highly qualified scientists and technicians can be met (Fig. 8.13). Within the global geography of individual TNCs there is often a division between the prosperous core regions and their monopoly of headquarter and R&D functions, and the economically less-developed regions, concerned almost entirely with routine manufacturing.

?

6 Study Figures 8.12 and 8.13. Describe and explain the global distribution of ICI's manufacturing activities.

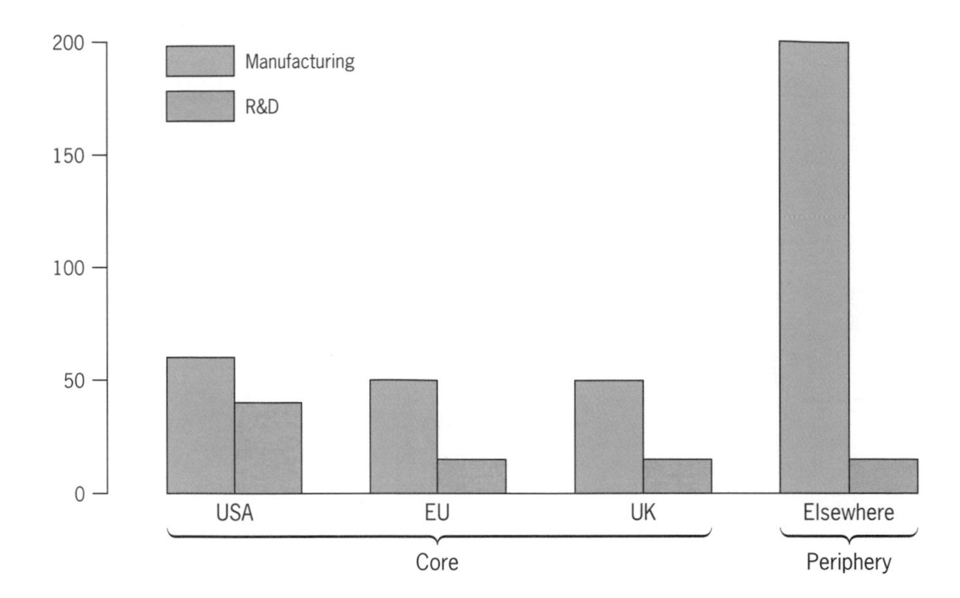

Figure 8.13 Distribution of ICI's manufacturing plants and R&D operations

Product life cycle theory

A **product life cycle** begins with the development of a product, and ends with its replacement by something better. Some products, such as high-quality porcelain, may have life cycles lasting one or two decades. Others, such as personal computers, may be obsolete after just three or four years. From a geographical point of view there is evidence that products at different stages of the cycle are associated with different locations (Fig. 8.14).

Stages in the product life cycle

Stage 1: Development Initial growth is slow. Efforts are made to improve quality and reliability, causing frequent changes in production processes and product design. Production relies heavily on scientific and engineering skills. Thus, production locates close to the firm's HQ and R&D operations.

Stage 2: Maturity The product has been perfected. Sales grow rapidly and efforts are directed towards lowering costs. The dispersal of production to peripheral branch plants begins.

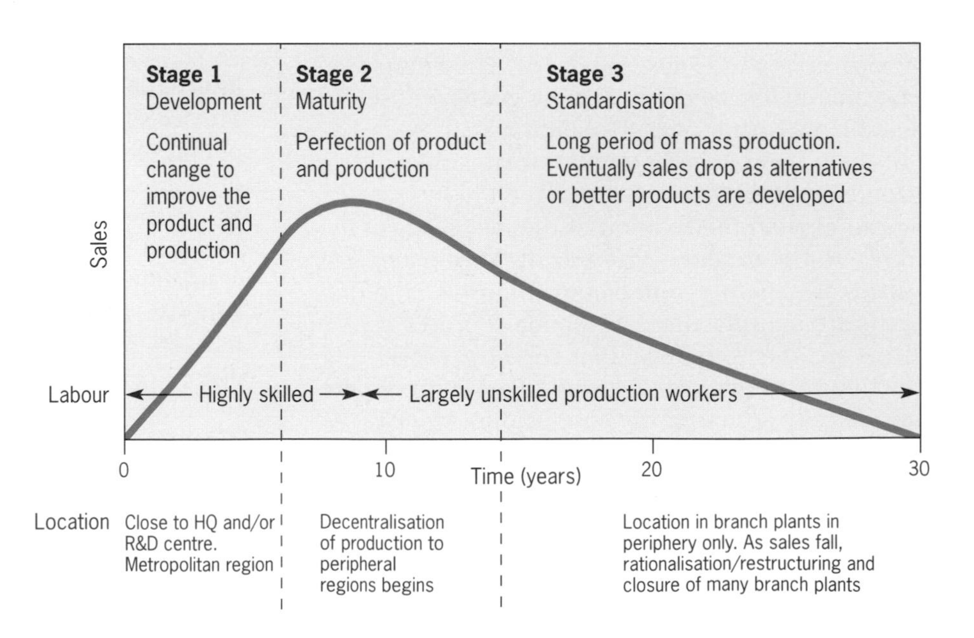

Figure 8.14 Product life cycle theory

Stage 3: Standardisation The importance of skilled labour and technical inputs diminishes. Costs are lowered by mass production and the replacement of skilled by semi-skilled (often female) labour. Locationally, the product is now more mobile. A locational preference for peripheral regions (for some products this may mean locating branch plants in LEDCs), where labour costs are low, is typical of this stage.

Because of the way multi-plant firms organise their activities geographically, some regions have a high proportion of manufacturing jobs in the first stage of the product cycle. This is seen in the electronics industry at the global scale. R&D is concentrated in the USA, Europe and East Asia. New products such as high-definition television, dependent on highly skilled, scientific staff, are developed in these core regions, and eventually production spreads out to the **periphery**. This means that most countries and regions are recipients rather than sources of new technology. They are also remote from the centres of decision-making in the core. Overall, this leaves the periphery in a position of **dependency** in relation to the core.

In the UK, the South-East is over-represented in the first stage of the product life cycle. By contrast, the manufacturing economies of peripheral regions like the North-East and Merseyside are dominated by branch plants engaged in stages 2 and 3.

?

7 Study the two theoretical product cycles in Figure 8.15.
a Describe the main differences between the two cycles.
b Explain how these differences are likely to influence industrial location.

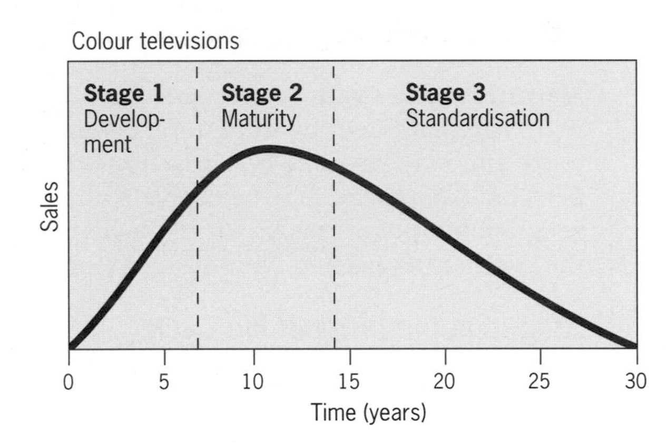

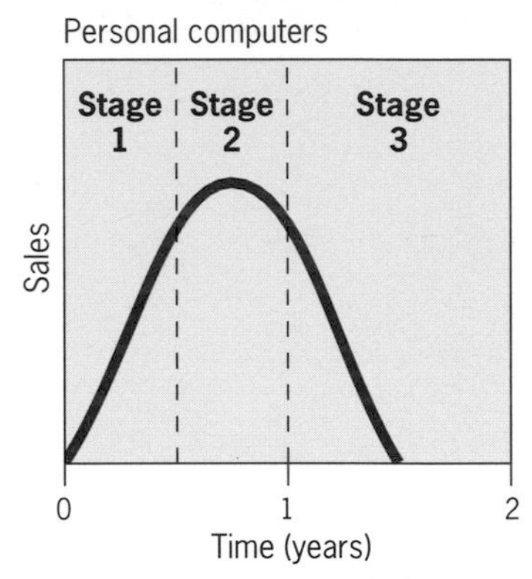

Figure 8.15 Theoretical product life cycles

The location of US branch plants in Mexico

US branch plants (or *maquiladoras*) have grown rapidly in Mexico since the 1960s (Fig. 8.16). They employ nearly one million workers, making goods whose value-added worth is over $7 billion a year. Almost half of them make mainly textiles or consumer electronics, mainly for the US market.

We can explain the location of the *maquiladoras* by reference to the product cycle model. Most *maquiladora* industries are 'mature' and labour-intensive, and their products are manufactured by standard processes. Thus US firms are able to reduce their costs by transferring production to a low-wage country like Mexico where wages are only a quarter of those paid in America. A further advantage is that the Mexican border region is a free-trade zone. This allows *maquiladoras* to import materials duty-free.

Locational choice

International scale

Mexico's low labour costs makes it one of the most popular locations for the labour-intensive branch plants of foreign firms. In the developing world only China has received more inward investment than the Mexican border region.

In addition to its low labour costs and duty-free zone, Mexico's proximity to parent companies and markets in the USA gives it an advantage over more distant countries. Its transport costs and delivery times to US markets are significantly lower than those of Fast Asian countries such as Thailand and the Philippines. Proximity to US manufacturers also means that the Mexican branch plants can be supplied easily with components.

Figure 8.16 Tijuana, home for many US *maquiladora* plants

?

8 Essay: Use the product life cycle theory (page 101) to explain the location of US branch plants in Mexico as shown in Figures 8.17 and 8.18.

——— Sole destination
–·–·– Primary destination (secondary shipments not shown)
- - - - - - Multiple destinations (with predominant direction of shipments indicated)
(44, 10) Number of Tijuana maquiladora with sole destinations, respectively, to Los Angeles (44, 10) or San Diego 34, 5)

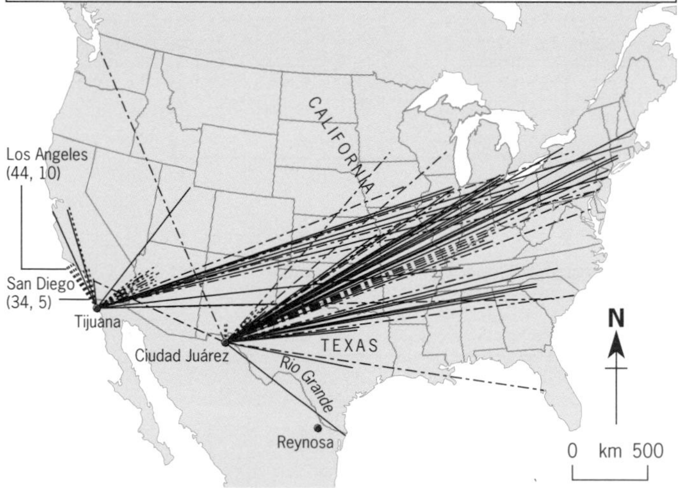

Figure 8.17 Destinations of *maquiladora* exports from Tijuana and Ciudad Juárez, 1986 (*Source:* South, 1990)

Regional scale

Within Mexico the *maquiladora* are concentrated close to the US border (Figs 8.17, 8.18). Such a location further reduces transport costs and delivery times to the USA. A border location also allows American branch plant managers to live in the USA and commute to Mexico.

Urban scale

Within the border zone, branch plants tend to locate in major urban centres, such as Tijuana and Ciudad Juárez, which are able to provide large pools of cheap labour. Industrial parks are also favoured. They provide essential infrastructure. Location on undeveloped sites can often lead to long delays and unexpected costs in LEDCs.

The Tijuana triangle

The area around Tijuana, with around 1,000 firms and 200,000 workers, has the largest concentration of *maquiladoras* in the border zone. Its main products are consumer electronics, notably television sets. Amond the Korean TNCs which have invested in Tijuana are Samsung, which makes television monitors, and Hyundai, manufacturing transport containers and lorry trailers. Japan is represented by Sony, Hitachi, Matsushita and Sanyo.

But Tijuana's labour intensive *maquiladoras* are already showing signs of change. There is a gradual movement away from dependence on low-wage, low value-added production to a sophisticated economy. Automation is increasing and more advanced products, such as motherboards for electronic equipment, are becoming important. The simple labour-intensive work has relocated to Latin American countries, such as Guatemala, which have even lower wages.

Environmental impact of the maquiladoras

Environmental and health and safety regulations are weaker in Mexico than in the USA or Japan, and are often ignored. Along the Rio Grande valley industrial growth and urbanisation have led to acute shortages of housing and water. Many *maquiladora* workers live in squalid shanty towns (*colonias*) without power or water. And as the demand for water has soared, farmers, who currently account for 90 per cent of water usage (for irrigation), have been badly hit. Meanwhile, water quality declines as chemical effluent and raw sewage are pumped into Rio Grande at Reynosa. Hazardous industrial waste presents another environmental problem. All toxic materials imported by *maquiladoras* from the USA should be returned to the USA for disposal. But in practice this rarely happens. Instead, many *maquiladoras* simply dump their toxic wastes at landfill sites in the desert.

However, some companies have a more responsible approach to environmental issues. Delphi Automotive

Systems (a vehicle components subsidiary of GM) employs nearly 75,000 workers in Mexico. The company has installed water treatment plants at most of its manufacturing sites, allowing water to be recycled. Where possible, the company also recycles plastics, cardboard, solvents and other chemicals. Meanwhile Delphi is pioneering a joint programme with the Mexican government to provide affordable housing for its workers.

Figure 8.18 Destinations of *maquiladora* exports from Nogales and Nuevo Laredo, 1986 (*Source:* South, 1990)

8.5 Strategies and policies of large firms

Branch plant closure

During periods of recession or times of severe competition, large firms may implement new policies in order to maintain their profitability (Fig. 8.19). Such policies are likely to have an effect on the geography of the firm.

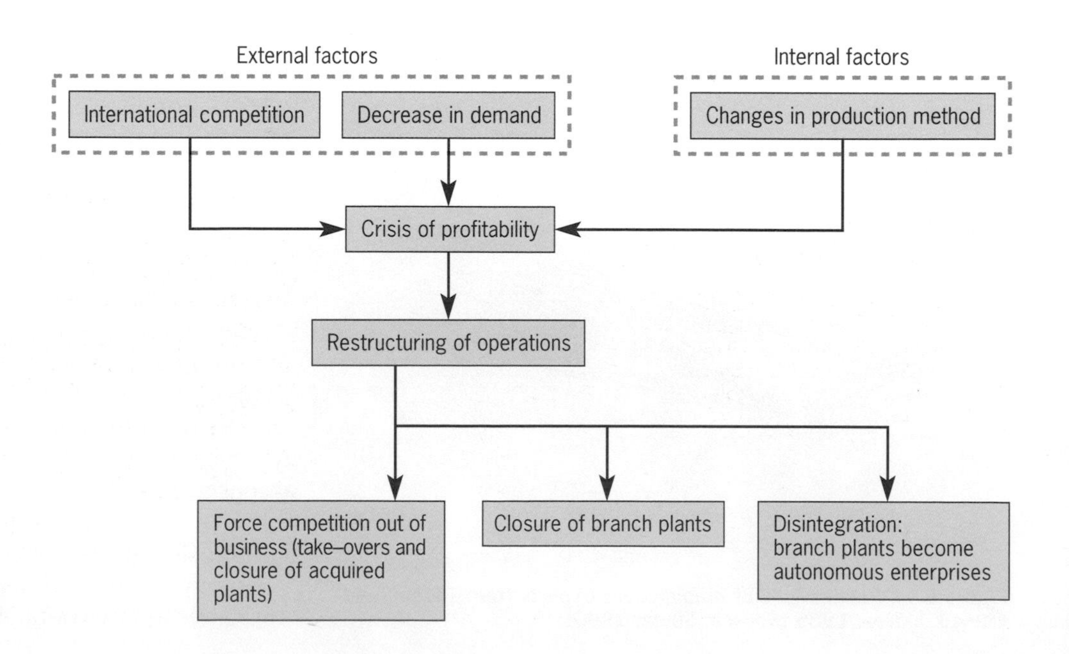

Figure 8.19 Response of multi-plant firms to crises of profitability

French jobs lost in EU lottery

FRENCH factory workers who are losing their jobs on a hi-fi assembly line, because their Japanese employer is concentrating production in Scotland, will tell a government minister today that they have been the victims of a European Union subsidies lottery.

A delegation of nine women, representing 243 workers who assemble mid-priced hi-fi systems for JVC in the industrial zone around the village of Villers-a-Montagne, in eastern France, claim the company is leaving because it has exhausted EU subsidies for their blighted former steel area and now intends to profit from hand-outs in Scotland.

The company – which manufactures televisions at East Kilbride, near Glasgow – last year expanded its Scottish operation to include the production of midi-systems previously made at the factory near the border with Luxembourg.

The French plant is now due to close at the end of January.

"When they have finished in Scotland they will return to the Far East, having spent 10 years benefiting from subsidies and securing a foothold in the European market," said Catherine Leblan, a 30-year-old production line worker.

Mourthe-et-Moselle, where the JVC factory opened in 1988, is one of dozens of European industrial regions – including parts of Wales, northern England and Scotland – which competed for the attention of hi-tech businesses during the 1980s.

In return for investing in former mining, steel and shipbuilding areas, the companies received billions of pounds in government and EU grants, and secured European markets.

For the likes of Aline Radosevic, whose father was a steel worker and was laid off at the age of 38, the departure of JVC is proof that eastern France is heading for its latest depression.

"JVC got a grant of Fr2 million (£225,000) to expand its television assembly plant outside Glasgow," she said. "Last year, they transferred part of our audio production there. JVC had already received EU money here; it's like robbing Peter to pay Paul."

But Mrs Radosevic, who will be laid off at the end of January unless a buyer to take over the factory is found, does not resent her Scottish colleagues, even though she believes they are paid considerably less than the Fr6,554 (£710) gross monthly minimum at Villers-la-Montagne.

Sitting behind a pile of documents which would humble an average financial director, Isabelle Banny explained Miss Leblan's prediction. "The market in hi-tech equipment – such as televisions, video recorders and hi-fi's – has slumped by 30 per cent in the past three years.

'When they finish in Scotland they'll return to the Far East, having spent 10 years benefiting from subsidies and securing a foothold in the European market'

"Glasgow's industrial history is similar to ours, but they must be warned that Far Eastern investment provides no future. Employers' costs may be low in Britain but still they cannot compete with Malaysia."

Miss Banny and her colleagues – some of them supporting their husbands and many of them single mothers – are far from resigned to their fate. Last week, they held hostage five of the company's directors for a day. On Thursday they plan a one-day strike.

The women – who are mostly aged under 35 because French employers receive grants to employ young people – are not drawing up plans for a buy out. "There is no market for hi-fi's," said Miss Leblan.

"The best we can hope for is that the factory will be sold and some of us re-employed."

Even if that happens, the women know they will be just the latest example of the transformation of an area where people once had father-to-son jobs for life into one of "relay factories".

This growing phenomenon sees companies move into a blighted area and receive grants, and then leave when the incentives or profits run out.

Before JVC's eight-year tenure of the factory, it was owned for two years by Thomson electronics, which employed 100 people. Yesterday, a compact disc manufacturer was looking around the site.

Miss Banny said: "We want our experience to serve as an example of how things should not be done. We shall tell the minister that when companies apply for grants in development zones they must promise not only to create jobs but to stay for 10 years."

The localised private sector uproar has come at a time of rising indignation of foreign investments and takeovers – seen as being agreed with few job guarantees. Last week, the French government sold part of the state-owned Thomson group to Daewoo, of South Korea.

Trade union activism is very limited in France's private sector – only 4 of JVC's 243 staff at Villers-la-Montagne are union members. National unemployment is 12.5 per cent, and higher in areas such as the east.

The company's personnel manager, Régis Spor, denied that JVC had let down its workers. "We brought eight years of employment to this area and in 1992 we employed 300 people.

"We have paid Fr26 million in taxes and Fr75 million in social charges."

He said the company had received Fr22.3 million in grants from the French state and from the EU, but had invested Fr80 million.

He conceded that the European audio market had been in decline since 1993, while denying that JVC was in the process of abandoning Europe.

But a JVC official added: "It would be suicide for anyone to open an audio plant anywhere in Europe now. All production is going back to the Far East."

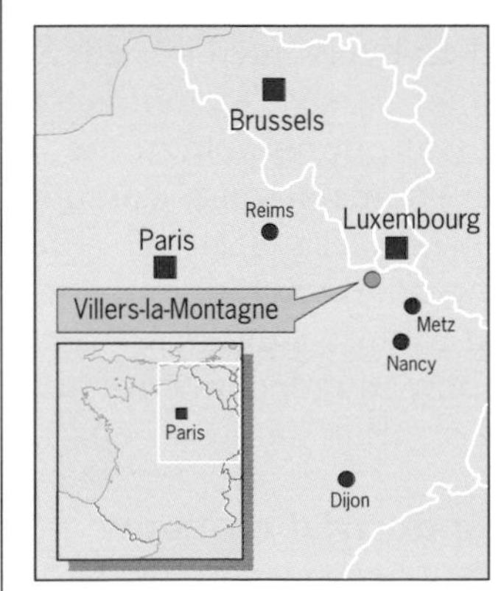

Reasons for relocating

Scotland's strong cards

- **Manufacturing productivity annual growth rate of 5.2 per cent;**
- **Less onerous restrictions on labour – average UK worker puts in two hours a week more than EU average;**
- **Days lost to industrial disputes in the UK were 24 per 1,000 workers, according to 1992 Eurostat figures;**
- **Success of inward investors attracts more companies. Inward investment in Scotland announced this year is over £2.7 billion.**

France's drawbacks

- **Manufacturing productivity annual growth rate of 2.7 per cent;**
- **Cost of labour up to 25 per cent higher than in UK;**
- **Days lost to industrial disputes: 37 per 1000 workers**
- **Total remuneration costs 37 per cent more than UK as benefits are more extensive;**
- **Wages councils under Social Chapter can set pay;**
- **Manufacturing output per hour about 80 per cent of UK level.**

Figure 8.20 Newspaper report of the closure of JVC's hi-fi factory at Villers-la-Montagne (*Source:* © *The Guardian*, 30 Oct. 1996)

?

When company profits fall, branch plants are the first to close. Such restructuring may have disastrous consequences for the economic and social well-being of the peripheral regions. Read the newspaper article (Fig. 8.20) on the closure of the JVC hi-fi factory at Villers-la-Montagne in eastern France.

9 Explain why the plant was closed. Give the views of the workers.

10 What does the article tell you about the transferability of capital by large multi-plant firms such as JVC? Why do you think the plant was relatively easy to close (consider labour skills, capital investment, resistance of workforce, etc.)?

11 What does the article suggest to you about branch plant economies?

12a Draw up a balance sheet to summarise the advantages and disadvantages of inward investment by foreign TNCs in a region.
b State and explain your views on foreign inward investment and the dependency it brings to regional economies.

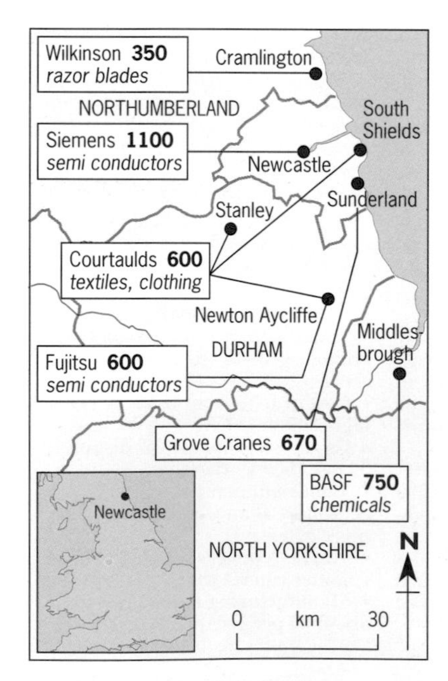

Figure 8.21 Major job losses in manufacturing industry in North-East England in 1998

Restructuring is one option available to firms. This usually means closing those plants which are least profitable. Branch plants, which are labour-intensive rather than **capital-intensive**, and which employ largely unskilled labour (which can easily be replaced when the economy improves) are most vulnerable to closure. As a result, regional economies which depend heavily on branch plants may be undermined, causing widespread unemployment. **Branch plant economies** have appeared in many peripheral regions of the UK since the 1970s. They have placed regions such as Merseyside and North-East England in a position of dependency: large sectors of their economies are controlled by decisions made elsewhere, often at overseas headquarters.

Branch plant economies: North-East England

Few regions in the UK were harder hit by deindustrialisation than the North-East. Between 1975 and 1990 staple industries such as shipbuilding and coal mining disappeared. Others such as steel, chemicals and heavy engineering survived, but only after shedding thousands of jobs.

Reindustrialisation

By the mid-1980s a revival of manufacturing was already under way. This revival was powered by inward investment by foreign TNCs. The regeneration of the North-East's industrial structure was led by the Japanese car giant Nissan, which opened its first European assembly plant at Sunderland in 1986. Several automotive parts manufacturers followed Nissan to the North-East, including Akeda and Yamato. Then in the 1990s came major investments in electronics. Fujitsu built a semi-conductor plant at Newton Aycliffe; Samsung embarked on a major investment programme to make consumer electronics on Teesside; and in 1995, Siemens, one of the world's largest electronics companies, chose North Tyneside for its new £1 billion semi-conductor plant. The North-East's success in attracting foreign inward investment transformed the region's industrial structure.

Even so, the restructuring of industry was not without cost. Foreign investment increasingly shifted control of the region's economy away from the North-East and the UK, to Japan, Germany, South Korea and the USA. The security of jobs in the region now depended more than ever on decisions taken in foreign cities such as Tokyo, Seoul and Frankfurt. The North-East, more than any other English region, had committed itself to the global economy.

Disinvestment

The implications for the North-East became apparent in the late 1990s. In 1998, in the space of two months, Fujitsu and Siemens closed their factories in the North-East. They blamed a worldwide overproduction of semi-conductors, plummeting prices and an overvalued pound. Siemens' state-of-the-art plant closed having operated for barely 18 months. On Teesside, Samsung's expansion programme was put on hold in the wake of the banking crisis in East Asia. Several other companies also closed factories or scaled down their operations in the North-East (Fig. 8.21). For the first time, people began to count the cost of relying on inward investment and foreign TNCs which had no long-term commitment to the region. The events also demonstrated how globalisation caused a new interdependence: that a financial crisis on the other side of the globe could result in job losses and factory closures in the UK.

The rise of the post-industrial economy

Despite the spate of factory closures in the North-East in 1998, the region emerged with a job surplus for the year. The loss of manufacturing jobs was more than offset by new jobs in services. This shift from manufacturing to services is a long-established trend. Today the service sector accounts for 70 per cent of employment in the North-East. So while the closure of Siemens' and

?

13 Investigate some of the ideas in this chapter by studying the organisation of a number of manufacturing firms in your local area. Obtain an up-to-date directory of manufacturers (often available from local planning departments) and select a random sample of firms. (If there is a sizeable industrial estate near by, this will make the task of data collection easier.) Devise a questionnaire like the one provided in Appendix A4. You will also find a list of hypotheses in Appendix A3. Select the ones which are likely to be relevant to the manufacturers in your area.

Figure 8.22 Siemens' semiconductor plant on North Tyneside

Fujitsu's semiconductor plants created 1,800 redundancies, a new banking call centre on Wearside, and a sales and customer call centre for an insurance company in Teesdale, created 2,250 jobs.

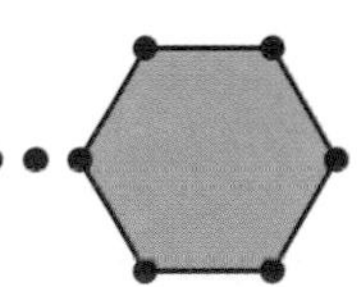

Structuralism

The theory of **structuralism** says that the location of industry and industrial change can only be explained by the structure of society. Thus it is argued that if we want to explain industrial location in the Western world we need to understand how the **capitalist system** works. It follows that the geography of capitalism is very different from that of a centrally planned or **command economy** like China.

The geography of capitalism

Increasing profit levels

The driving force behind capitalism is the creation of profit. Firms must raise money for investment if they are to compete successfully in a capitalist system. Profits are increased through a number of strategies.

One strategy is to shift production to lower-cost locations. Thus, a Japanese TNC might transfer production from Japan to a low-wage country in South-East Asia, such as Indonesia or Malaysia.

Another strategy is to pass on part of the costs of manufacturing to local people and the environment. Steel mills and chemical works pollute the atmosphere and may damage the health of people living near by. Pulp mills pollute lakes and rivers, destroying habitats and wildlife. If firms were to bear all the costs for which they were responsible, their profits would fall. This in turn would influence their choice of location and drastically alter the industrial geography of Western countries.

Regional inequality

Under capitalism, capital moves freely to places where it can make the greatest profit. As a result, economic growth becomes spatially uneven. This effect is apparent at different scales. At the global scale, capital is mainly concentrated in the MEDCs. Within these countries it is attracted to metropolitan core areas, which grow at the expense of peripheral regions. And within the metropolitan core, capital is increasingly abandoning inner-city areas in favour of the suburbs or small towns.

Control and dependency

Industrial production and control in the capitalist system often occupy geographically separate locations. Today, TNCs, which are largely controlled from headquarters in major world cities, can transfer capital easily between countries and continents. This enables them to restructure their operations in periods of economic downturn. Often this means closing branch plants and transferring production to more profitable locations. With control far removed from the centres of production, peripheral regions are placed in a position of dependency. In their quest for profit, TNCs give only a low priority to the impact of their decisions on the economic and social life of local communities. This conflict of interest between capital and labour is at the heart of the structuralist approach.

14 Study the maps in Figures 8.8, 8.12 and 12.12. Using the ideas of structuralism, try to explain the spatial distribution shown on these maps.

Summary

- Large manufacturing firms are often multi-plant, multi-locational and transnational.
- Transnational corporations (TNCs) are very large firms with manufacturing activities in several countries.
- The main functions of multi-plant firms – headquarters, R&D, branch plants – often have a distinctive spatial organisation, with headquarters and R&D located in core regions, and branch plants on the periphery.
- Overseas investment by TNCs since the mid-1970s has been dominated by Japanese firms.
- TNCs are a major influence on the global economy: much of the world's production and trade are accounted for by TNCs.
- Very large TNCs have a distinctive geography of their own.
- Inward investment by TNCs is often undertaken either to avoid trade barriers or to exploit cheap sources of labour and/or materials.
- Governments compete to secure inward investment by offering loans, grants, etc., to foreign TNCs. However, investment is usually linked to performance requirements such as sourcing parts and materials locally.
- Large manufacturers like TNCs are likely to undertake thorough searches for potential locations before committing themselves to investment.
- The product life cycle has an important influence on the location of manufacturing industry.
- During periods of economic downturn, large firms often close branch plants and restructure their operations. This process generally hits peripheral areas harder than core areas.
- Structuralist theories contend that the hidden structures of the capitalist system explain uneven regional development, branch plant closures and environmental pollution caused by industry.

9 Regional disparities and industrial change

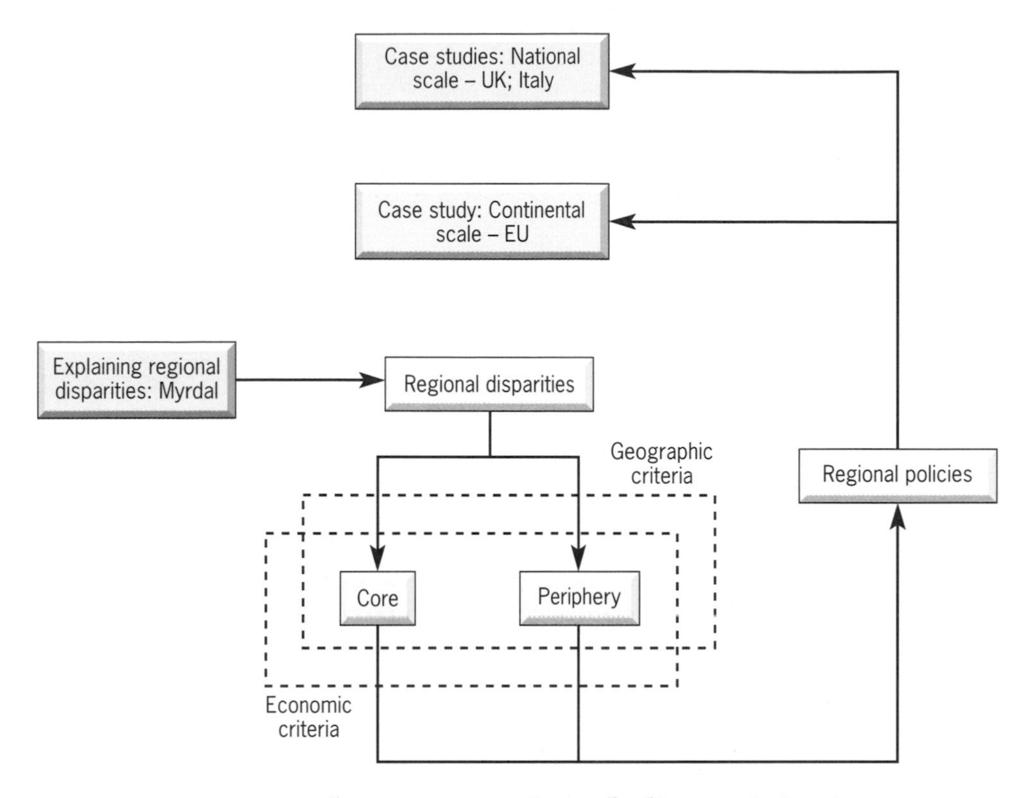

9.1 What are regional disparities?

Table 9.1 shows that there are considerable differences in wealth both between and within EU countries. We refer to the latter as **regional disparities**.

Table 9.1 Regional disparities in the EU (1998)

	GDP per capita (EU average = 100)	Most prosperous region (EU average = 100)	Least prosperous region (EU average = 100)
Austria	112	127	90
Belgium	112	173	89
Denmark	119	–	–
Finland	97	119	97
France	104	160	85
Germany	108	192	61
Greece	68	77	65
Ireland	97	–	–
Italy	103	133	66
Luxembourg	169	–	–
Netherlands	107	115	93
Portugal	70	71	50
Spain	79	101	59
Sweden	101	–	–
UK	100	140	81

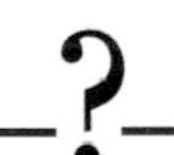

Is it possible to say that in the EU the more prosperous a country is, the greater its regional disparities? Test this idea for yourself using the information in Table 9.1 and completing the following exercises.

1 Calculate an index of regional disparity for each country. To do this, divide the GDP index for a country's most prosperous region by the GDP index for its least prosperous region.

2 Calculate the Spearman rank correlation coefficient between GDP per capita and your regional disparity index.

3 Describe and explain the relationship between GDP and regional disparity.

Regional industrial growth in a country seems to occur unevenly: some regions grow rapidly and prosper, while others lag permanently behind. Often this inequality has a geographical dimension: the prosperous regions are situated in a central or core area, and the less prosperous on the outskirts.

9.2 Causes of regional disparity

The Swedish economist Gunnar Myrdal explained regional disparities in his model of cumulative causation (Fig. 9.1). He said that industrial growth starts in a region because of its **initial advantages**. (These advantages might include any of the locational factors we considered in Chapters 3–6.) When development is under way, it triggers a series of virtuous growth cycles. The process is a cumulative one: a snowball effect leads to growth which reinforces further growth.

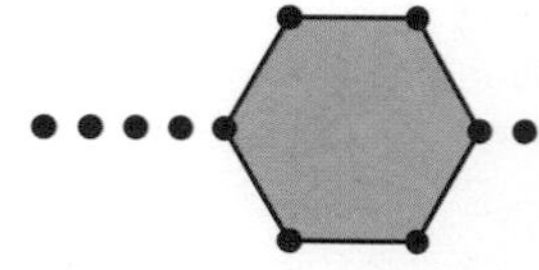

Cumulative causation

If you study Figure 9.1 you will see that the starting-point in the cycle is the location of a new industry in a region. Its effect is to increase local employment and create a pool of skilled labour. Eventually other industries are attracted into the area. As they expand they provide markets for parts suppliers and encourage localisation economies (see section 6.1). Finally, rising prosperity attracts producer services and investment in transport and social infrastructure (schools, hospitals, housing, etc.). The result is the emergence of a core region with powerful agglomeration economies (see section 6.1).

Meanwhile, cumulative causation works against peripheral regions. Investment is diverted from the periphery; better job prospects in the core encourage the more enterprising workers to leave; and local industries may collapse under competition from the core. These are Myrdal's **backwash effects**. Together with cumulative growth in the core they explain regional differences in prosperity.

However, Myrdal was not entirely pessimistic about the economic prospects for the periphery. He believed that the prosperity of the core would eventually spread (or 'trickle down') to the periphery. But so far this has only happened where peripheral regions have some resource advantage, such as Almería's favourable climate for agriculture, and eastern Scotland's access to North Sea oil.

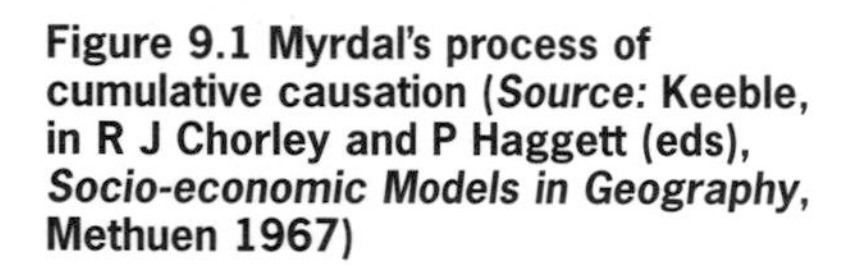

Figure 9.1 Myrdal's process of cumulative causation (*Source:* Keeble, in R J Chorley and P Haggett (eds), *Socio-economic Models in Geography*, Methuen 1967)

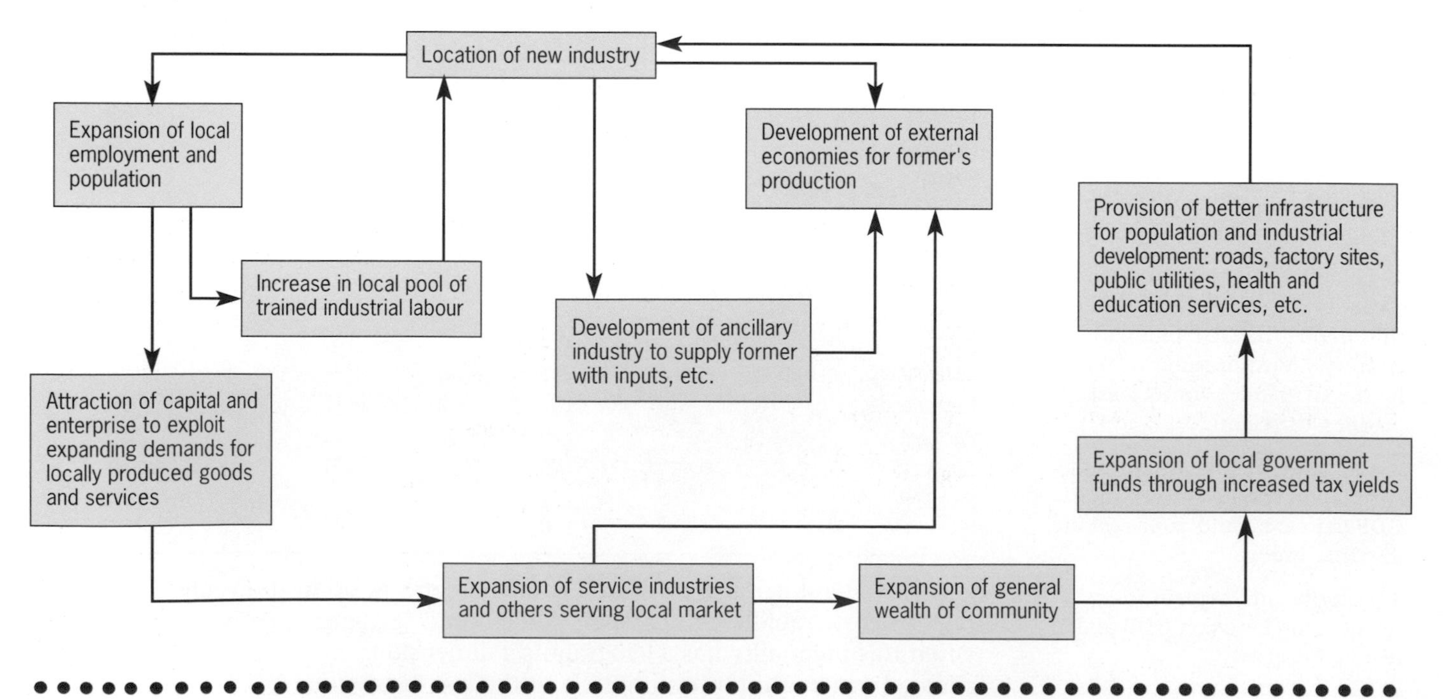

?

4a Draw an annotated diagram to show how a cycle of decline can affect the periphery.
b Around your diagram add the possible responses to regional decline.

5 Identify a range of values and attitudes that influence responses to regional decline.

9.3 Tackling the regional problem

Most governments think that wide regional disparities are undesirable and have devised policies to reduce regional imbalance. For the most part these policies have two common features: a package of financial incentives (loans, grants, tax concessions, labour subsidies, etc.) to promote investment in the periphery; and measures to restrict development in the core.

Economic arguments

The case for regional policy is economic, social and political. Economic arguments focus on the best use of resources. For instance, policies which create more jobs in the periphery should, in theory, increase output, and make the whole nation better off. On the other hand, excessive growth in the core can lead to labour shortages, forcing wages up, and hampering economic growth. Most often the prosperity of the core brings diseconomies (see section 6.2) such as traffic congestion, sharp rises in rents for offices and factories, and soaring house prices.

All this contrasts starkly with conditions in the periphery, where unemployment is often high and wages are depressed. The response is out-migration. In remote rural regions this may cause depopulation. This sets in motion a vicious cycle of decline. As population levels fall, schools, hospitals, transport and other services are gradually withdrawn. Further population decline then becomes inevitable.

Social and political arguments

Most governments accept that regional disparities in income and employment are unacceptable for social and political, as well as economic, reasons. Where regional disparities are wide, a sense of social injustice may force governments to take action. Where they coincide with linguistic differences (as in Belgium, where the French-speaking South has been markedly less prosperous than the Flemish-speaking North) regional policies may help to unify the state. The political factor also figures strongly where governments fear that ignoring the regional problem threatens their chances of re-election.

Regional disparities at the continental scale: the European Union

Major regional disparities exist at a continental scale within the European Union (EU) (Fig. 9.2). In the 1980s and 1990s these disparities widened: economic recession hit peripheral regions hardest; and the enlargement of the Union incorporated several very poor regions in southern Europe and eastern Germany.

Unequal development in the EU has produced three types of region: **core**, **semi-periphery** and **periphery**. These terms can refer to both geographical location and economic performance (Hudson, 1988). The difference between them is explained below.

Geographical regions

The geographical core of the EU is roughly the triangle formed by London, Hamburg and Milan. Peripheral regions are remote from the core, in the Mediterranean

Table 9.2 The EU's least-prosperous regions

Region	GDP per capita: EU average = 100	% GDP from agriculture	industry	services
Azores (Portugal)	50	11.8	34.5	61.4
Madeira (Portugal)	54	3.8	18.2	78.0
Kentriki Ellada (Greece)	58	24.6	27.2	48.2
South (Spain)	59	8.6	22.8	68.6
Mecklemburg-Vorpommern (Germany)	61	2.9	29.8	67.3
Sachsen-Anhalt (Germany)	61	2.1	37.0	60.9
Thuringen (Germany)	61	1.9	36.1	62.0
Sachsen (Germany)	64	1.2	36.9	61.9
Vareia Ellada (Greece)	65	22.1	28.8	49.2
Sicily (Italy)	66	6.2	19.1	74.7
Campania (Italy)	66	3.8	20.1	76.1
South (Italy)	67	7.4	20.2	72.4
North-East (Spain)	67	7.1	30.4	62.5

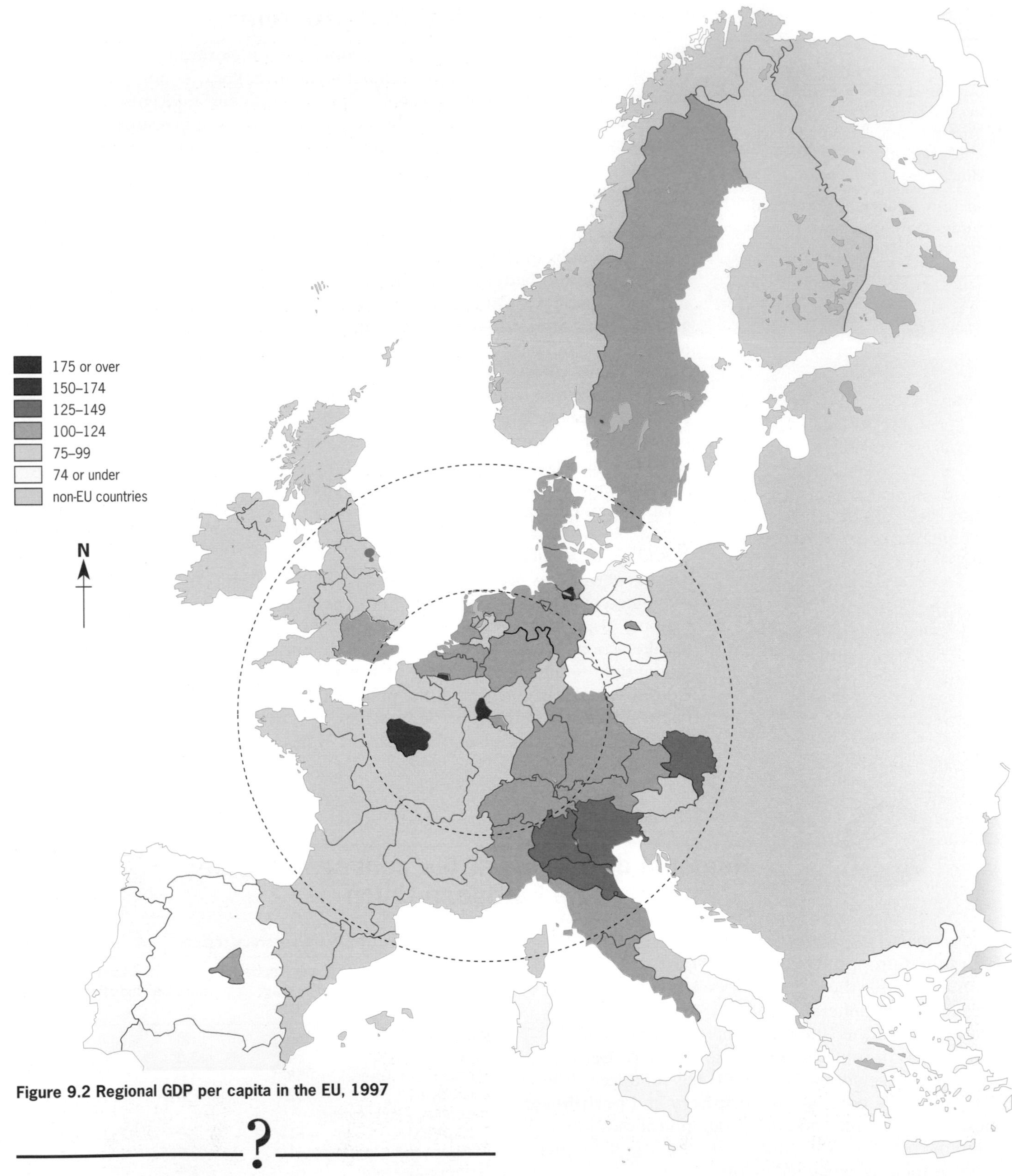

Figure 9.2 Regional GDP per capita in the EU, 1997

?

6 Find the regions in Table 9.2 on the map of the EU (Fig.9.2). What do the locations of most of these have in common?

7 Suggest how the economic structure of the regions in Table 9.2 might explain their lack of prosperity.

deep south (Spain, southern Italy and Greece), on the western Atlantic fringe (Scotland and Ireland), eastern Europe (east Germany) and Scandinavia. In a geographical sense, semi-peripheral regions, such as northern England and southern France, are situated between the core and the periphery.

Economic regions

We also use the terms core, periphery and semi-periphery in a non-geographic sense to describe the relative prosperity of the EU's regions. The most prosperous regions belong to the core. They are centres of economic and political power. Most are focused on major cities like Paris, Brussels, London and Frankfurt, which play key roles in the global economy. They attract the headquarters of big TNCs; have important producer services (financial, legal, advertising, marketing and consultancy); and have a large share of high-tech industry, especially R&D.

Semi-peripheral regions are less prosperous, and more widely scattered. They include former economic core areas like the Ruhr in Germany, Lorraine in France and central Scotland. These regions were once the leading industrial centres of Europe, but they have suffered through their overdependence on declining industries like steel, textiles and coal mining. The semi-periphery also includes a number of up-and-coming rural regions. Almería, in the Andalucia region of southern Spain, is an example. It has prospered recently thanks to innovative agriculture and a switch to export-oriented crops, such as early flowers and vegetables. There are other fast-growing regions which have moved from the economic periphery to the semi-periphery. Among them are newly industrialising regions like Emilia–Romagna in Italy, and tourist regions such as Languedoc in France.

The economic periphery includes some of the poorest regions in the EU. Often handicapped by high transport costs, and isolation from information and markets, many peripheral regions lack modern service industries or technologically advanced industries. These regions, in large measure dependent on low-productivity agriculture (e.g. Mediterranean Europe and northern Europe), often have high levels of underemployment and out-migration.

Regional policies in the EU

The EU is committed to reducing regional disparities within its member states by providing assistance to regions whose development is lagging behind. The main instruments of policy are the EU's Structural Funds (principally the European Regional Development Fund and the European Social Fund). Spending on the Structural Funds increased from 3.7 billion ECU in 1985 to 33 billion in 1999. This represents about a third of total EU spending.

The EU recognises five types of assisted area (Fig. 9.4). However, only two – Objective 1 and Objective 2 areas – include urban and industrial problem regions. Table 9.3 summarises the main characteristics of Objective 1 and Objective 2 areas. Assisted areas in the UK currently receive twice as much financial aid from the EU's Structural Funds than from the UK government's own regional policy.

In spite of much increased spending, the EU's regional policies have achieved only modest reductions in regional inequalities. However, it is not easy to assess the effectiveness of the EU's regional policies. The enlargement of the EU since the 1980s has embraced many poorer regions in southern Europe. Meanwhile the reunification of Germany incorporated the economically-backward regions of the former East Germany into the EU.

?

In this exercise we shall assess the extent to which regional economic prosperity in the EU is influenced by geography. In Figure 9.2 the centre of the EU is assumed to be Luxembourg. The two circles centred on Luxembourg in Figure 9.2 mark arbitrary boundaries between the geographic core (less than 400 km), the semi-periphery (400–800 km) and the periphery (more than 800 km).

8 Copy Table 9.3. Using the information on GDP per capita and distance from Luxembourg in Figure 9.2, enter the number of regions in each of the 12 classes shown in Table 9.3.

Figure 9.3 La Défense, Paris: an out-of-centre business complex and the headquarters of several major French TNCs.

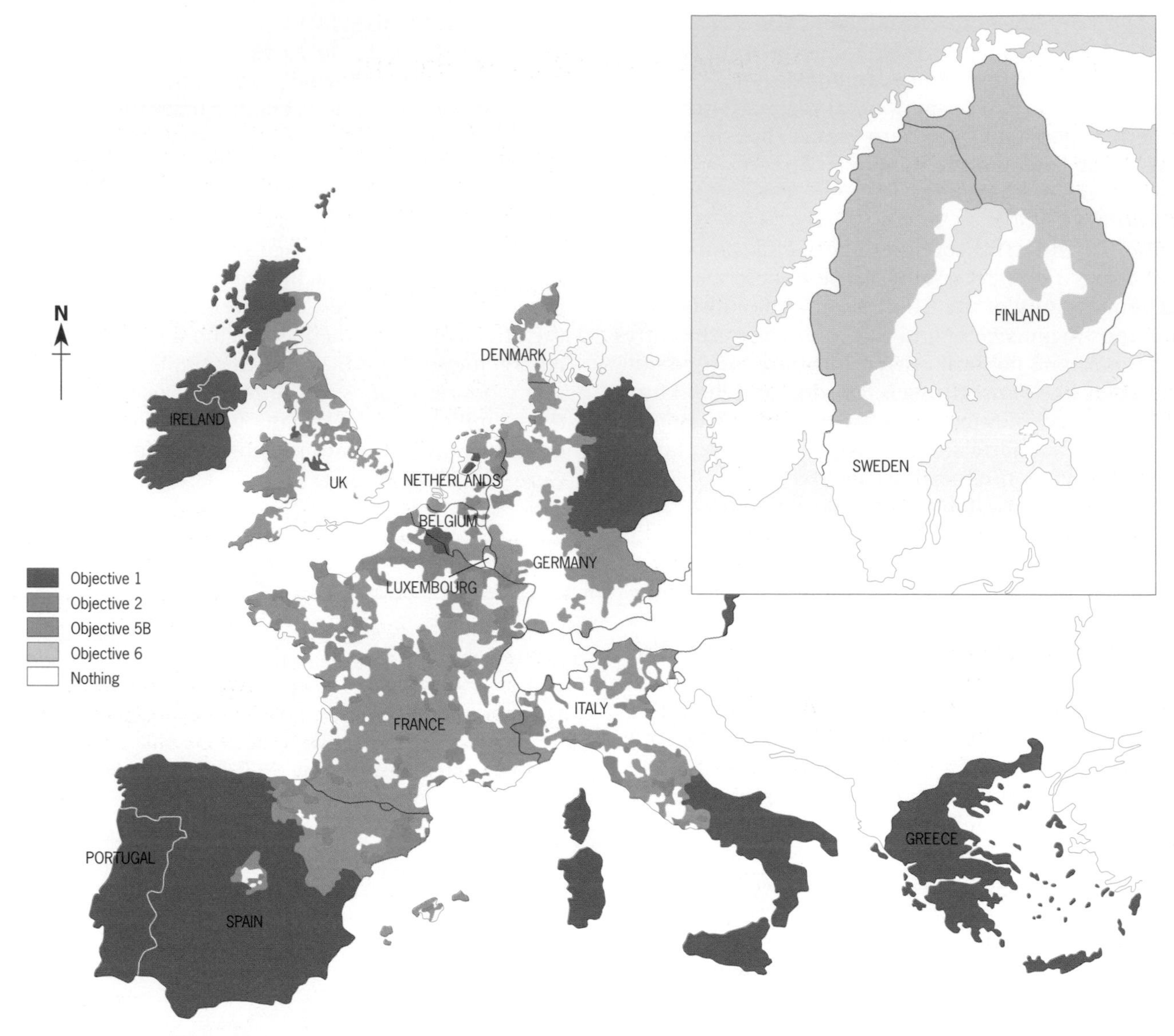

Figure 9.4 EU regions eligible for Structural Funds (*Source:* R Hudson, *Geography Review* (Philip Allan Updates), March 1998)

?

9 Describe the pattern of geographical and economic regions in your table.

10 Are the patterns in the table statistically significant? You can test for significance using the chi-squared statistic (see Appendix A2) from the data in your table. This will tell you whether the distribution of economic regions within the EU is random or related to distance from the core.

11 Set out your conclusions in a paragraph or two. How accurate is the geographical core–periphery concept as an indication of regional prosperity?

12 Test the hypothesis that the more peripheral the location of a region, the greater the spending from the ERDF. Use the information in Figure 9.4 and follow the methods used in exercises 8–11.

Table 9.3 Geographic and economic core–periphery regions

		Geographic regions		
		0–400 km	**401–800 km**	**>800 km**
Economic regions	>124			
	100–124			
	75–99			
	<75			

Table 9.4 Regional inequalities in the EU: 1983–97 (GDP per capita; EU average = 100)

	1983	**1997**
Richest: Hamburg (1983 and 1997)	184	192
Poorest: Extremadura (1983)		
Azores (1997)	43	50

Regional disparities in the UK

Regional policy

The origins of the UK's regional policies date back to the 1930s and the Great Depression. Successive British governments have pursued policies with varying commitment, aimed at reducing regional inequalities particularly those in unemployment. But since the early 1980s spending on regional policy has fallen steeply. Indeed by the 1990s the cost of regional policy was less than 0.1 per cent of GDP. Details of the current policy are given in Table 9.5. A major review of regional policy including a new map of assisted areas was completed in 2000.

Today the UK's assisted areas are the same as those eligible for the EU's Structural Funds. Regional assistance is now targeted with greater selectivity at the most needy areas. The list of assisted areas in the UK varies from the county and district scale to individual wards in cities.

In 1999 the UK government introduced a new initiative to help reduce regional inequalities: it set up Regional Development Agencies (RDAs) for the English regions. Previously only Scotland, Wales and Northern Ireland had their own development agencies. As a result they received more generous funding than the English regions. The RDAs will have a funding of £1 billion a year. They will have responsibility for closing the wealth gap between the regions, as well as building factories and clearing derelict land.

Regional policies: an audit

Despite more than 60 years of regional policies, there is little evidence that regional inequalities have narrowed in the UK. Does this mean that regional policies have failed? Experts find it hard to agree on this point. Yet regional policies have had a positive impact by:

- creating thousands of jobs;
- helping to diversify regional economies and improve infrastructure (e.g. by building new roads, factories etc.);
- by securing inward investment, including high levels of FDI in Scotland, Wales, Northern Ireland and northern England;

One estimate suggests that between 1960 and 1981 regional policies created 800,000 new jobs in assisted areas. Regional policies in the 1960s and 1970s also help to protect and renew the manufacturing base of

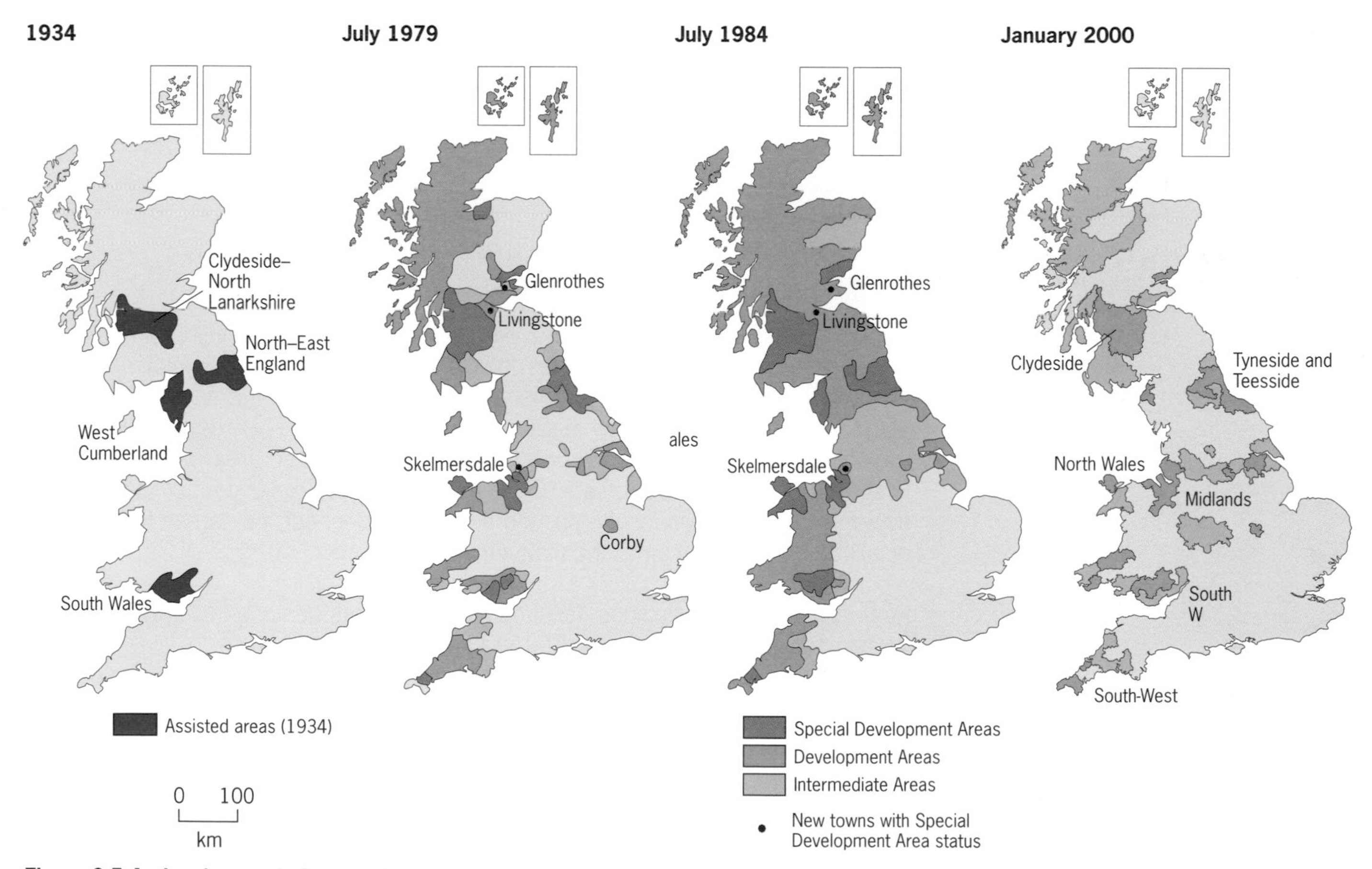

Figure 9.5 Assisted areas in Britain (*Source:* Armstrong and Taylor, 1988)

?

13 What evidence is there in Figure 9.5 to suggest that regional policy between 1934 and 1992 had little impact in reducing regional disparities?

Figure 9.6 Seal Sands chemical works in the assisted area of Teesside: large amounts of regional aid were spent on capital industries that created few jobs

many old industrial regions. Indeed in the absense of regional policy it could be argued that regional disparities would be even greater than they are today. However, regional policies have come at a price. Each new job created in assisted areas in the 1970s cost the government £35,000 (at 1981 prices)! Furthermore, while there were healthy increases in manufacturing firms locating in assisted areas, much of this increase was in capital-intensive industries and externally-controlled branch plants. Capital-intensive industries such as chemicals provided relatively few jobs; and the success of branch plants (as we saw in section 8) depends on the economic fortunes of individual companies and the world economy.

Table 9.5 UK regional policy: assistance available

- During the 1990s two types of area qualified for assistance: development areas and intermediate areas. Both had problems of above-average unemployment and economic decline, though the problems were more acute in the development areas. In January 2000 a new assisted areas map (Fig. 9.7) brought the UK's regional policy into line with EU policy. Two tiers of assisted area were defined. First-tier areas have a per capita GDP that is less than 75 per cent of the EU average. The definition for the second tier is less clear-cut, i.e. 'areas that are disadvantaged in relation to the national average'.

- Regional selective assistance (RSA) consists of grants designed to attract new investment and safeguard jobs. RSA grants cover up to 15 per cent of the cost of new investment in plant, machinery, buildings and land preparation. Firms do not have to be located in an assisted area to qualify for a grant, but it is easier to get a grant in assisted areas.

- Regional enterprise grants (REGs) are available to very small firms. They provide money for capital investment and financial assistance up to 50 per cent of the costs of projects.

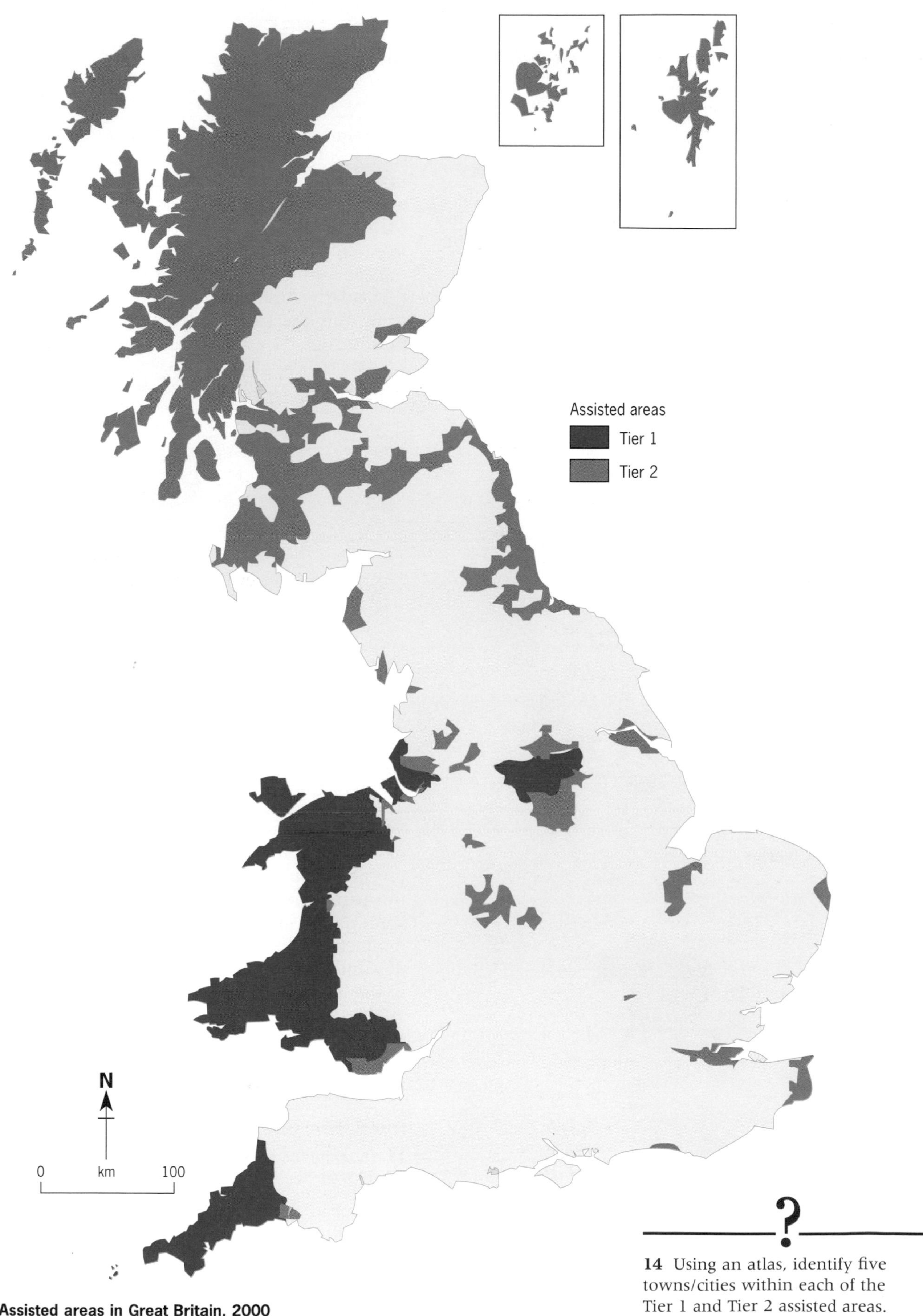

Figure 9.7 Assisted areas in Great Britain, 2000 (*Source:* Department of Trade and Industry)

?

14 Using an atlas, identify five towns/cities within each of the Tier 1 and Tier 2 assisted areas.

Regional disparities in Italy

Regional disparities in Italy are among the highest in the EU. Despite the best efforts of governments over the past 50 years, the gap between the rich North and poor South (or Mezzogiorno) remains as large as ever. In the South, GDP per capita is only 70 per cent of the Italian average. Poverty and unemployment resulted in a massive out-migration from the South for most of the last century. Between 1955 and 1970 nearly 2 million headed north to find work and at least as many emigrated overseas. Since 1970, however, there has been no net migration from South to North. Indeed, because of the high cost of living in the North, southerners may be better-off remaining in the South, even if they are jobless.

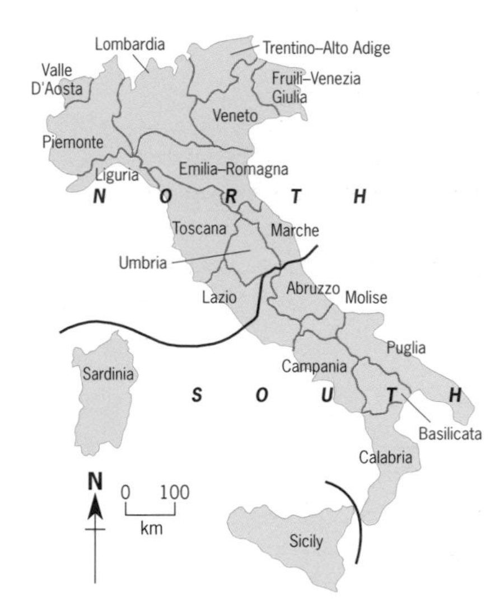

Figure 9.8 Italy's regions

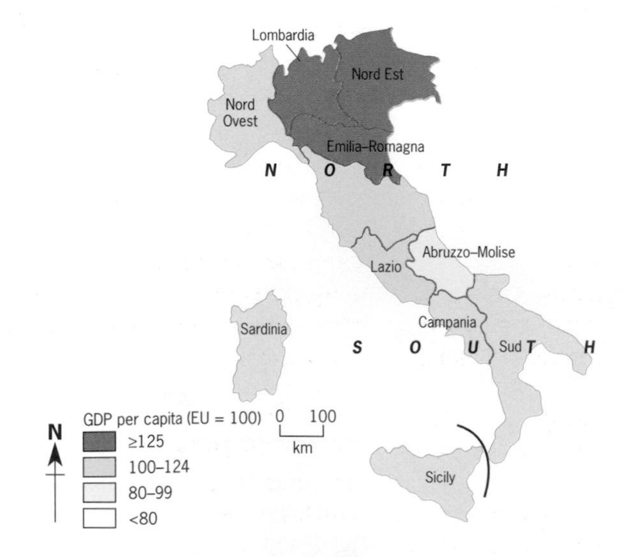

Figure 9.9 Italy: GDP per capita, 1997

The North

Economically, northern Italy is a different country to southern Italy. Whereas the South has suffered chronic unemployment for decades, in the late 1990s the North experienced full employment and labour shortages. Many industries in the North rely on immigrant labour from LEDCs. A growing number of firms are relocating outside northern Italy. Eastern Europe has proved particularly attractive. Between 1995 and 1997, 100 firms from the Treviso (Veneto) area set up operations in Romania.

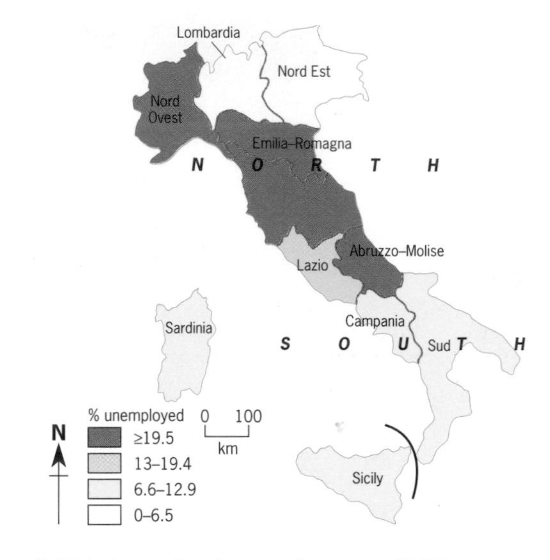

Figure 9.10 Italy, regional unemployment, 1997

Of Italy's 20 regions, three in the North – Lombardy, Veneto and Piedmont – generate 40 per cent of the nation's GDP. These regions are among the most prosperous in the EU. The strength of manufacturing in this part of Italy is based on small- and medium-sized firms. Firms specialising in export-oriented 'mature' products, such as textiles, clothes and furniture, form dynamic interlinked clusters. A measure of the success of these enterprises can be gauged from the fact that in 1996, 55,000 small businesses around Treviso produced export goods worth more than the industrial output of the whole of southern Italy.

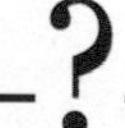

15 Describe the regional pattern of GDP per person in Figure 9.9.

16 Compare the regional distribution of unemployment (Fig. 9.10) in Italy with GDP per person.

17 Explain why the spacial patterns of GDP and unemployment have little influence on migration between northern and southern Italy.

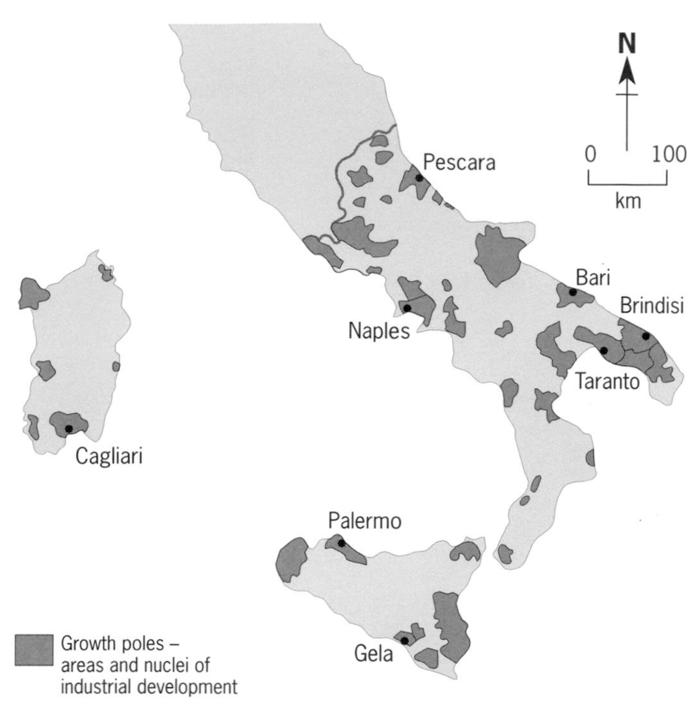

Figure 9.11 Areas of industrial development in southern Italy (*after Mountjoy*, 1982)

?

18 Essay: With reference to Myrdal's model of cumulative causation (Fig. 9.1 and section 9.2) explain how the concentration of investment in growth poles was likely to lead to economic development in the South.

Figure 9.12 The Taranto steelworks in Puglia, a classic example of state controlled heavy industry.

The South

Southern Italy is one of the poorest regions in the EU. GDP per capita over much of the South is barely two-thirds of the EU average. Between one in four or one in five workers are unemployed, and agriculture still accounts for around 12 per cent of the workforce. Why does the South lag so far behind the North (and most of the EU)? The reasons are complex but include:

- Italy's high labour costs (the social costs of pensions, national insurance, etc. are twice those in the UK), which in the South (unlike the North) are not offset by high levels of productivity;
- organised crime (the Mafia) and an image of the South as violent and corrupt;
- a poorly developed infrastructure, especially road and rail services, with little integration;
- a location on the periphery of Europe; regions such as central and eastern Europe are closer to the EU core and have lower labour costs.

Regional policies and the Cassa per il Mezzogiorno

Italy has tried to reduce regional disparities by promoting the economic development of the South. The Cassa per il Mezzogiorno, set up by the government in 1950, was given this task. After 1957 it placed more emphasis on industrial development. Grants, loans and tax exemptions were provided to attract private investment and state-controlled industries were forced to make 40 per cent of their investments in the South.

Growth pole policy

Investment was channelled into special areas or growth poles which had potential for successful industrial development. (Fig 9.11)

'Cathedrals in the desert'

Investment in the South in the 1960s and 1970s was dominated by state-controlled heavy industries, such as petrochemicals and steel. Following the growth pole principle, it was hoped that these industries would act as a focal point for industrial development and attract new businesses and light industries to the area. These capital-intensive (rather than labour-intensive) 'cathedrals in the desert' absorbed huge amounts of government money but created few jobs.

Meanwhile the lack of industrial tradition and technical know-how, discouraged many firms from moving south. Also with so many growth poles, investment was thus too thinly spread to generate cumulative growth.

Changing policies in the 1980s

The Cassa was finally wound up in 1981, its expensive policies having failed to stimulate 'take-off'. The large state-owned enterprises which had been forced to locate in the South remained unprofitable and attracted few other industries.

During the 1980s policies shifted towards investment labour-intensive industries to generate new jobs in the South.

After 1986, regional institutions and local government became responsible for regional policy. The incentives on offer remained generous and included grants, interest rate relief, tax exemptions and social security concessions. The new policies gave small firms the biggest subsidies. Aid was also extended to services, including tourism.

Regional policies: the future

Decades of regional policy have failed to reduce the inequalities between northern and southern Italy. Huge sums continue to be spent on policies to create employment, but with a few notable exceptions, success has been limited. Thus, incomes in the South are increasingly subsidised by the North.

Now there is growing recognition that attempts to transplant industry in the South are unlikely to succeed. The region's future could lie in tourism – which until recently has been curiously underdeveloped. This is surprising, given the region's obvious physical attractions (warm sunny climate, coastline) and cultural/historical heritage. For many politicians and businessmen the promotion of tourism has become a top priority. However, massive investment in essential tourism infrastructure – hotels, roads, airports, etc. – will be needed. Some of this is already under way: golf courses, a new opera house in Palermo, and a new airport at Naples are just the beginning.

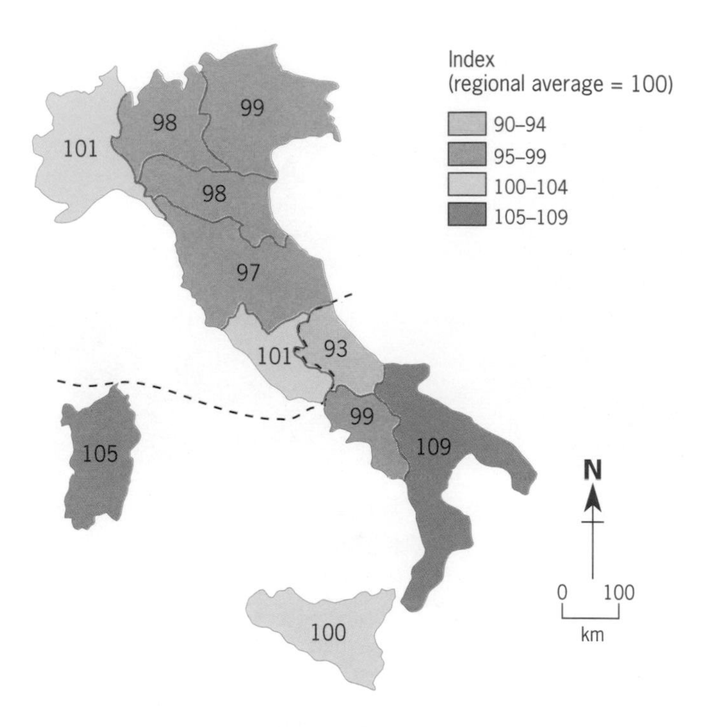

Figure 9.13 Average earnings of manual workers in mechanical engineering in Italy

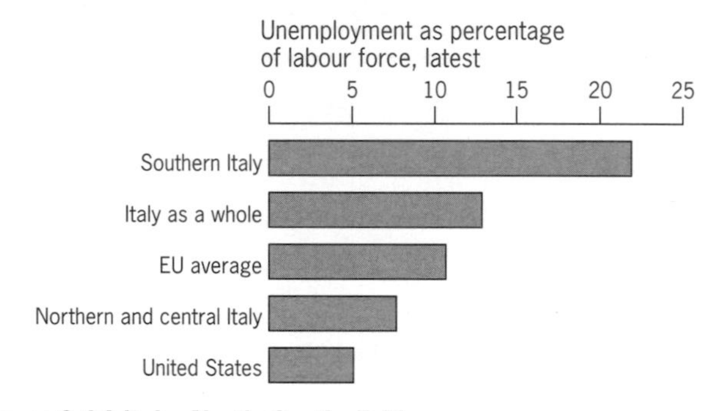

Figure 9.14 Italy: North–South divide

Regional disparities in Thailand

Rising prosperity: growing regional disparity

> Anyone who has spent a day in Bangkok could well have spent half of it in traffic. The clogged streets of the nation's bustling capital support the saying that on the one hand there is Bangkok, and on the other Thailand. Economic development has hit Bangkok with a vengeance: it is a long way from spreading throughout the country. (*Financial Times*, 5.12.1988)

The above quote is a reminder that regional disparities are not just found in developed countries. Wherever industrial development occurs, it is associated with regional inequality and imbalance.

Thailand is one of the world's fastest-growing economies. It belongs to the second generation of emerging Asian economies, and may soon join the ranks of South Korea, Taiwan, Singapore and Hong Kong as a newly industrialising country (NIC).

However, while living standards in recent years have improved for all Thais, the benefits of industrial growth have not been shared equally. Bangkok, the capital and primate city, has emerged as the centre of a highly prosperous core region (Fig. 9.15). Thirty per cent of Thailand's GDP is produced in Bangkok, which contains just 15 per cent of the total population. Average incomes are 10 times greater than in the North-East, Thailand's poorest region.

So far, industrialisation has widened the economic gap between Bangkok and the rest of the country. This 'backwash' effect has also placed severe strains on the capital's infrastructure. Bangkok's roads are choked with traffic (average speed of traffic is only 8 km an hour); its seaport at Klong Toey is badly congested; and overcrowding and pollution are severe and persistent problems.

Will the poorer regions benefit from national growth?

The Thai government has committed itself to balanced regional growth. First, it wants to ensure that the 65 per cent of the population still employed in agriculture can benefit from economic development. And second, it needs to solve Bangkok's urgent problems. The most likely development model is therefore a mixed agricultural–industrial economy. This would not only take advantage of Thailand's large agricultural base, but also provide more balanced growth in areas away from Bangkok.

The last two development plans (1981–86 and 1986–91) aimed to encourage development in the more backward regions. Would-be investors were offered a package of financial incentives. These included an 8-year tax 'holiday' in regions outside Bangkok (compared with 3 years in Bangkok itself). However, the government, which believes strongly in the free market, is not prepared to force firms to locate in peripheral regions by imposing development controls.

The most impressive attempts to decentralise industry have been those connected with the Eastern Seaboard Development Zone (ESDZ). Two zones, 80 and 120 km south-east of the capital, were launched in 1981. The aim is to attract US$4 billion of investment, and draw off industry from Bangkok. Planned to be fully operational by 2001, the ESDZ is designed to use the natural gas reserves of the Gulf of Thailand. Its first phase includes a gas-separation plant. The emphasis is on heavy, capital-intensive industries. Industrial estates and an export processing zone have been established at Laem Chabang. The second phase is based around a new petrochemical plant at Rayong. It includes an array of 'downstream' operations, including the making of synthetic rubber, polythene and polystyrene. This complex will serve a domestic market which includes over 2000 manufacturers of plastics. In addition, new ports, industrial parks and a fertiliser plant are planned.

On the peninsula in the south, a similar growth area is taking shape as a counterbalance to Bangkok. It aims to transform southern Thailand into an industrial region, diversifying away from its traditional dependence on tin mining, plantation agriculture and tourism. The Southern Seaboard Development Project will include two new ports: Ao Tha Len on the Andaman Coast, and Khanom on the Gulf of Thailand. Large areas of land have been purchased around the port sites to create industrial zones. It is hoped to establish an economic corridor between the ports by building a pipeline, a new highway and a railway.

Also proposed is a free trade zone in the provinces nearest Malaysia, and the establishment of three Andaman Sea provinces as an international tourism zone.

Finally, the latest development plan stresses the importance of small-scale industrial developments in rural areas. There is a recognition that Thailand is ideally suited to labour-intensive industries, especially in rural areas where unemployment is much higher than the official 7 per cent figure.

?

19 Look back to section 9.2 and Myrdal's model of cumulative causation (Fig. 9.1). With reference to the model, can you explain how Bangkok has grown at the expense of the rest of Thailand?

20 What does the model suggest about the future for Thailand's peripheral regions?

Figure 9.15 Bangkok, the industrial core of Thailand

?

To complete this Section on regional disparities, try the following role play and values exercise.

A large TNC has decided to locate a new electronics factory in the UK. It has short-listed a site in the core region near London. However, there is strong opposition from MPs from northern England who feel that the factory should be located in an assisted area.

Imagine that you are either (*a*) a Labour MP representing a former mining constituency in northern England, or (*b*) a Conservative MP from a prosperous commuter area in South-East England.

21 State your attitude towards the proposed factory site. Then argue your case in the form of a letter to the Secretary of State for Trade and Industry. Make clear your values and priorities (economic, social, political, environmental), and beliefs (your views about the benefits/disbenefits of location in the two areas).

22 Clarify your own views on regional policy. Are you in favour or against? Think about your own values and priorities, and your beliefs concerning the effectiveness of such policies. Prepare a brief so that you can discuss your views with others in class.

Summary

- One outcome of economic growth in capitalist countries is regional disparities in the distribution of wealth.
- Myrdal's model of cumulative growth provides an explanation for regional disparities.
- Prosperous regions within a country are termed the 'core'; less prosperous are known as the 'periphery'.
- Core and periphery have a geographic as well as an economic meaning.
- Regional policies have been adopted by most economically developed countries to reduce regional disparities.
- Within the EU, regional policies are applied at a continental scale.
- The arguments in favour of regional policies are economic, social and political.
- Regional policies usually include a package of incentives (loans, grants, tax concessions) to attract economic activities to the periphery, and limits on development in the core.

10 Manufacturing in the peripheral regions

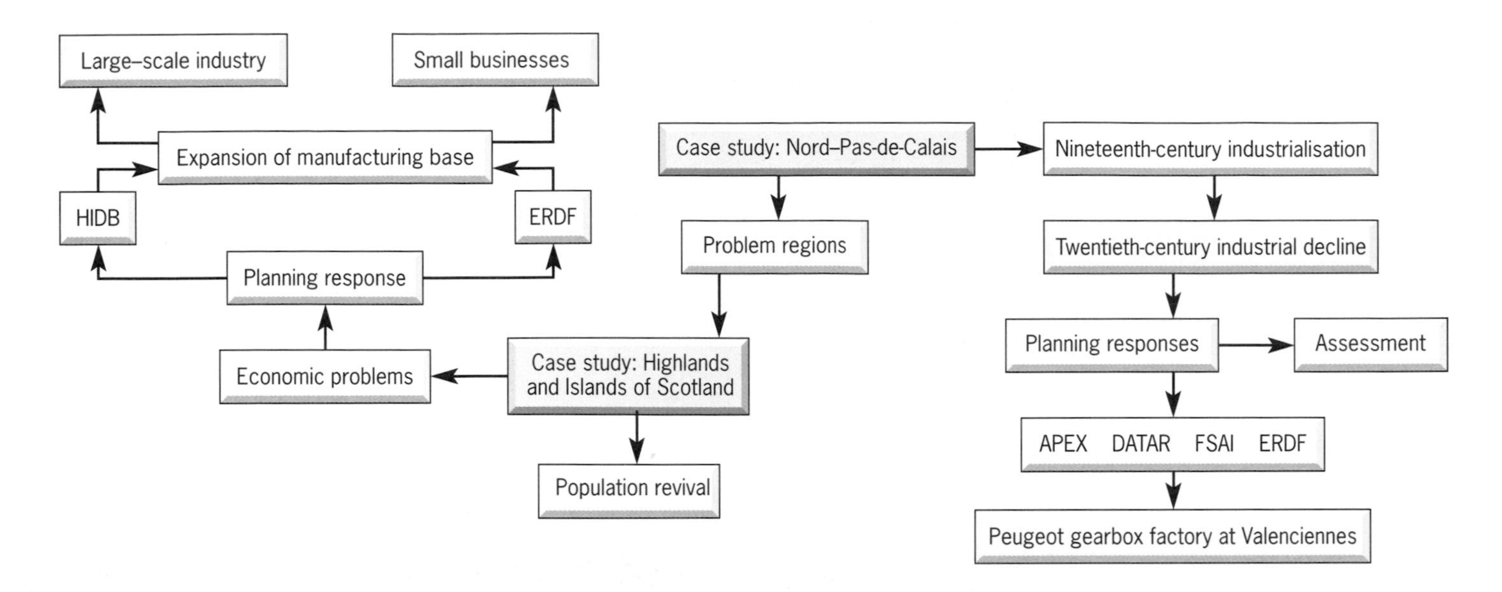

10.1 Introduction

In the last chapter we focused on regional disparities within countries. This discussion is taken a step further in this chapter as we look at manufacturing in two peripheral regions: the Highlands and Islands of Scotland, and Nord–Pas-de-Calais in France.

The Highlands and Islands is a peripheral region in both a geographical and economic sense (see the Case Study on page 111). Nord–Pas-de-Calais, although geographically part of the core, has suffered prolonged economic decline. Like many other deindustrialised regions in the EU, in economic terms the region is today part of the semi-periphery.

The Highlands and Islands of Scotland

The Highlands and Islands of Scotland is a remote rural region on Europe's periphery. The region presents considerable obstacles to manufacturing. Although it covers nearly 15 per cent of the area of the UK, it accounts for less than one per cent of the total population. Thus, its labour force and market are small.

More serious is its peripheral location which makes it remote from markets in the UK and the EU. Neither is transport within the region easy. There are over fifty inhabited islands as well as the most extensive upland area in the UK (Fig. 10.1). These problems are reflected in the region's economy and employment structure (Fig. 10.2). They can be summarised as overdependence on agriculture, fishing and tourism; and a GDP per capita which is just 78 per cent of the UK average. Although the Highlands and Islands recorded a small population increase between 1991 and 1998, in remote areas out-migration is severe enough to cause depopulation.

The planning response

In 1965 the UK government acknowledged the special problems of the region by setting up the Highlands and Islands Development Board. Its successor, Highlands and Islands Enterprise (HIE), aims to: (*a*) assist people in the Highlands and Islands to improve their economic and social conditions, and (*b*) enable the Highlands and Islands to play a more effective part in the economic and social development of the UK.

The HIE (along with the Development Board for Rural Wales) is the only rural development agency in the UK. Its value can be gauged from the number of new business start-ups in the Highlands and Islands, which throughout the 1990s was well above the average for the rest of Scotland.

A key objective of the HIE is to expand the region's manufacturing base. Its policies aim to provide more jobs by attracting inward investment, strengthening the employment structure, increasing incomes and encouraging more people to stay in the region. Grants and low-interest loans are provided to a wide range of businesses, including manufacturing. The HIE also promotes the region through advertising; provides an advisory service; and conducts its own research into local economic and social problems. However, its main expenditure is on building industrial premises, especially advance factories.

The Highlands and Islands also receive assistance through the EU's Structural Funds and the UK's regional policies. With a GDP per capita that is only 76 per cent of the EU's average, the Highlands and Islands qualify for assistance from Europe as an Objective 1 region. Through the UK's regional policy, the Highlands and Islands are also eligible for regional selective assistance.

The geography of manufacturing

A useful distinction can be made between large-scale and small-scale manufacturing industries in the region.

Large-scale industries

The largest manufacturing enterprises in the Highlands and Islands are resource-based. The natural resources which attract firms are North Sea oil, hydro-electric power, timber, water, fish and deep-water harbours. The HIE's policy has been to concentrate heavy industries in the eastern Highlands, especially around Inverness and the inner Moray Firth.

The discovery of North Sea oil in the 1970s attracted on-shore activities such as rig and platform construction. These were directed to deep-water sites at Nigg Bay, Ardersier, Kishorn and Stornoway (Fig. 10.1).

Outside the oil sector, heavy processing industries locate in the inner Moray Firth, Caithness and Lochaber. In response to government pressure and HIE promotion, the British Aluminium Company built a smelter at Invergordon in 1972. However, the plant operated for just ten years before high energy costs forced its closure with the loss of 1000 jobs. The government, concerned about the level of unemployment, designated the area an **Enterprise Zone** (EZ), and offered special incentives to attract investment (see section 11.5).

Industry in Caithness depends heavily on the UK's nuclear industry, and in particular on the Dounreay

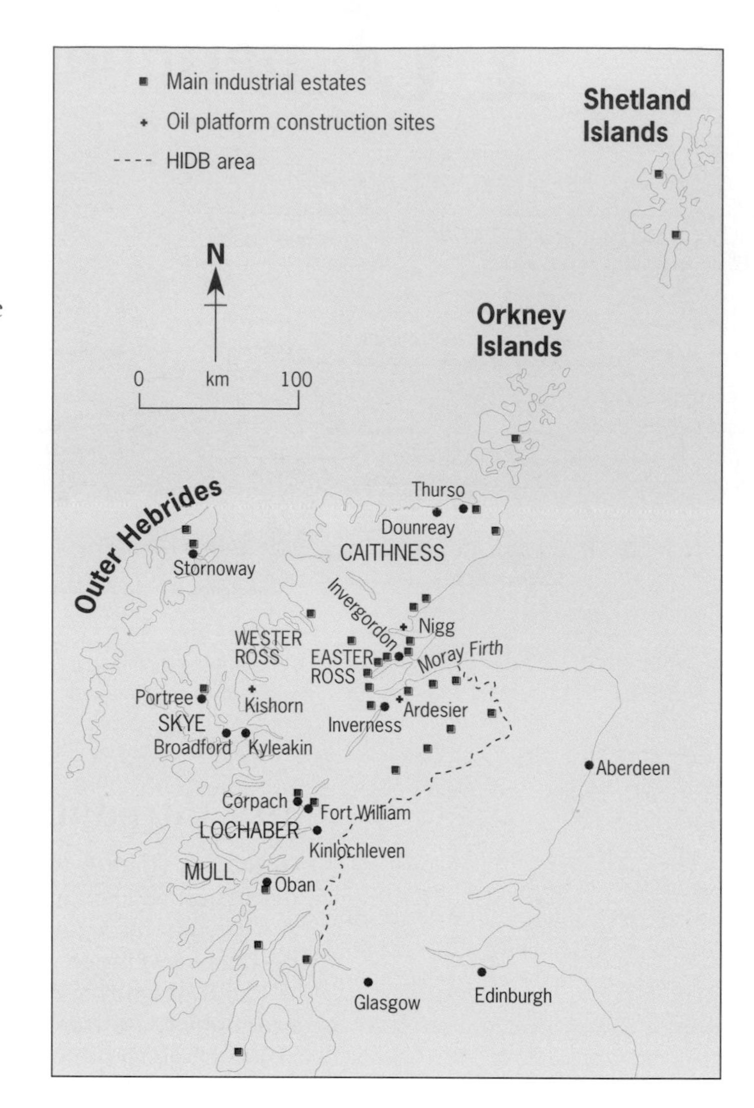

Figure 10.1 Highlands and Islands region of Scotland

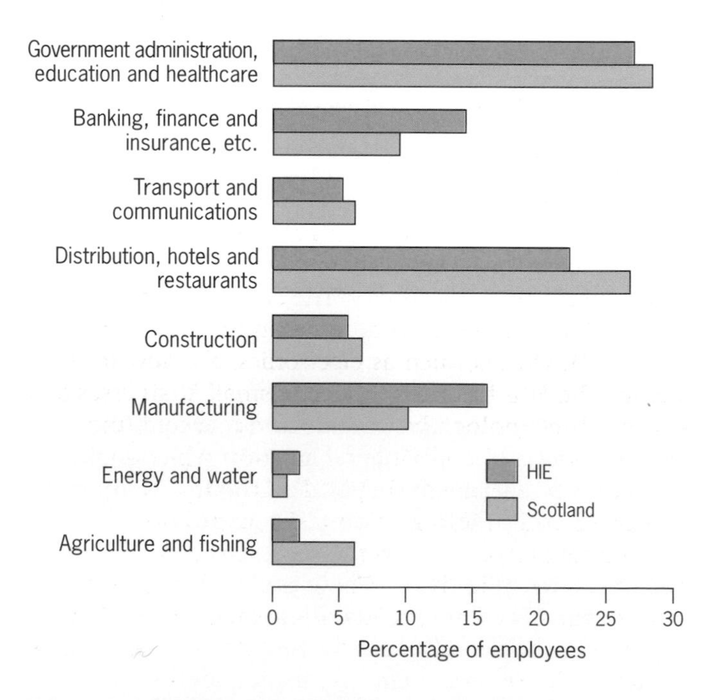

Figure 10.2 Distribution of employment in the HIE area

Figure 10.3 The British Alcan aluminium plant at Fort William

reactor near Thurso. Although the government announced the closure of Dounreay in 1998, the impact on employment will be minimal. It will take until 2095 to dismantle the plant – work that will provide employment for hundreds of people in the area.

There are three major industrial employers in Lochaber: the Fort William and Kinlochleven aluminium smelters, and the pulp and paper mill at Corpach. Closure of the pulp mill and increased automation of the Fort William smelter created serious unemployment in the area in the 1980s. None the less, paper manufacturing based on imported pulp has survived at Corpach. A deep-water harbour adjacent to the plant and abundant supplies of water are important locational factors. Aluminium smelting, attracted by cheap hydro-electric power, was established at Kinlochleven and at Fort William early this century and currently employs around 500 workers (Fig. 10.3).

Small-scale industries

In terms of employee size, the structure of manufacturing in the Highlands and Islands is dominated by small to medium-sized enterprises. Some small businesses are traditional activities such as whisky distilling, food processing, knitwear and textiles (Fig. 10.4). Others, such as electronics, are new to the region. The HIE has tried to attract small businesses and advanced-technology-based industries. Indeed, the region has one of the highest rates of small-business formation in manufacturing in the UK (Fig. 10.5). Improvements in information technology and telecommunications are steadily offsetting the disadvantages of isolation. For some small businesses any problems of communications are outweighed by the high quality of the physical environment and the attractive lifestyle on offer.

The population turn-round

Since 1971 there has been a revival of population growth in the Highlands and Islands, reversing a trend of continuous decline since the mid-nineteenth century (Fig. 10.5). Between 1971 and 1998 the population grew by 13 per cent. Although growth slowed between 1991 and 1998 (to just 0.3 per cent), it was still comparable with that in Scotland as a whole. Most of the growth of the last thirty years has been due to in-migration.

Population revival in rural areas is part of a wider process in MEDCs known as **counter-urbanisation**. This process has often been accompanied by an urban–rural shift of manufacturing (see section 11.1). However, in the Highlands and Islands the strong population growth of the 1970s was largely due to North Sea oil developments, especially in Shetland, Easter and Wester Ross, and Lochaber. Elsewhere growth was modest, and in the Outer Hebrides depopulation and rural decline continued.

As the oil boom of the 1980s passed, many migrant workers left the region and growth slowed. For instance, the Shetlands experienced 53 per cent growth between 1971 and 1981. This became a 15 per cent decline between 1981 and 1987 (Fig. 10.5). During the 1990s the population stabilised with a modest gain of 1.6 per cent. Even so, there were signs of a counter-urbanisation in some areas. On Skye, for example, depopulation was reversed, as young self-employed migrants were attracted to the island. Significantly, growth took place in the larger settlements of Portree, Broadford and Kyleakin, though this was partly at the expense of remoter areas in the north and west.

Figure 10.4 The Highland Park whisky distillery in the Orkneys

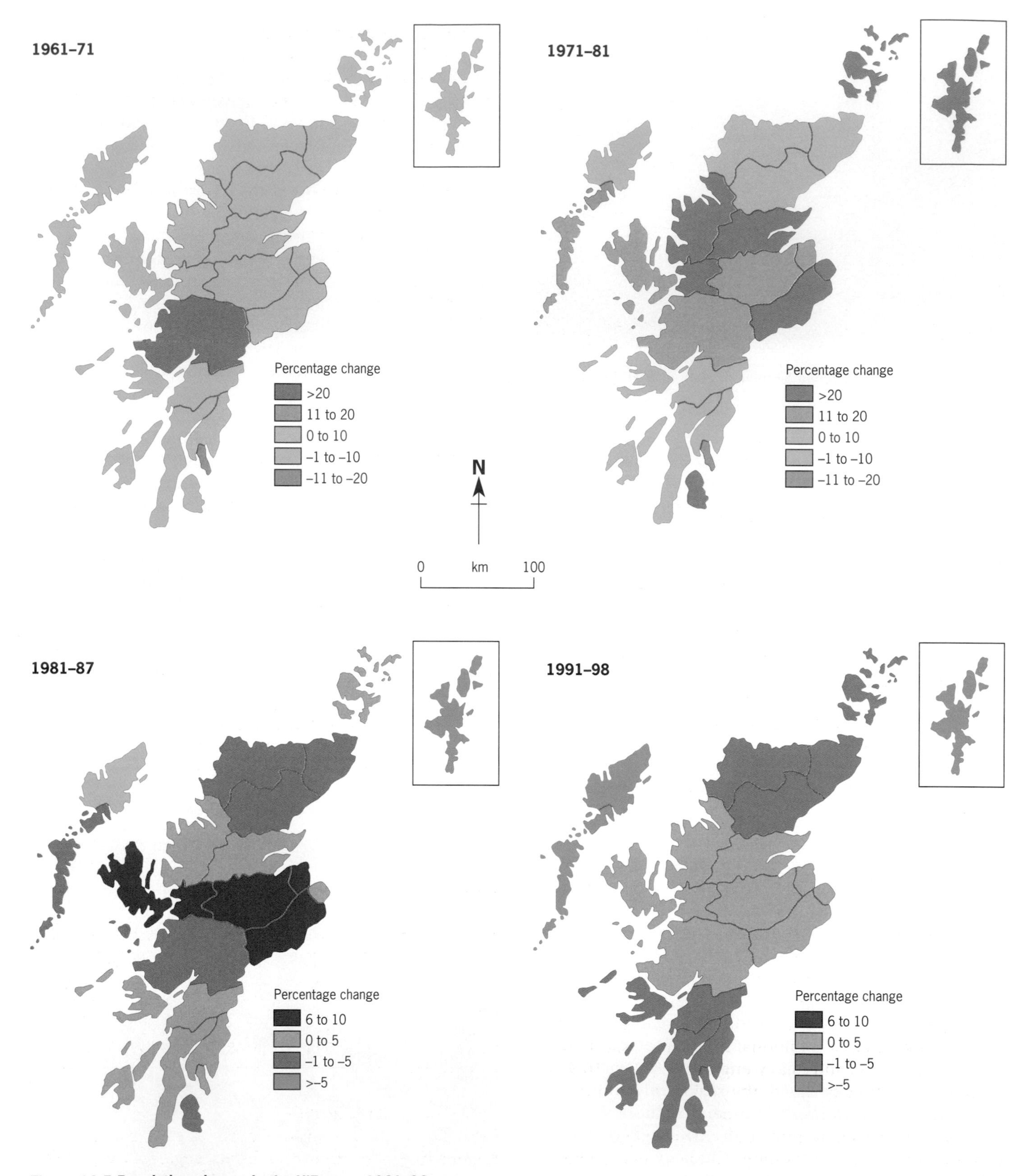

Figure 10.5 Population change in the HIE area, 1961–98

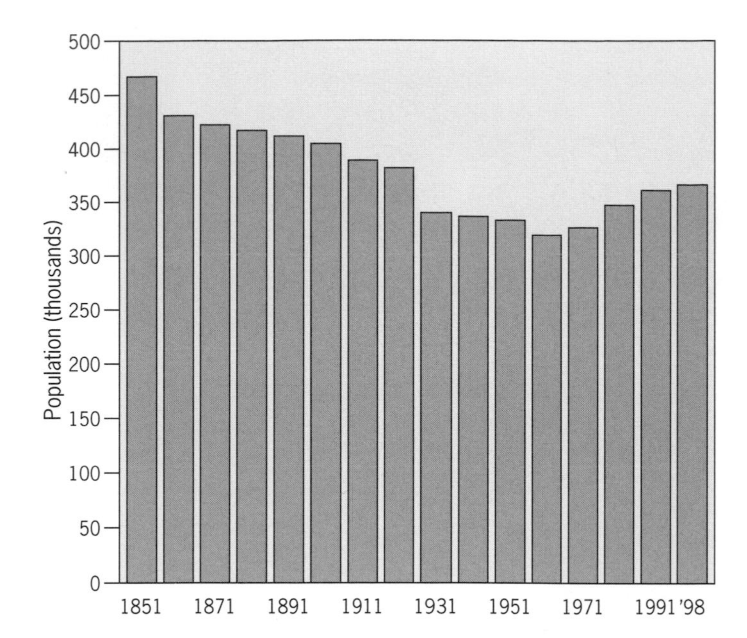

Figure 10.6 Population change in the Highlands and Islands, 1851–1998

The main reason for the population turn-round was the attractiveness of the physical environment. Employment opportunities proved less important. Indeed, many incomers created their own jobs, often in service activities. Generally, this group was prepared to trade off a lower income for a better quality of life. Isolation and lack of services, far from being thought of as disadvantages, were seen as positive attractions. Conventional retirement migration, typical of that in South-West England, is almost entirely absent in the Highlands and Islands.

The overall contribution of manufacturing to counter-urbanisation has been modest. Though many new businesses and technologically-based firms have appeared, their impact on employment has been small. Despite this, the widely dispersed locational patterns of small businesses has been an advantage, particularly when compared to the spatially concentrated oil industries of the 1970s.

Table 10.1 Employment structure of the Highlands and Islands and the UK: 1998

	Highlands and Islands	The UK
Primary	6.2	2.0
Manufacturing	3.8	18.0
Construction	6.5	5.0
Services	83.5	75.0

1 Describe and explain the main differences between the employment structures of the Highlands and Islands and the UK.

2 Although the Highlands and Islands is a peripheral region both within the EU and the UK, it has some attractions for manufacturing industry. What are they, and what types of industry have they attracted?

3 Explain why the remoteness of the Highlands and Islands is becoming less of a handicap to manufacturing industry.

4 Describe and explain the pattern of population change in the Highlands and Islands of Scotland between 1951 and 1998 (Fig. 10.5). To what extent have industrial developments influenced population change?

Nord–Pas-de-Calais

The coalfield area of North-East France comprising the départements of Nord and Pas-de-Calais (Fig. 10.7) belonged to Europe's prosperous economic core until early in the twentieth century. Like other nineteenth-century industrial regions, its fortunes were closely tied to a narrow range of traditional industries, especially coal, steel, textiles and heavy engineering (Fig. 10.9). As these industries declined, the region slipped into the less prosperous, economic semi-periphery.

Since the 1960s determined efforts have been made to widen the region's economy. Their success can be seen in Table 10.2: the employment structure of the Nord–Pas-de-Calais region in 1998 was more or less comparable with the rest of France. However, Nord–Pas-de-Calais remains a problem region. It has the highest levels of long-term unemployment and the lowest GDP per capita in the whole of France.

Table 10.2 Employment structure of Nord–Pas-de-Calais and France

	Nord–Pas-de-Calais		France	
	1990	1998	1990	1998
Agriculture	4.5	1.4	7.6	2.4
Industry	36.4	30.1	31.2	27.4
Service	59.1	68.5	61.2	70.1

Industrial development

Nord–Pas-de-Calais was one of the first industrial regions in France. Industrialisation began early in the nineteenth century and was based on local reserves of coal and iron ore. Later the region acquired the added advantages of a well-developed transport infrastructure and a large labour force. Prosperity centred around the coal industry, with production peaking at 32 million tonnes in 1939. Iron- and steel-making was located

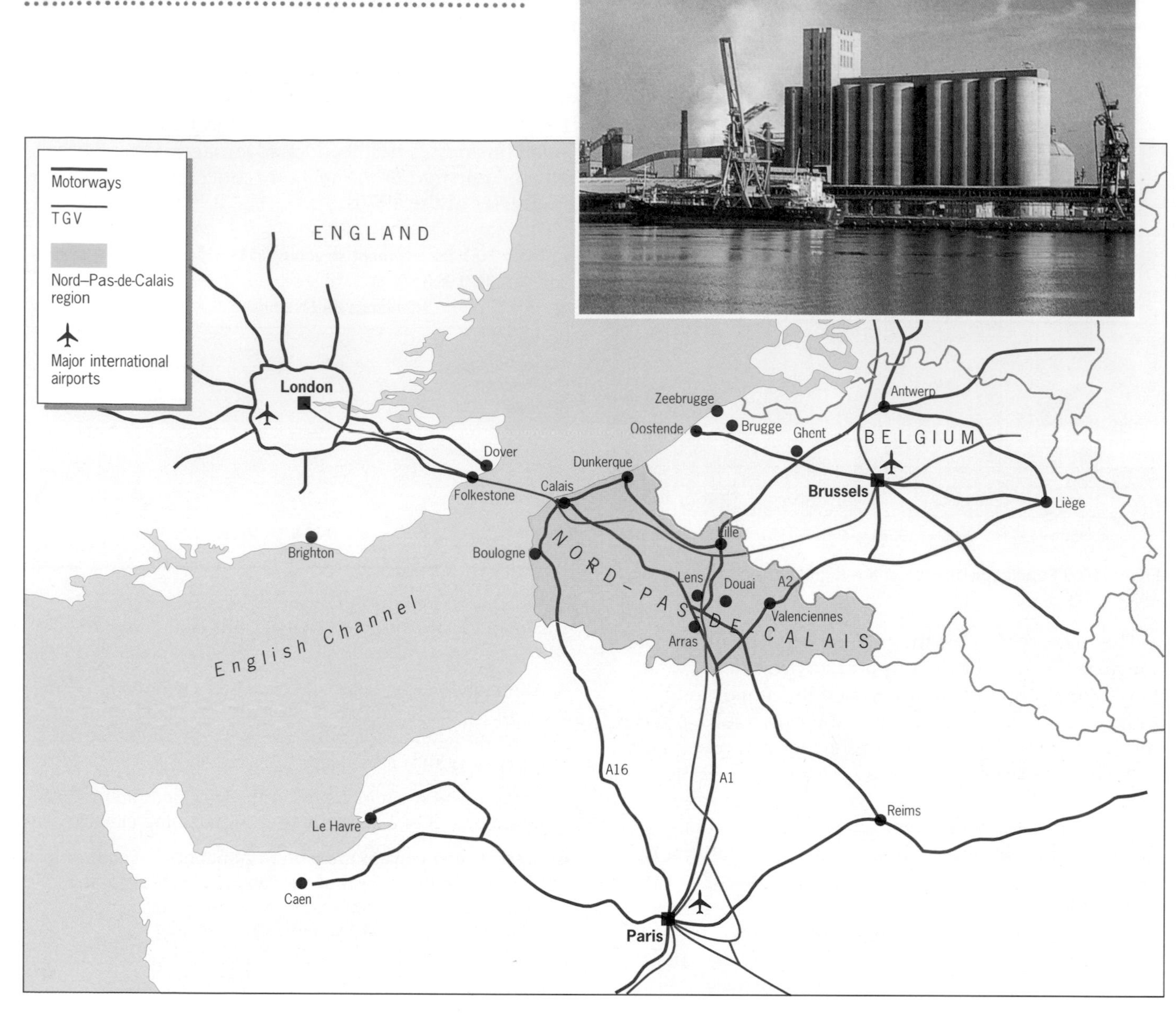

Figure 10.7 The Nord–Pas-de-Calais region

Figure 10.8 Dunkerque, as a deep-water location, has attracted development capital away from the old industrial centres

close to coal and iron reserves at Denain and Valenciennes, and in the Sambre Valley around Maubeuge. Textiles (wool and cotton) was the third staple industry. Originating in an earlier craft industry, its main centre was the Lille, Roubaix–Tourcoing and Armentières area. By 1900, Nord–Pas-de-Calais was one of the most densely populated, highly urbanised and prosperous regions in France.

Economic decline

Since 1960 the region has suffered from the structural decline of its traditional industries. Deindustrialisation occurred on a massive scale between 1975 and 1985, when 10 per cent of jobs in industry were lost. The coal industry (Fig. 10.7), with its low productivity and high costs, cut over 150 000 jobs between 1947 and 1975. It also suffered the loss of its markets: the French economy increasingly moved towards nuclear energy in the 1970s, and many local steelworks closed. The last colliery finally closed in 1991, ending 250 years of mining in the region.

Rising international competition contributed to the decline of the region as a steel-making centre. Its steel plants, occupying inland sites, were small, outdated and inefficient. The collapse in demand for steel after the oil shocks of 1973–74 led to overcapacity and the closure of several older plants. The Denain–Valenciennes area lost 7000 steel jobs between 1975 and 1980, as unemployment rose to 20 per cent. Meanwhile, investment in the steel industry was directed to coastal sites like Dunkerque (Figs 10.7, 10.8) and Fos-sur-Mer near Marseille.

The region's third staple industry, textiles, also experienced steep decline. At the end of The Second World War, the textile industry employed 125,000 workers in the Nord region alone. Today the industry provides work for only 25,000 and traditional textile towns like Roubaix and Tourcoing have experienced high levels of unemployment. The main cause of decline was international competition from South and East Asia. Many firms relocated their production overseas to benefit from lower labour costs. Closures reflected a lack of competitive edge caused by years of neglect and underinvestment.

?

5a Describe the evidence for industrial change in Figure 10.9.

b Explain how the geographical situation of Nord–Pas-de-Calais has assisted its economic regeneration.

Reviving the regional economy

In the late 1960s the French government devised a detailed plan to regenerate Nord–Pas-de-Calais and other deindustrialised regions such as Lorraine. It established a national agency for regional development (Délégation à l'Aménagement du Territoire à l'Action Régionale or DATAR) and an agency for the Nord–Pas-de-Calais region (Nord–Pas-de-Calais Développement or NPCD).

Figure 10.9 The landscape between Lens and Douai

Through these agencies improvements were made to the region's physical infrastructure, and financial incentives were offered to attract inward investment and promote the region. Priority was given to improving transport routes linking Nord–Pas-de-Calais to Paris and other industrial centres in Belgium and Germany. New autoroutes to the Belgian border and to Le Havre were constructed, and after completion of the Channel Tunnel in 1993, the region became the gateway to Britain. A new high-tech university was established at Lille, aimed at introducing advanced technology to local manufacturing. Further infrastructural help is given to relocated firms, which are provided with utilities, sewerage and other services in designated industrial zones.

A wide range of financial support is available to companies moving to the Nord–Pas-de-Calais region (Table 10.3).

Table 10.3 Financial incentives available to firms locating in Nord–Pas-de-Calais

Local incentives:	rent reductions on business premises;
	special reductions on some infrastructural costs (connections to utilities, parking facilities, etc.).
Regional incentives:	cash grants from regional authorities and from the ERDF;
	training programmes for employees recruited in northern France.
National incentives:	special state grants (prime d'aménagement du territoire).

NB The level of financial assistance is based on the size of the investment and the number of jobs created.

The impact of regional policies

Nord–Pas-de-Calais attracted 13.5 per cent of all foreign direct investment (FDI) to France in 1998. Given that Nord–Pas-de-Calais accounts for only 5.75 per cent of the country's employed population, this was a remarkable achievement. Even more remarkable, Nord–Pas-de-Calais attracted more FDI than any other French region, including Ile-de-France and the Paris Basin. This success is not a one-off: it has been sustained for several years (Fig. 10.10). Indeed between 1989 and 1998 the NPCD attracted investments that brought 30,000 new jobs into the region. Yet despite these successes, unemployment remains stubbornly high. In 1998, only the Mediterranean region had higher levels of unemployment than Nord–Pas-de-Calais.

The leading foreign investors in Nord–Pas-de-Calais are the USA, Germany, the UK, Belgium and Japan. Among the major TNCs that located in the region in the 1990s were Caterpillar, Toyota and Delphi.

Caterpillar (USA) makes heavy earth-moving equipment, employing 66,000 worldwide. Its plant at Arras manufactures gearboxes for its production facilities at Peterlee (UK), Grenoble (France) and Gosselies (Belgium). Toyota (Japan) chose Valenciennes as the site of its second European assembly plant in 1997. The attractions of Nord–Pas-de-Calais included its excellent infrastructure, its geographical situation, its proximity to several car parts manufacturers and generous grants. Production of a new small model of car for the European market begins in 2001. Delphi (USA) is one of the world's largest car parts manufacturers. Its presence in the region (at Douai) is explained by the large concentration of car assembly plants in Nord–Pas-de-Calais (Renault, Peugeot, Toyota) and nearby Ile-de-France and Belgium. The strong technical skills and flexibility of the workforce were additional advantages.

?

6 Essay: Both the Highlands and Islands of Scotland and Nord–Pas-de-Calais have been described as 'problem regions'. Compare and contrast the problems faced by each region, the efforts of governments to generate economic growth and their degree of success.

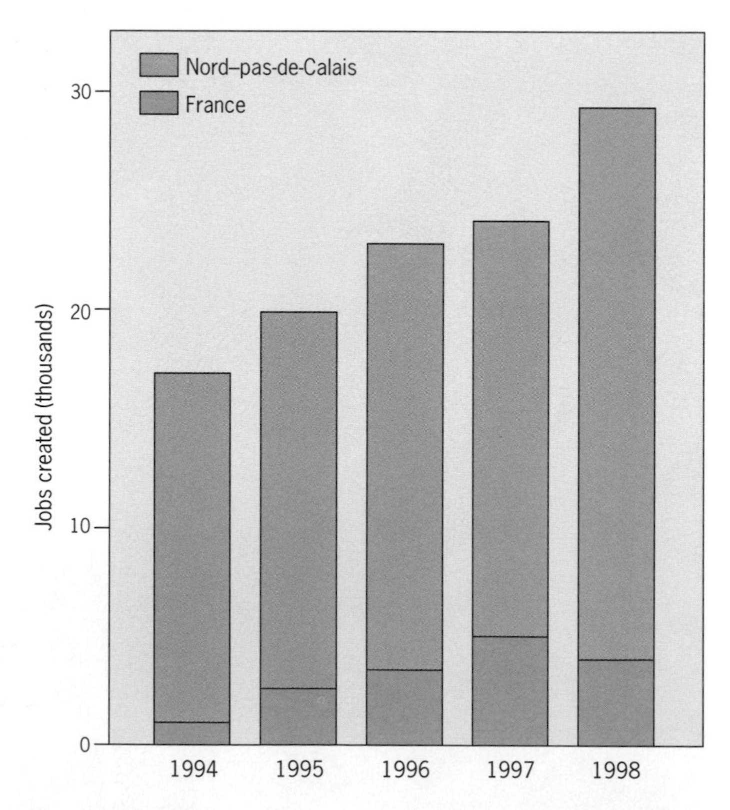

Figure 10.10 Jobs created by foreign inward investment, 1994–98

Summary

- The less prosperous regions in the EU are the so-called peripheral regions.
- The Highlands and Islands of Scotland is a remote rural region. It is peripheral both in a geographic and an economic sense.
- Nord–Pas-de-Calais is situated in the geographic core of the EU, but long-term industrial decline in the twentieth century places it in the economic semi-periphery.
- There are major problems for manufacturing industry in the Highlands and Islands: these include the remoteness of the region, its limited supplies of labour and its small market.
- The UK government has tackled the problems of the Highlands and Islands by setting up a rural development board (Highlands and Islands Enterprise) with responsibility for the region.
- Large-scale industries are concentrated around the Moray Firth and in Lochaber. They are mainly resource-based, making use of deep-water terminals, HEP and timber.
- Small businesses are widely dispersed. Many are based on advanced technology and local resources and are attracted by the quality of life on offer in the Highlands and Islands.
- Since 1971 there has been a revival of population over much of the Highlands and Islands.
- Nord–Pas-de-Calais is a nineteenth-century industrial region whose prosperity was based on coal, textiles, steel and heavy engineering.
- The decline of the Nord–Pas-de-Calais industrial base has created large-scale unemployment.
- The French government has attempted to solve the region's problems through various agencies (DATAR, NPCD) whose task has been to attract inward investment and improve the regional infrastructure.
- Major inward investments have been attracted into the Nord–Pas-de-Calais region, and there has been significant economic regeneration.

11 Urban industrial change

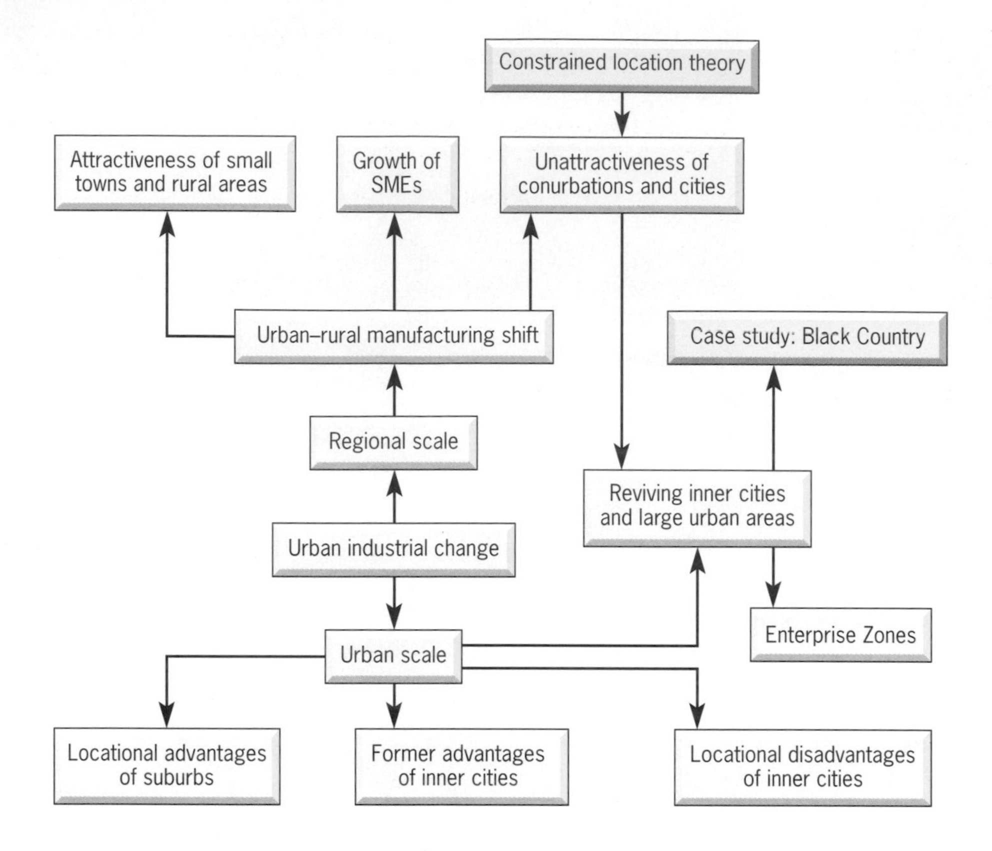

11.1 Introduction

In MEDCs old industrial cities and conurbations have proved increasingly unattractive to modern industries. This failure to attract significant new investment has had unfavourable results for many large urban areas. Fast-growing industries based on advanced technologies are often poorly represented; huge job losses have occurred as plants have contracted and closed; and levels of unemployment have risen sharply compared with smaller towns and rural areas.

The geographical impact of these changes has been felt at two scales. Regionally there has been an urban–rural shift of manufacturing industry. Within cities, a redistribution of industry has taken place from the inner to the outer suburbs. In this chapter we shall investigate these trends and look at the attempts of government to revive manufacturing industry in the largest urban areas.

11.2 The urban–rural shift

Since 1960, the UK's largest cities have lost manufacturing jobs at a faster rate than the country as a whole. This has led to an urban–rural shift which is acknowledged as 'the single most powerful trend in manufacturing location in the UK since the 1960s' (Keeble, 1987).

Between 1960 and 1981 manufacturing employment fell by a quarter. Rates of change differed sharply between the conurbations and major cities, and small towns and rural areas. London, for example, lost half its manufacturing employment (700 000 jobs) during this period. But in rural areas the trend was reversed. There, manufacturing employment rose by nearly a quarter (or 100,000 jobs). As a general rule, the bigger the settlement the greater the decline. Thus all the conurbations lost at least one-third of their manufacturing jobs, and only two of the seventeen free-standing cities experienced any increase (Fig. 11.1).

The urban–rural shift was not confined to the UK. It was closely paralleled in other industrialised countries, especially in the EU and the USA.

Table 11.1 Manufacturing employment change in UK conurbations, 1952–81 (relative to UK average in thousands per year)

	1952–60	1960–66	1966–73	1973–78	1978–81
London	-9.1	-16.5	-34.1	-17.2	+2.9
Conurbations	-14.4	-21.1	-29.3	-12.1	-31.1

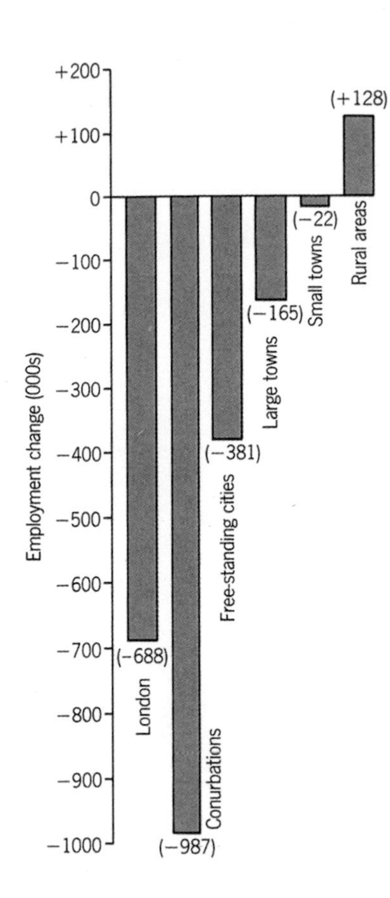

Figure 11.1 Manufacturing employment change by type of area, 1960–81

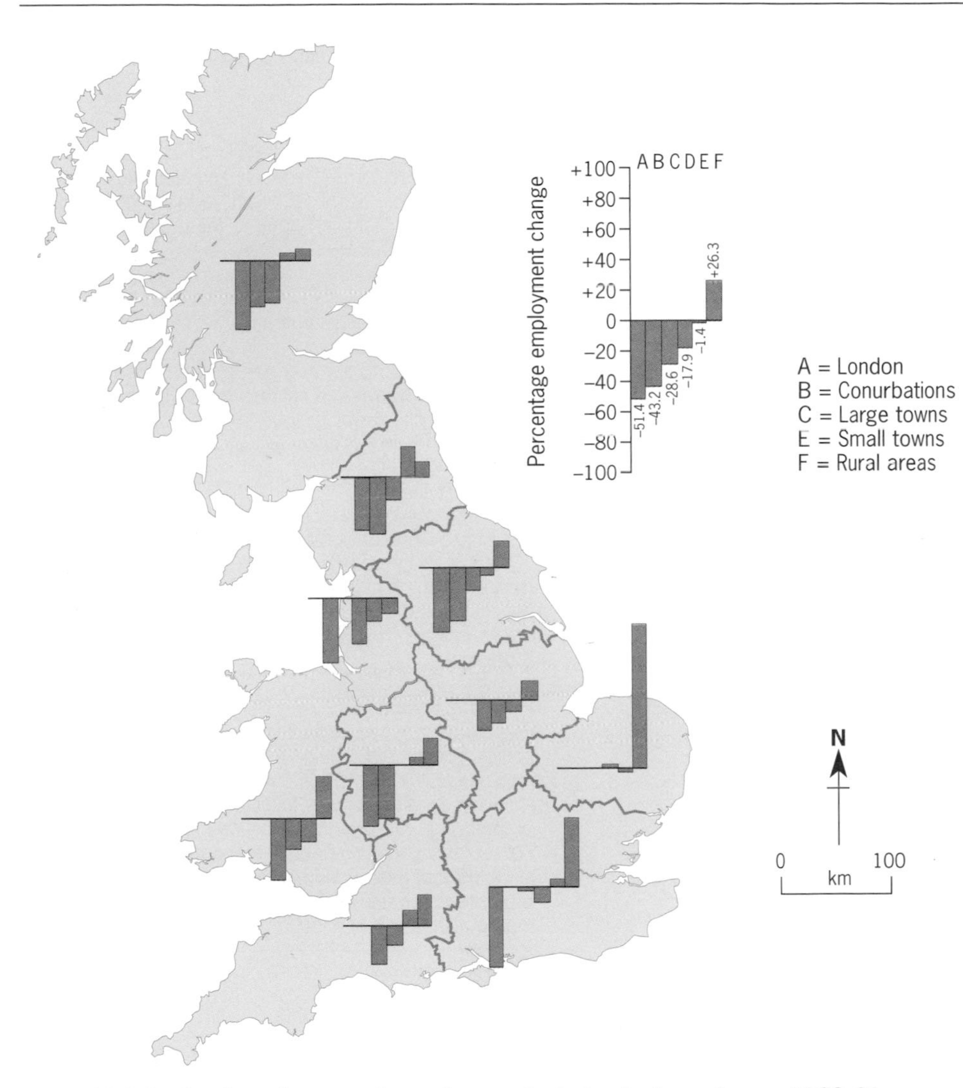

Figure 11.2 Regional employment change in manufacturing by type of area, 1960–81

?

1 Analyse the regional pattern of the urban–rural shift in the UK by studying Figure 11.2. You can pick out the main trends by completing a table like the one below. For example, there are eight regions with conurbations, and each conurbation experienced a net loss of jobs.

	Conurbations (including London)	Large cities	Large towns	Small towns	Rural areas
Job gain	0				
Job loss	8				

2 Describe the pattern of employment change shown in your table. How widespread was the urban–rural shift in the UK as a whole?

3 Compare the patterns of employment change in East Anglia and the North-West. How do they differ from the average?

11.3 Explaining the urban–rural shift

There is no simple explanation for the urban–rural shift. However, one thing we can be sure of: the urban–rural shift does not result from the closure of factories in cities and their transfer to small towns and rural areas. The creation of new jobs in small towns and rural areas is due largely to the expansion of existing factories.

4 Study Table 11.2. Which explanation of the urban–rural shift seems best to you? Give reasons for your choice.

Several hypotheses have been advanced to explain the urban–rural shift (Fothergill, Kitson and Monk, 1985). You will find a brief explanation and analysis of each one in Table 11.2.

Table 11.2 Explanations of the urban–rural shift

Industrial structure
Conurbations and large free-standing cities have large concentrations of old, declining industries. As a result, rates of factory closure are high in these areas.
There is no conclusive evidence for this. London, which had a favourable industrial structure, lost jobs more rapidly than any other conurbation.

Residential preference
People prefer to live in small towns and rural areas.
In multi-plant companies, locational decisions are rarely influenced by the residential preferences of decision-takers. Decisions are taken at boardroom level by HQ executives who will not have to work at the plant in question. But the residential preferences of key workers may have some significance if their skills are in short supply. The location of high-tech industry in southern UK and California has been influenced by the environmental preferences of highly qualified scientists, managers and technicians (see the Case Study on page 69).

Labour
Conurbations, as traditional centres of trade union organisation and strength, discourage investment.
Only two conurbations (Clydeside and Merseyside) experienced an above-average number of days lost through industrial action (after adjusting for industrial structure). There is, however, some evidence that labour turnover is higher owing to the greater employment opportunities available in large centres of industry.

Production costs
Manufacturing is declining in conurbations and large cities because they are uneconomic locations with high production costs.
In the UK, costs do not vary greatly from place to place. Labour costs show little regional variation (see section 5.2). Wages and salaries for multi-plant firms are determined by national pay bargaining, and bear little relationship to local labour market conditions. Apart from industrial land prices and property rentals, which are high in cities, costs vary little between urban and rural areas.

Transport costs
Twentieth-century advances in transport technology and telecommunications have decreased the advantage of cities as places where firms could benefit from easy access to local supplies and markets.
Transport costs do little to hinder industrial development in rural areas. Today, 75 per cent of UK manufacturing has transport costs of less than 3 per cent of gross output. If small towns and rural areas have a transport costs advantage (access to motorway junctions, absence of congestion), it is likely to be a small one. The urban–rural shift was evident in many parts of the UK before the development of motorways.

Planning policies
Regional policy, new and expanded towns, and local planning control (planning permission, land-use zoning, green belts) could, in theory, encourage an urban–rural shift.
Regional policy before 1980 made no distinction between urban and rural areas.
New and expanded towns favoured small centres and rural areas; they accounted for only a small proportion of the growth in manufacturing away from the UK's cities.
Local planning has often discouraged growth in small towns and rural areas for environmental reasons. In large urban areas, green belt policies have had an impact. They have limited urban sprawl, but have often prevented new industry from locating on the edges of cities. To this extent, physical planning policies have played a role in encouraging the urban–rural shift.

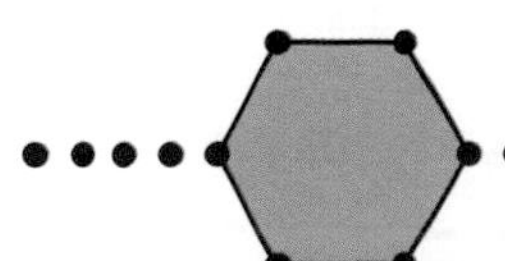

Constrained location theory

Few of the factors listed in Table 11.2 are thought to have had much influence on the urban–rural shift. Most of the evidence points to the shortage of industrial land in cities as the driving force behind the trend. Lack of room for expansion has become critical as manufacturing has moved towards greater capital intensity and increasing space needs per worker employed. Thus we can point to physical constraints on industrial expansion in congested cities as the main force behind the urban–rural shift of manufacturing industry (Fig. 11.3).

?

5 Study the aerial photographs of central Glasgow (Figs 11.3a and 11.3b). The CBD is the area of high-rise buildings top centre left.
a What evidence is there that in 1969 (Fig. 11.3a) the inner city was already an unattractive location for manufacturing industry?
b Compare Figures 11.3a and 11.3b, and describe the main land-use changes that occurred between 1969 and 1988. Suggest an explanation for these changes using the ideas of the urban–rural shift.

We refer to this explanation as the **constrained location theory**. What exactly are these physical constraints?

Factory buildings

Industrial expansion in large urban areas is often limited by an outdated stock of factory buildings. Many factories built in the nineteenth and early twentieth

Figure 11.3 The changing face of the inner city

***a* Glasgow, 1969**

***b* Glasgow, 1988**

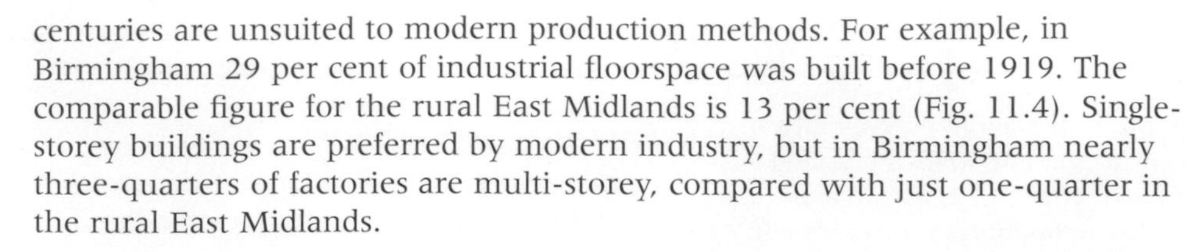

centuries are unsuited to modern production methods. For example, in Birmingham 29 per cent of industrial floorspace was built before 1919. The comparable figure for the rural East Midlands is 13 per cent (Fig. 11.4). Single-storey buildings are preferred by modern industry, but in Birmingham nearly three-quarters of factories are multi-storey, compared with just one-quarter in the rural East Midlands.

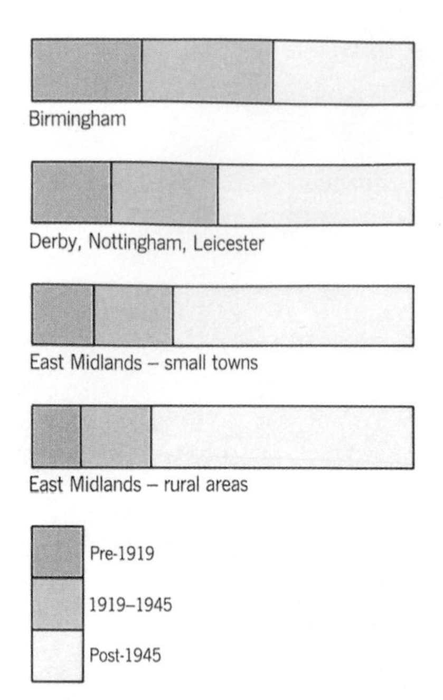

Figure 11.4 Age of industrial buildings by type of area (percentage of floor space built in each period)

On-site space for expansion

Factories in cities have less room for on-site expansion. In rural areas, sites next to a factory are often used for farming, and permit expansion. In cities, potential factory sites are likely to be occupied by other urban users, making expansion impossible.

Availability of sites

Small towns and rural areas have substantially more industrial land available than cities. Despite years of industrial decline, the UK's largest cities have only small amounts of land available for industrial development. Where large sites are available, they are often derelict and need costly reclamation. Difficulties of this type are not encountered on greenfield sites in small towns.

Size of sites

Most industrial sites in cities tend to be small. They are unable to accommodate either large factories or industrial estates. The latter are especially important, because they attract speculative factory building in advance of demand. Without large sites, new factory development is necessarily held back.

Land prices

Land prices are higher in cities than in surrounding rural areas (Table 11.3). Urban land which could be used by industry could also be used for housing, offices and other purposes. Thus, competition for limited space pushes up prices. For firms which have no compelling reason to locate in the city, land prices will encourage investment in new factories in small towns and rural areas.

Table 11.3 The price of industrial land in England and Wales, 1981–82 (£ thousand per ha)

	London	Conurbations	Major cities	Towns and rural areas
North		62	62	38
Yorks and Humberside		116	88	73
North-West		108		86
East Midlands			145	87
West Midlands		153	113	135
East Anglia				116
South-East	748		427	426
South-West			275	134
Wales			87	41

6 Compare the land prices in Table 11.3 with the regional employment changes in Figure 11.2. To what extent can we explain changes in employment in conurbations (including London) and large cities by variations in land prices?

7 Essay: Read through the Case Study of conflict between manufacturing industry and the green belt in Bradford, West Yorkshire (page 171). To what extent does this example support constrained location theory as the main force behind the urban–rural shift?

Branch plant decision-making

The process of decision-making in the location of new branch factories favours an urban–rural shift. First, a firm chooses a suitable area – e.g. the South-East, or within 50 kilometres of London. Second, a search for a suitable site within the chosen area is made. Some firms stop their search as soon as they find somewhere satisfactory. Others do a thorough evaluation of alternative sites. Site selection is strongly affected by the availability of land, so most sites are in small towns and rural areas. If firms have no special preference for location in an urban area, the simple facts of land availability are likely to attract most new factories to small towns and rural areas.

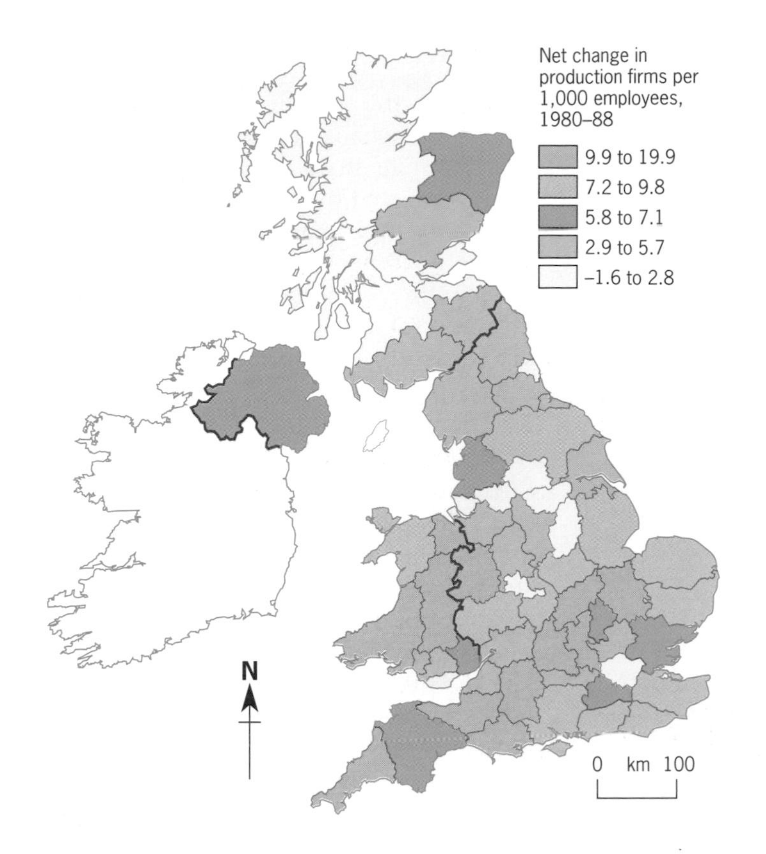

Figure 11.5 The geography of small-business development in production industry, 1980–88 (*Source:* Keeble, 1990a)

Small manufacturing enterprises and the urban–rural shift

Much of the job growth in the manufacturing sector during the 1980s and 1990s in the UK occurred in small manufacturing enterprises (SMEs) employing fewer than 100 workers. A study of SMEs in London, Hertfordshire and rural northern England showed that SMEs in rural areas grew faster than SMEs in towns and conurbations during in the period 1979 to 1990 (Table 11.4). This urban–rural shift of employment continued, though at a slower pace, throughout the 1990s.

The trends in Table 11.4 have been confirmed by similar surveys in other parts of the UK. Two factors explain the more rapid growth of rural firms:

- Rural SMEs have not experienced the same problems of recruiting labour with appropriate skills as SMEs in large urban areas. Thus, while rural SMEs could expand their output by taking on more workers, many urban SMEs have been forced to adopt other strategies (e.g. more automation).
- Clusters of linked economic activities are less common in rural areas. Thus rural SMEs, compared to their urban counterparts, have fewer opportunities to sub-contract work to other businesses. As a result, rural SMEs create more jobs in-house. For example, the study by Smallbone and North showed that while nearly half of SMEs in London used sub-contractors, in rural areas the proportion was less than one-third.

8a Using a percentage scale, represent the data in Table 11.4 as divided bar charts.
b Summarise the main differences in employment change between London, the Outer Metropolitan Area, and rural areas in northern England.

Table 11.4 Employment change in a sample of SMEs in different environments, 1979–90 (*Source:* Smallbone and North).

	Inner London	Greater London	OMA*	Rural
1979 employment	1,682	1,929	1,634	1,067
1990 employment	1,729	2,123	2,013	1,606
% change	3	10	23	51

**Outer Metropolitan Area.*

11.4 Intra-urban manufacturing change

Since 1945, the location of industry in cities has undergone considerable change. This change has followed a consistent theme: the outer suburbs have increased their share of industry at the expense of the inner areas. This has usually occurred through the contraction of industry in the city centre, and its expansion in the suburbs. For example, the inner core of Greater Manchester lost nearly 40 per cent of its jobs in manufacturing between 1966 and 1975. However, in some cases suburban growth reflects a relative shift. This happens when central areas either contract faster than the suburbs, or the suburbs expand faster than central areas.

The historic advantages of the inner city

The inner city in the UK is the area within one or two kilometres of the CBD, which was built largely before 1914 (Fig. 11.6). In the nineteenth century capital flowed into this part of the city, attracted by its access to materials, labour and local markets (Fig. 11.7). Rail termini for the collection and shipment of products were located there. In port cities like Liverpool and Glasgow, docks penetrated into the inner city and were sources of imported materials for manufacturing.

By the mid-nineteenth century the inner city had emerged as a mixed zone of low-class housing and factories, where workers lived within walking distance of their jobs. Before the development of efficient urban transport systems, inner-city locations gave firms the best access to labour supplies. Meanwhile, the rapid expansion of urban populations in the nineteenth century meant that cities became large and important markets in their own right.

A Early industrial concentrations
B CBD-orientated industries
C Heavy and port-based industry, late 19th and 20th century
D Light industry, 20th-century surburban
E Light industry, 20th-century planned estates

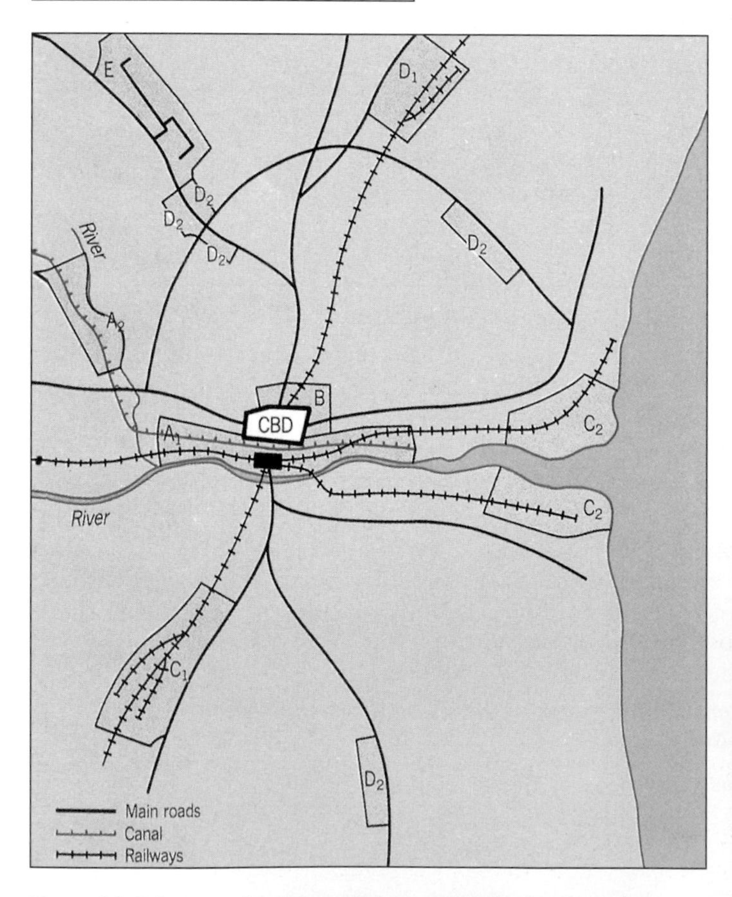

Figure 11.6 A generalised descriptive model of manufacturing areas in cities (after Herbert and Thomas, 1990)

Figure 11.7 The inner city and its transport links attracted investment for both commerce and industry in the late nineteenth and early twentieth centuries

?

9 Study the aerial photograph of central Glasgow in 1969 (Fig. 11.3a). Using the evidence of the photograph, describe the possible advantages to nineteenth-century manufacturing industry of location in this area.

Later, these initial advantages were added to by agglomeration economies which further reduced costs. Large concentrations of industry were interlinked by complex flows of finished and semi-finished products and information. Given its advantages, it is not hard to see how the inner city became the least-cost location for so many manufacturing activities in the nineteeth century.

The modern city: declining core and expanding suburbs

The twentieth century has seen the steady decline of the advantages of the inner city. Capital has deserted the area as multi-plant firms have closed factories, attracted by more profitable locations in the suburbs, small towns and rural areas. Today the inner city (without government assistance) is a high-cost and unattractive location for most manufacturing activities.

The decline of the inner city is due to several factors. Some of these – obsolete factory buildings, lack of space for expansion and congestion – have been discussed earlier in this chapter as part of the urban–rural shift.

Suburbanisation of the population

The suburbanisation of populations, divorcing inner-city industries from their workforce, has contributed to inner-city decline. This process has been carried furthest in US cities. Detroit, for example, lost half its population to its suburbs between 1950 and 1990. Because suburbanisation operates selectively, it leaves behind the less skilled, the poorly educated and the old. This leaves inner-city firms with only a narrow choice of labour skills. In addition, increased mobility of the workforce through private car ownership has helped to free industry from the inner city and its dependence on the public transport system.

Accessibility

As accessibility has declined, so too has the locational value of inner-city sites. Instead of being the least-cost transport locations, inner cities are now among the most costly. Multi-storey factory buildings, occupying cramped sites, and surrounded by narrow, congested streets, cannot be served efficiently by road transport. In addition, rail sidings, marshalling yards and docks, once so attractive, have little relevance to modern industry. By comparison, suburban locations are geared to road transport. They are served by ring roads and expressways, and have easy access to motorway junctions. Greenfield sites on the edge of the city also offer space for single-storey factories, container handling and parking.

Disruption of inter-firm linkages

As inner-city factories have closed, agglomeration economies through inter-firm linkages have been disturbed. The effect has been particularly damaging where industries with high levels of linkage (such as mechanical and electrical engineering) have closed. Often the result has been to undermine the profitability of the remaining firms, inducing a spiral of decline (Fig. 11.8).

Urban renewal

In the UK, planning policies have also contributed to inner-city decline. The need to clear large tracts of slum housing in the inner cities in the 1960s and early 1970s inevitably resulted in the demolition of many old factories as well. One effect of this urban renewal policy was to undermine business confidence and reduce the flow of investment funds to the inner city.

Deteriorating social and physical environments

Lastly, the deteriorating social and physical fabric of the inner city has given a further push to manufacturing decline (Fig. 11.9). In many US cities the suburbs are independently governed. Declining population levels in the older,

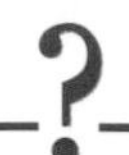

10 Study the vicious circle of manufacturing decline in inner cities (Fig. 11.8). Using the ideas in Figure 11.8 and the information in section 11.4 headed 'The modern city', construct your own virtuous circle of manufacturing growth in the outer suburbs.

11 Make a sketch of the industrial estate in Figure 11.11. What locational advantages for industry are evident from the photograph? Add these as annotations to your sketch.

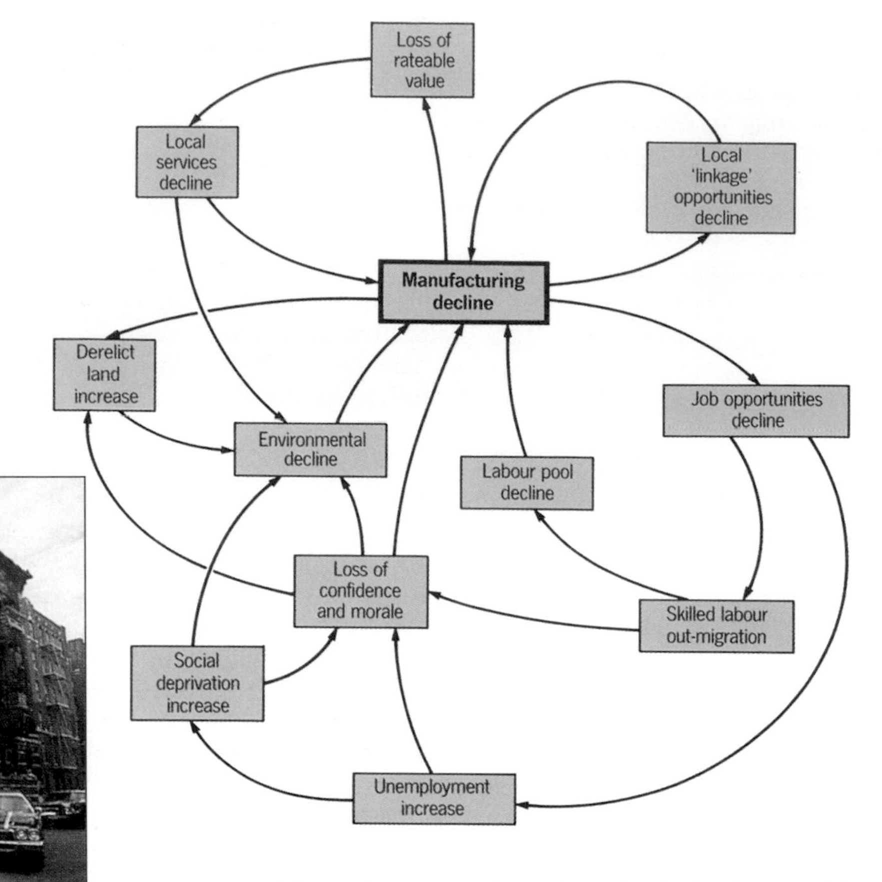

Figure 11.8 The vicious circle of manufacturing decline in inner cities

Figure 11.9 Inner-city dereliction in the Bronx, New York

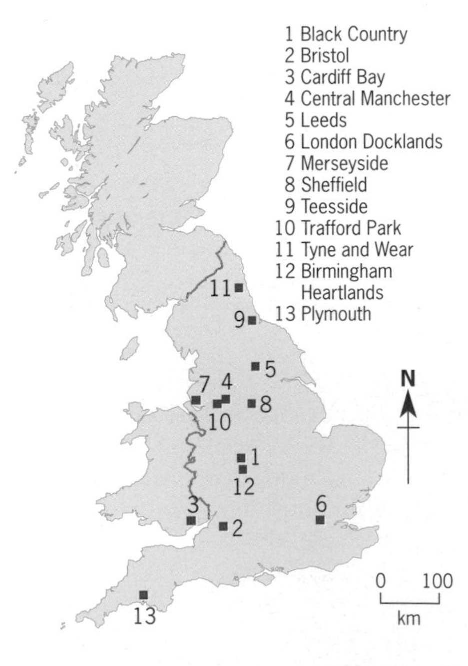

Figure 11.10 Urban Development Corporations

inner areas, have undermined tax bases, producing great inequality in service provision. Segregation leaves the inner cities dominated by poor, black populations. High rates of crime, slum housing and an aged infrastructure, provide an unattractive environment for firms. Today's prestige locations are industrial estates and science parks in modern suburbs where other expanding firms are found.

11.5 Regenerating urban and industrial areas

In the 1980s the UK government shifted the focus of its spatial policy from the regions to areas of urban dereliction. The economic regeneration of these areas was based on two instruments of policy: **Urban Development Corporations** (UDCs), and **Enterprise Zones** (EZs).

Urban Development Corporations

The main thrust of government policy for the revival of the UK's inner cities were the UDCs. Thirteen UDCs were set up between 1981 and 1993 (Fig. 11.10). They covered the worst examples of inner urban decay (Table 11.5). Appointed by government, the UDCs were given wide powers. These included derelict land reclamation, land assembly, environmental improvement and the provision of infrastructure. Financial assistance was available to private firms as loans and grants. This 'demand-led' approach aimed to create new local economies. Relatively modest public funds were intended to attract much larger private investment, eventually leading to self-generating growth.

Table 11.5 UDC output measures: cumulative achievements (*Source: Geography Review*, 12, 3, Jan. 1999)

	Land reclaimed (ha)*	Housing units*	Non-housing floorspace* (000 m²)*	Infrastructure roads (km)	Jobs (gross)*	Private investment* (£m)	Grant-in-aid (£m) Lifetime target
London Docklands	728.4	19,844	2,283.9	244.7	66,683	6,277.5	1,860.3
Merseyside	363.2	2,875	555.0	84.0	16,595	461.0	385.3
Black Country	314.7	2,914	826.4	28.3	15,517	833.0	357.7
Teesside	434.4	1,187	362.2	26.1	10,086	928.9	350.5
Trafford Park	151.7	283	572.1	37.5	21,063	1,012.8	223.7
Tyne and Wear	485.7	3,639	844.5	33.2	23,473	937.3	339.3
Bristol	69.0	676	121.0	6.6	4,825	235.0	78.9
Central Manchester	35.0	2,583	138.6	2.2	4,944	372.8	82.2
Leeds	68.0	571	374.0	11.6	9,066	357.0	55.7
Sheffield	239.8	0	358.2	12.7	12,747	577.2	101.0
Birmingham Heartlands	75.6	603	165.2	19.9	2,253	174.7	39.7
Plymouth	10.8	0	3.0	4.4	25	0.5	44.5
Total	2,976.3	35,175	6,604.1	511.2	187,277	12,167.7	3,919.0

**To 31 March 1996.*

Figure 11.11 The London Docklands Development area

The Black Country

The Black Country is a sub-region within the West Midlands conurbation. It consists of four districts: Dudley, Sandwell, Walsall and Wolverhampton. With a population of 1.1 million, the Black Country accounts for one-fifth of the population of the West Midlands region.

As a coalfield region with local deposits of iron ore, the Black Country was at the forefront of the industrial revolution. Its main specialities were iron and steel and metalworking. By 1850 the region was the UK's leading iron-making centre (Fig. 3.31). Industrialisation led to the expansion of the Black Country's small towns. Urban growth created a poly-nuclear urban structure with towns such as Wolverhampton, Dudley, Walsall and West Bromwich merging together. As a result, the Black Country has no dominant commercial focus.

Economic and environmental problems

Early industrialisation and regional industrial specialisation created a legacy of economic and environmental problems in the Black Country. Among the problems that became particularly acute in the second half of the twentieth century were the region's excessive dependence on a declining industrial base; relatively high levels of unemployment; widespread dereliction (Fig. 11.12); poverty and social exclusion.

Planning responses 1987–98

Assisted areas

The Black Country is recognised as an assisted area, with businesses eligible for regional selective assistance funded by the UK government.

EU assistance

All four Black Country boroughs have EU Objective 2 status and receive assistance from the European Structural Funds. Indeed European money has been the largest single source of external funding and has provided a major boost to the regeneration of the Black Country. The Wolverhampton Science Park and the Midland Metro rapid transit system are examples of major projects in the region supported by EU money.

Black Country Development Corporation (BCDC)

The BCDC had responsibility for regenerating 10 square miles of the Black Country between 1988 and 1998. Its main tasks were to reclaim derelict land for industry,

Figure 11.12 Industrial dereliction in the Black Country

commerce and housing; improve infrastructure; attract private investment; and market the region. The achievements of the BCDC and other urban development corporations are summarised in Table 11.5.

City Challenges

City Challenges were set up in 1992. Urban authorities were invited to bid competitively for funds targeted at business promotion and social/environmental improvements. Sandwell, Walsall and Wolverhampton all received substantial funding from the scheme, which covered urban renewal, land reclamation, training, education and community projects. Like the BCDC, City Challenges came to an end in 1998.

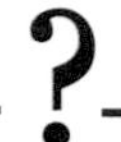

12a Using the information in Table 11.5, plot private investment (y) in UDCs against grant aid (x) as a scattergraph (excluding London Docklands).
b Draw a best-fit line on your graph and label those UDCs below the line as 'least successful' and those above the line as 'most successful'.
c Make two league tables of UDCs, using the following criteria: private investment divided by grant-in-aid; and jobs created divided by grant-in-aid.
d From the evidence of your scattergraph and the two criteria in (c), comment on the relative performance of the Black Country Urban Development Corporation.

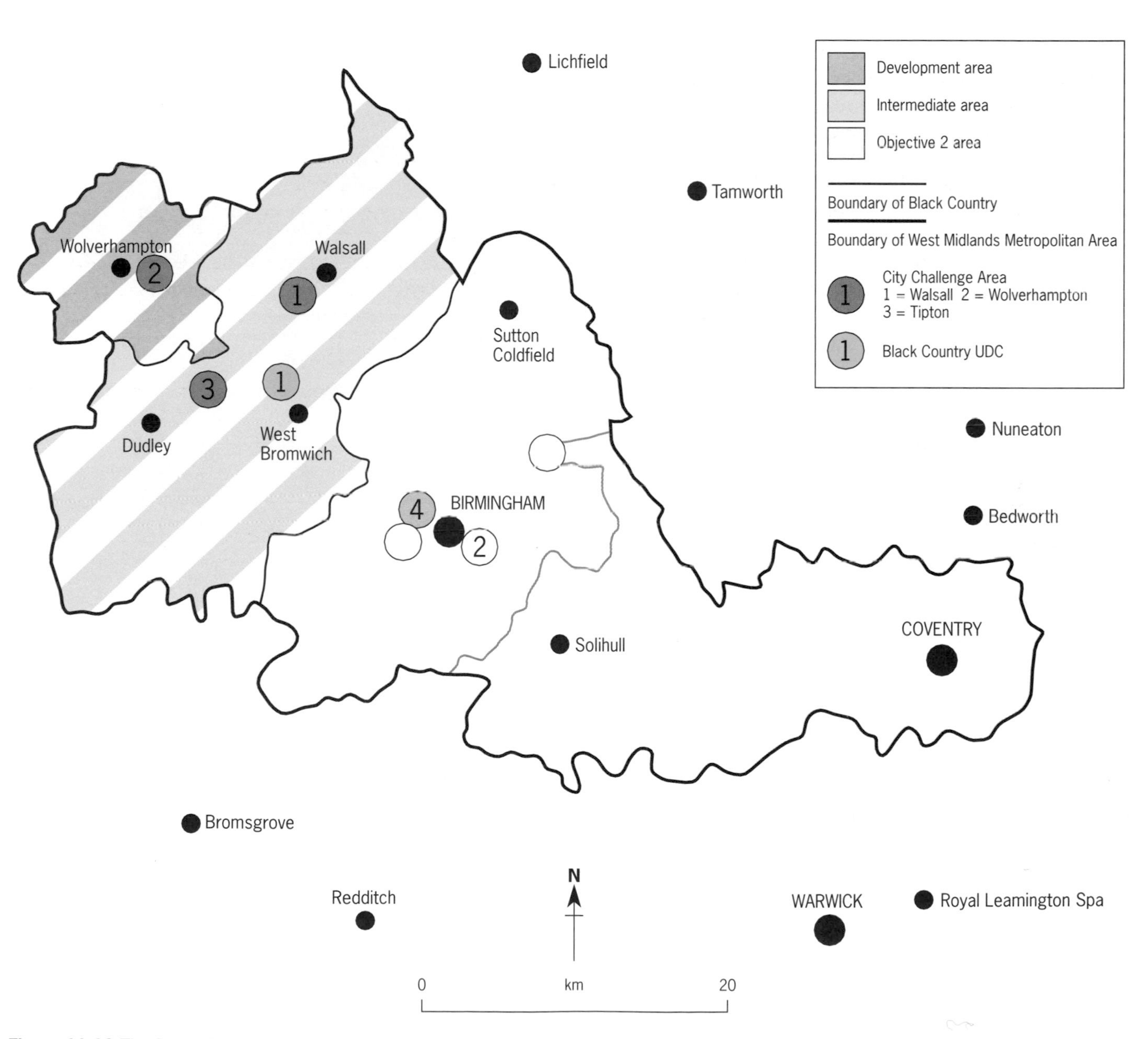

Figure 11.13 The Black Country: economic areas

Single Regeneration Budget (SRB) Challenge Fund

Introduced in 1994, the SRB Challenge Fund attracted four successful Black Country bids worth £57 million. The schemes focused on small areas and ran for five to seven years. They covered urban forestry, economic growth, education, training, improvements in housing, tackling crime and reclaiming derelict land.

Outcomes

Although substantial regeneration of the Black Country occurred between 1987 and 1998, a number of problems remained. The most significant were:

- an economy overdependent on manufacturing (33 per cent of employment compared with 18 per cent nationally), and in particular on metalworking (e.g. forging, stamping, pressing, etc.);
- low levels of investment by SMEs;
- inappropriate skills, owing to the region's dependence on its traditional industrial base;
- large-scale environmental dereliction, with over 80 per cent of derelict land in the West Midlands conurbation located in the Black Country;
- a poor image, making it difficult to attract inward investment;
- few development sites of sufficient size and quality to attract inward investment – a significant proportion of the region's industrial land supply is on derelict land;
- few large industrial premises available – only 39 buildings of more than 4000 square m were available in 1997;
- low earnings, with concentrated pockets of high unemployment and deprivation.

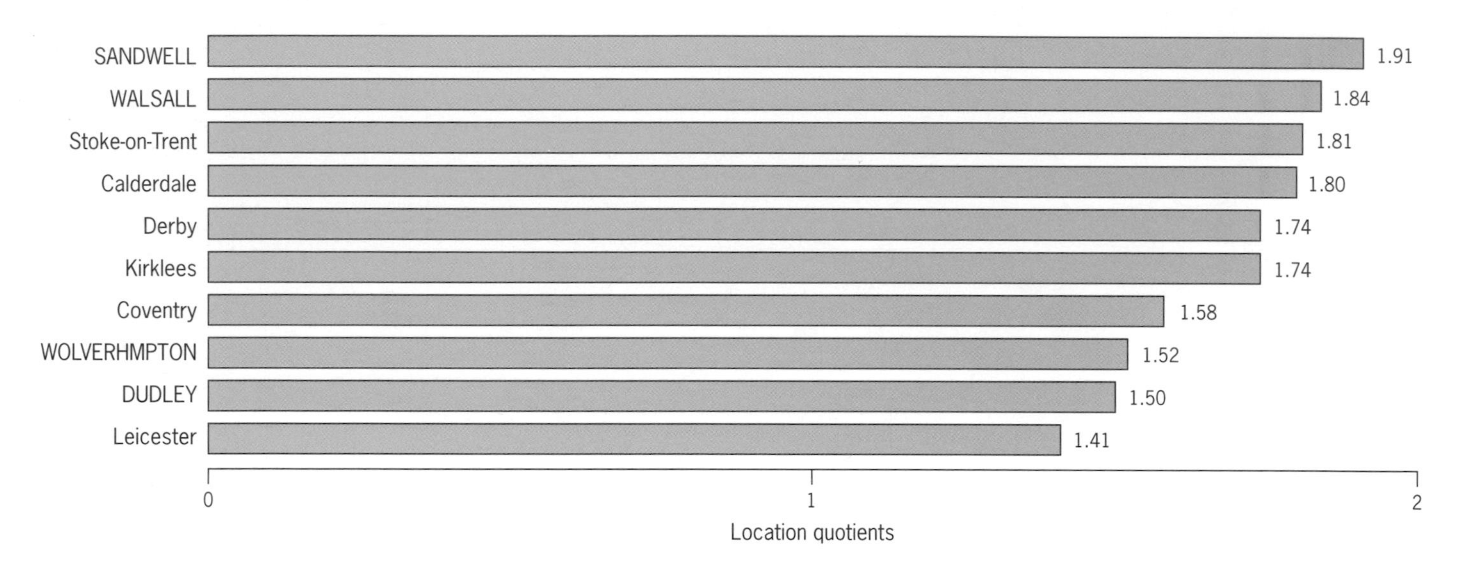

Figure 11.14 Location quotients of top 10 UK manufacturing centres

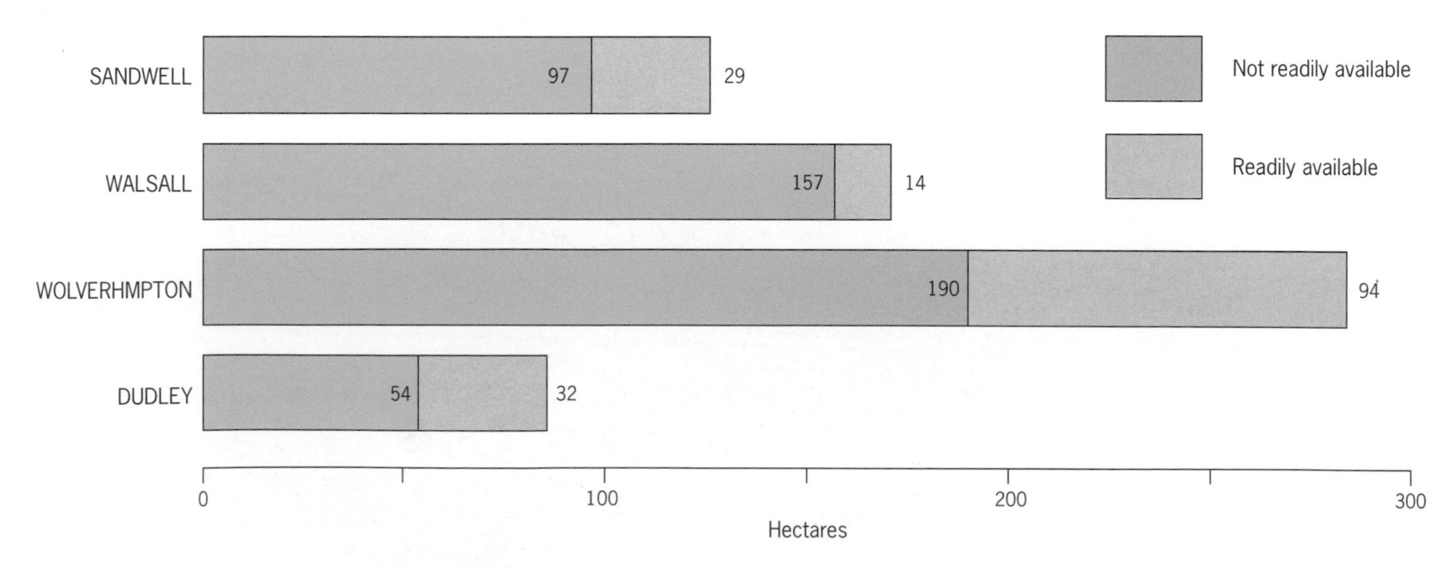

Figure 11.15 Industrial land supply in The Black Country

Table 11.6 Economic profile of the Black Country in the late 1990s

	Black Country	UK
GDP per capita	92.40	100.0
Value added in manufacturing	79.70	100.0
Metalworking/fabrication (LQs)	5.45	1.0
Unemployment % (1997)	8.00	6.7

13a Using the information in Table 11.6 and Figure 11.14, write a brief but critical assessment of the economic profile of the Black Country.
b Study Figure 11.15 and comment on the availability of sites in the Black Country for industry.

11.6 Enterprise Zones

Enterprise Zones (EZs) formed the government's second major policy drive designed to tackle the problems of the UK's inner cities and areas severely hit by deindustrialisation in the 1980s. They were launched in 1981, and by 1984 25 EZs had been created (Fig. 11.16). The idea was to encourage private investment by offering financial incentives and simplifying planning controls and other kinds of bureaucracy. Firms locating in EZs received a rates 'holiday', exemption from development land tax and 100 per cent capital allowances for industrial building for a period of 10 years. EZs in assisted areas also qualified for regional aid. Further EZs were designated after 1984.

The EZ idea has been widely adopted in the USA and other EU countries. Although scaled down in the UK, EZs were in operation at 10 separate locations in 1999 (Fig. 11.18). Most of these locations have been badly hit by either deindustrialisation or the decline of the coal industry.

The effectiveness of Enterprise Zones

The UK government evaluated its Enterprise Zone policy in 1995. Based on 22 of the 25 original EZs, its main findings were as follows.

- By 1990, 22 EZs employed nearly 126,000 people. Of these, it was estimated that nearly half were in new jobs that would not otherwise have been created in the local areas.
- The cost per job created was £2,100 per year.
- Between 1981 and 1993, the total cost of the 22 EZs was estimated at between £798 million and £968 million.
- The most important benefit attracting firms to EZs was tax relief; also important were capital allowances, the relaxation of planning controls and the availability of factory premises.
- Around 80 per cent of the available land in EZs was developed during the EZs' lifetime.
- More than £2 billion of private capital (at 1994/95 prices) was invested in property on the 22 EZs.
- There have been considerable environmental improvements on EZs through the reclamation and removal of derelict land.

While the achievements of the UK's EZs appear considerable, the EZ programme was largely abandoned once the lifetime of the 1984 EZs had expired. The ever-rising costs of the programme, measured against the jobs created, were high. Furthermore, there is no doubt that many businesses relocated in EZs from elsewhere in the locality in order to qualify for benefits. EZs attracted both manufacturing and service activities. Arguably the two most impressive investments in EZs were in retailing: the regional shopping centres at Gateshead (Metro Centre) and at Dudley (Merry Hill). Without EZ status, neither Gateshead nor Dudley would have attracted such large-scale investment.

14 Compare the EZs in Figure 11.16 with assisted areas in Figure 9.7.

a What proportion of EZs are located in assisted areas?

b To what extent would you say that EZs are merely a more spatially-focused version of regional policy?

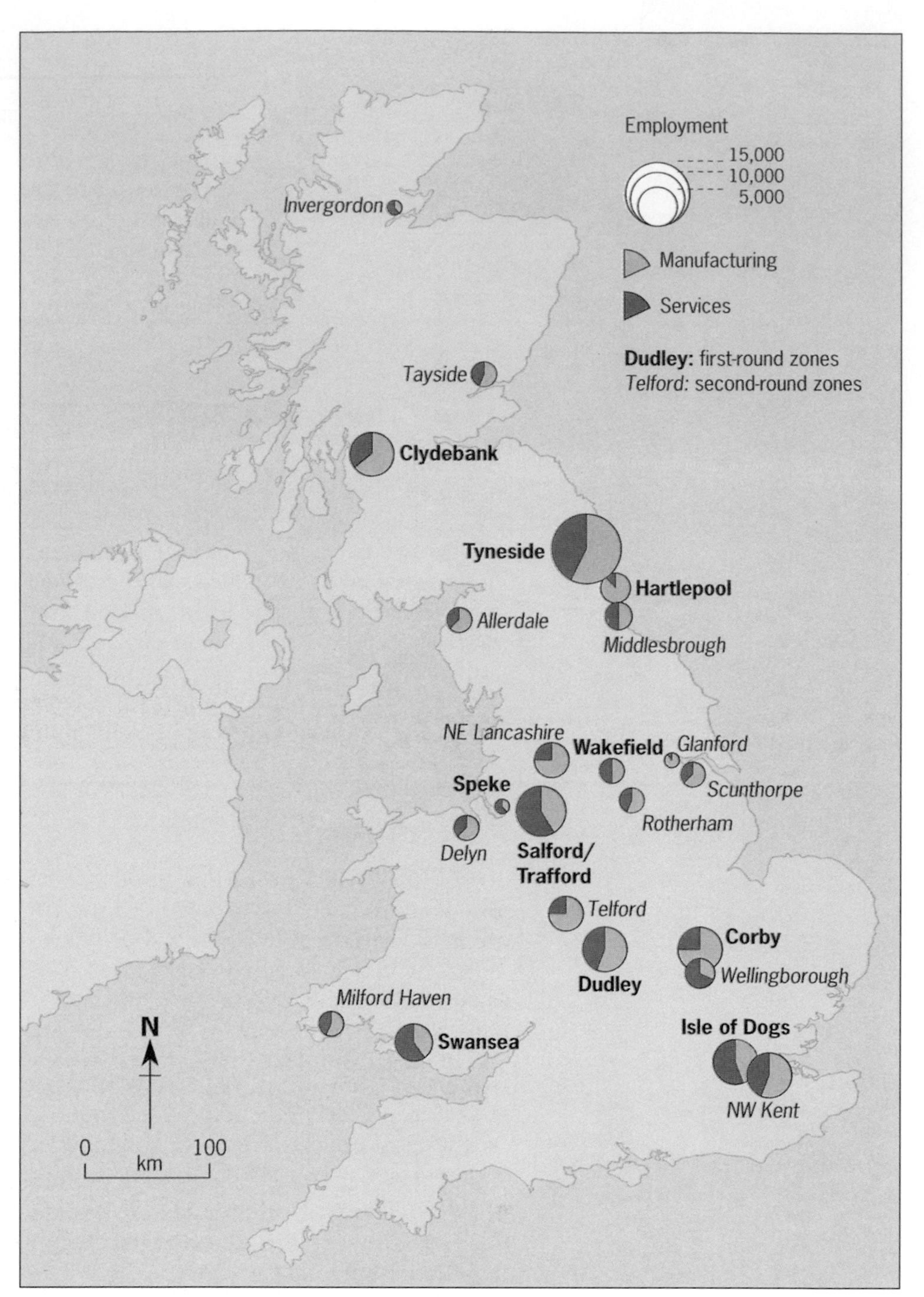

Figure 11.16 First- and second-round Enterprise Zones designated between 1981 and 1984

Figure 11.17 Development opportunities in the Tyneside Enterprise Zone

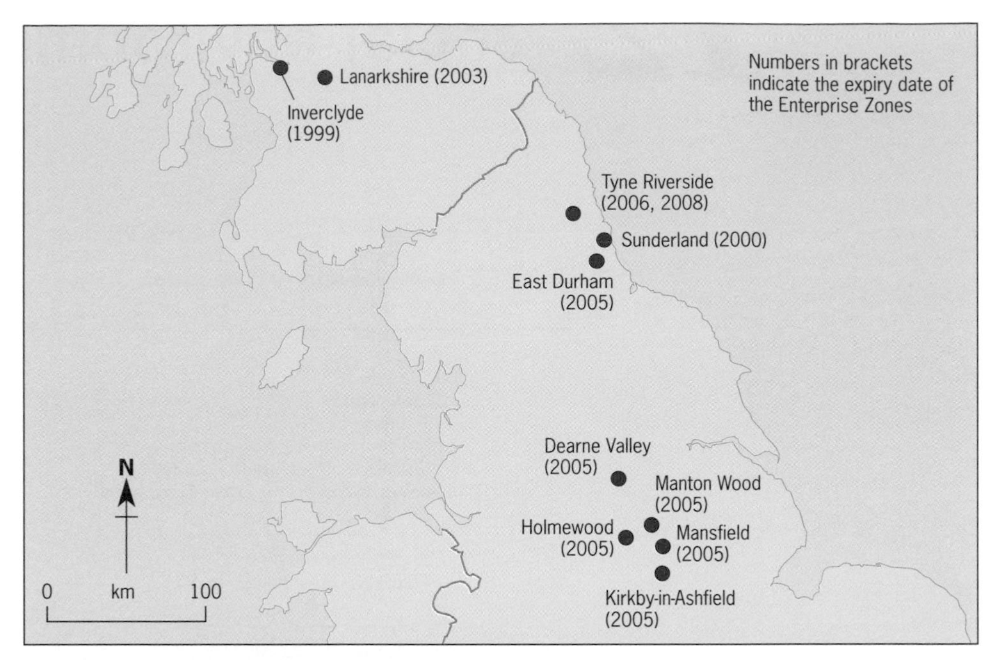

Figure 11.18 Enterprise Zones in the UK in 1999.

Summary

- Old urban industrial centres in MEDCs are increasingly unattractive to modern manufacturing industry.
- Industrial decline and the failure to attract new manufacturing activities to these centres has led to serious unemployment.
- The severity of job losses in old urban industrial centres has been greatest in conurbations and large cities.
- The effect of manufacturing decline in urban areas, and growth in small towns and rural areas, has set in motion an urban–rural manufacturing shift throughout the economically developed world.
- The urban–rural shift is not caused by a transfer of firms from urban to rural areas.
- In the nineteenth century, inner urban areas were often optimal locations for manufacturing industry.
- The decline of manufacturing in large urban areas is due principally to lack of space for development or expansion – the so-called constrained location theory.
- The main intra-urban locational trend is the relative decline of manufacturing in inner-city areas and growth in the suburbs.
- Congestion, lack of suitable sites, obsolete buildings, etc., combine to make the inner-city locations highly unattractive to most manufacturing firms.
- A large part of employment growth in the UK since 1980 has occurred in small businesses.
- Rural SMEs have grown faster than urban SMEs, and have contributed to the urban–rural shift of manufacturing.
- The UK government and the EU have attempted to regenerate many run-down urban and residential areas through the creation of Urban Development Corporations, Enterprise Zones and programmes such as City Challenge and the Single Regeneration Budget Challenge Fund.

12 The physical and social environment

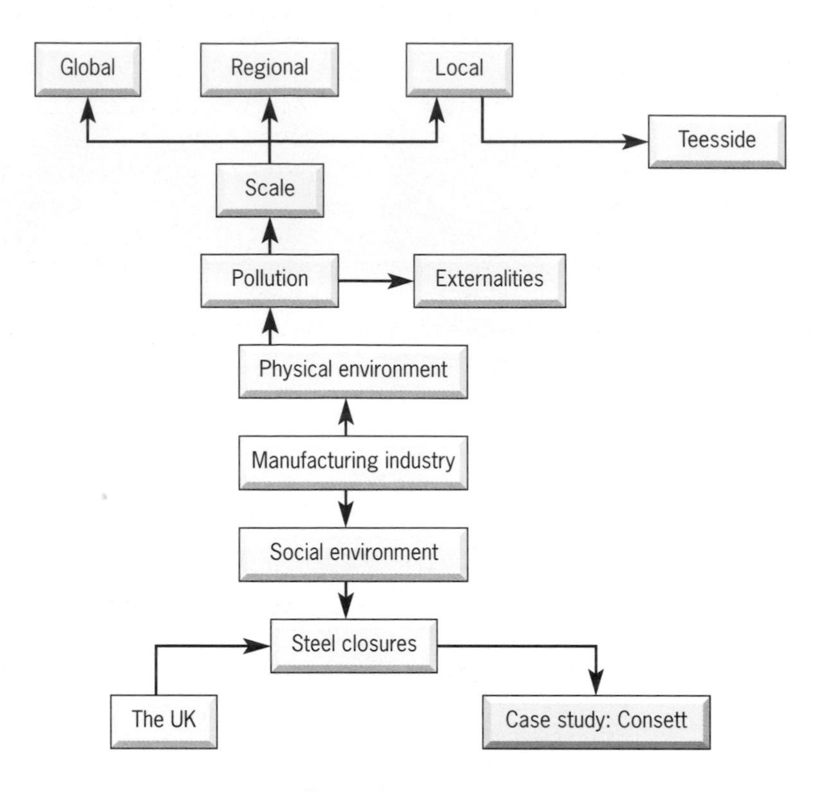

12.1 Introduction

In previous chapters we have emphasised the relationship between manufacturing and the economic environment. In the remaining chapters we shall broaden the discussion to include the impact of manufacturing industry on the physical and social environments.

12.2 Environmental pollution

Manufacturing industry is a major source of environmental pollution. As geographers we are interested in two particular aspects of this problem. First, the fact that pollution is felt at different scales, from the global to the local. And second, that the disbenefits of pollution fall unevenly on the population, being strongly influenced by where people live.

Pollution can be seen as an unwanted but inevitable by-product of economic activity. The environmental resources most easily degraded are those whose ownership is in common, such as air, soil and water. Unlike mineral ores or energy supplies, industry treats these resources as though they were free. The real costs of using them are passed on to the environment and to the people.

Table 12.1 Air pollution (CO_2) by manufacturing industries

	Therms to produce £100 output
Iron and steel	40.34
Chemicals	18.69
Mineral products	18.60
Non-ferrous metals	13.42
Paper, printing, publishing	5.35
Food, drink, tobacco	4.67
Vehicles	3.90
Textiles, leather, clothing	3.84
Mechanical engineering	3.71
Electrical engineering	2.13

The concept of externalities

Manufacturing industries generate a range of benefits and disbenefits known as **externalities**. Externalities are the side-effects of manufacturing which are felt beyond the factory site, but are not reflected in costs and prices. Positive externalities include the jobs and prosperity that industry provide. Negative externalities describe the adverse environmental effects of industries, and their impact on human well-being. These include ill health caused by polluted air

1 Refer back to your answers to Questions 12 and 13 in Chapter 2, or do these questions now.

Figure 12.1 shows that industrial pollution originating in St Louis spreads downwind over a large area.

2 What are the two main sources of pollution from St Louis and how far downwind can they be detected?

3 At what distances downwind are the following at a maximum:
a ozone pollution,
b light scattering caused by dust particles?

4 Which plume patterns in Figure 12.1 are most likely to cause the downwind pollution shown in Figure 12.2?

5 In general terms, where would you locate the four main polluting industries in Table 12.1 if your aim were to minimise pollution levels? How might such locations conflict with economic demands?

and contamined water, and expensive programmes for cleaning up polluted environments funded by tax payers. Hitherto, only a tiny fraction of these costs have been borne by the polluters themselves.

Who are the main polluters?

The major polluters tend to be large-scale processing industries, especially coal-burning power stations, chemicals, iron and steel, and non-ferrous metals (Table 12.1). Some of them have strong inter-industrial linkages and operate most efficiently in agglomerations (see section 6.1). This localises pollution problems, which are most severe around oil-refining and petrochemical complexes, like Teesside in North-East England, and Europoort in the Netherlands.

However, the effects of pollution are not just confined to their immediate source. In the UK, south-easterly winds often bring low-quality air and poor visibility as dust, smoke and other pollutants are blown across the Channel from industrial areas on the continent, just as south-westerly winds blow polluted air from the UK to Scandinavia.

Scales of pollution

Global

At the global scale industry's main polluting effect is on the atmosphere. Pollution from industry contributes significantly to global warming, the thinning of the ozone layer and acid rain. It is estimated that around one-quarter of all greenhouse gases (especially carbon dioxide and methane) come from manufacturing industry. Industry's use of CFCs also adds to global warming, as well as being largely responsible for damage to the ozone layer. Acid rain is caused by the burning of fossil fuels, and the emission of sulphur dioxide, nitrous oxides and other gases.

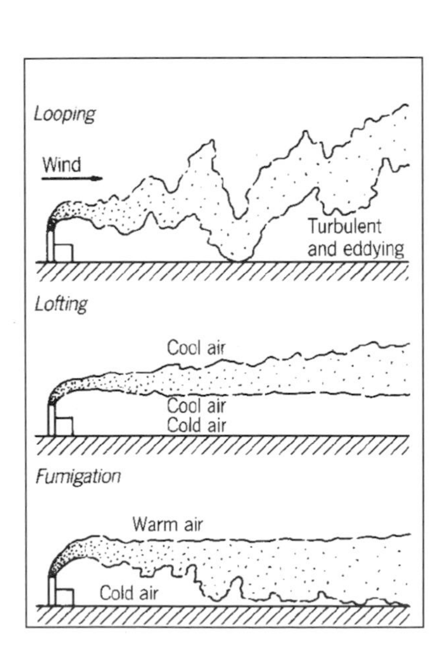

Figure 12.1 Plume patterns under different atmospheric conditions

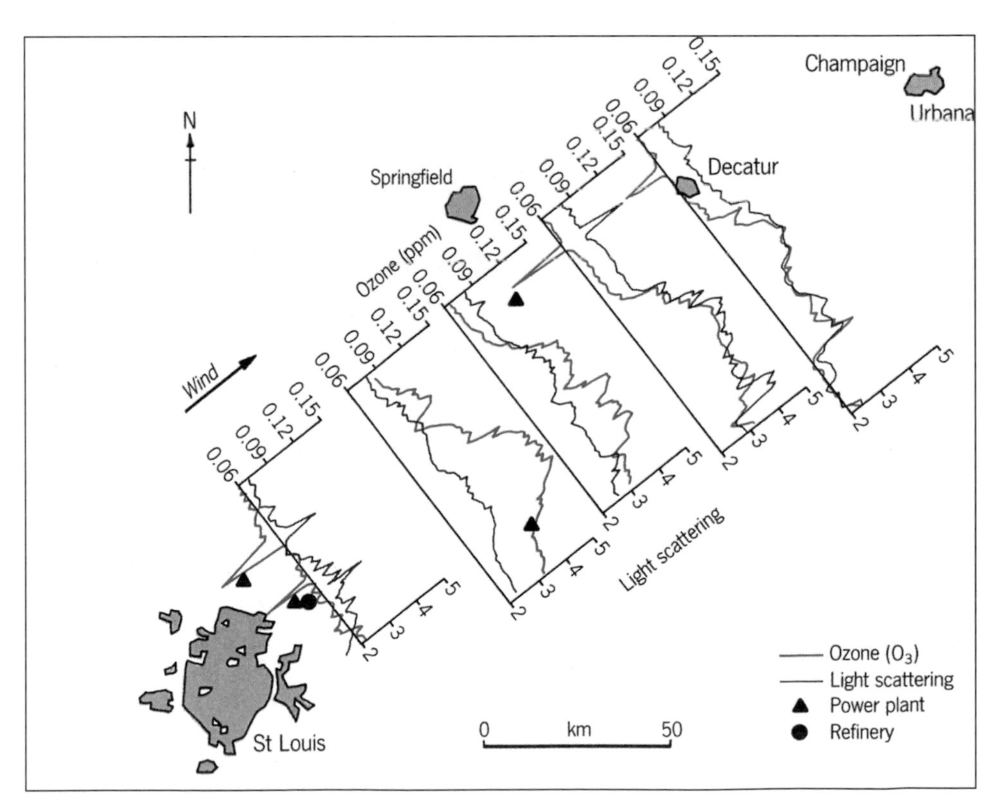

Figure 12.2 The pollution plume of St Louis, USA, 18 July 1975 (*Source:* T R Oke, *Boundary Layer Climates*, Methuen, 1978)

Regional

In addition to atmospheric pollution, pollution of rivers and coasts by industry becomes important at the regional scale. Europe's largest chemical companies (BASF, Hoechst and Bayer) routinely (and legally) discharge effluent containing heavy metals like mercury, zinc and cadmium into the River Rhine. Similar pollution takes place in the UK. ICI, for example, has permission to discharge more than 120 tonnes of sulphuric acid and 45 tonnes of cyanide into the sea each day. Some chemical pollutants remain toxic for many years. Others are concentrated in food chains, where they pose a direct threat to human health.

The tragic case of Minamata in Japan in the 1960s shows how serious this type of industrial pollution can be. A plant manufacturing vinyl and other plastic products discharged liquid wastes containing mercury into the sea. The mercury was passed along the food chain until it became concentrated in lethal amounts in fish. As fish were a major item in the diet of the local people, the effects on the community were devastating. Over one hundred people died from mercury poisoning, and hundreds of others were permanently disabled.

Local

At the local scale several other forms of pollution are added to the list of negative externalities. The dumping of toxic waste materials presents a hazard where leaching carries the wastes into streams, rivers and groundwater supplies. Visual disamenity is caused by waste tips, and unsightly factory structures. New factory buildings may result in loss of amenity, as urban sprawl extends into greenbelt and agricultural land. For residents living close to industry, noise and traffic congestion may add to the list of disbenefits. Most serious is the threat of accidents at plants making chemicals, illustrated by the tragedies at Flixborough in 1974 (Fig. 12.3) and Bhopal (India) in 1984.

Figure 12.3 Devastation at the Flixborough chemical plant

?

6 Figure 12.4b shows a factory where there is no overlap between positive and negative externality fields. Explain the likely attitude of local residents to pollution caused by this factory.

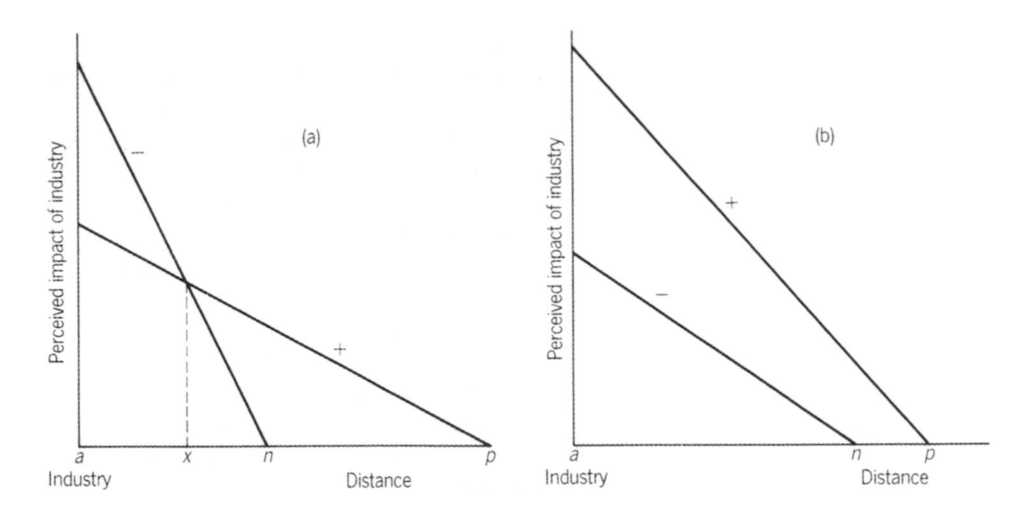

Figure 12.4 Externality gradients (after Bale, 1978)

Externality fields

Externalities also have a spatial dimension. They impact most on the geographical space close to a factory. As distance increases they follow a gradient of decline. Figure 12.4a shows two imaginary externality fields: a negative field *an*, and a positive field *ap*. The positive externality field describes the advantages of employment provided by the plant. Between *ax* there is a net negative externality field, because the disadvantages of pollution outweigh the advantages of access to employment. However, at *xn*, although the factory still creates some nuisance, this is now small enough to be outweighed by its positive effects, producing a net positive externality field.

The effect of negative externality fields caused by industry has some influence on who lives where in urban areas. Higher-status groups will choose residential areas where the negative externalities of industry are minimised. In many Western cities there is evidence that these groups prefer housing areas upwind of polluting industries. Meanwhile, lower-status groups often have little choice but to occupy housing in disadvantaged areas near to industry.

Industrial pollution and ill health

The spatial inequality of industrial pollution

Modern manufacturing processes produce waste products such as gases, effluents and solid materials. Disposal of these products in the atmosphere, in rivers, in seas and landfill sites may threaten human health and wildlife and damage the environment. When this occurs we refer to these waste products as pollutants.

Table 12.2 Most polluting industries

Industrial sector	Examples
Fuel and power	Power stations, oil refineries
Mineral industries	Cement, glass, ceramics
Waste disposal	Incineration, chemical recovery
Chemicals	Pesticides, pharmaceuticals, organic and inorganic chemicals
Metal industries	Iron and steel, smelting, non-ferrous metals
Others	Paper manufacture, timber preparation, uranium processing

The cheapest way for industry to dispose of pollutants is to release them into the environment. When this happens, the true costs of manufacturing, in the form of pollution, are passed on to society as negative externalities. This pollution burden falls unevenly on society.

Industrial pollution invariably hits the poor hardest. A report published in 1999 by Friends of the Earth on pollution and poverty in England and Wales exposed this inequality. It found that polluting factories were most concentrated in neighbourhoods with the lowest household incomes. Hence the poorest families with annual incomes of less than £5,000 are twice as likely to have a polluting factory in their locality than families with incomes of more than £60,000. This disbenefit often comes on top of others, such as high levels of morbidity (ill health), unemployment and neighbourhood crime, as well as poor access to education and medical services. Thus pollution is just one other contributor to social exclusion among the poor.

Table 12.3 Household incomes and the geographical distribution of the most polluting factories in England and Wales, 1996

Average annual household income by post-code sector	Proportion of post-code sectors (%)	Actual number of polluting factories	Expected number of polluting factories
<£10,000	5.12	104	67.6
£10,000–14,999	29.17	558	
£15,000–19,999	34.94	461	
£20,000–24,999	21.49	168	
£25,000–29,999	7.28	24	
£30,000–39,999	1.97	5	
>£40,000	0.04	0	

Industrial pollution and human health

In 1996, factories in England and Wales released 1.28 million tonnes of sulphur dioxide, 0.65 million tonnes of nitrogen oxide, 0.17 million tonnes of carbon monoxide and 0.08 million tonnes of sulphur oxides. These chemicals (together with 13 other major emissions) threaten human health. Some, such as dioxins, are cancer-causing agents and others are known to harm the development of young children (e.g. PCBs) and disrupt the hormonal systems of people and animals. People living close to polluting factories are at greatest risk from both routine emissions and accidents (see Figure 12.6).

As we have seen, poor people face the greatest risk from health-threatening chemicals because they are most likely to live in neighbourhoods with

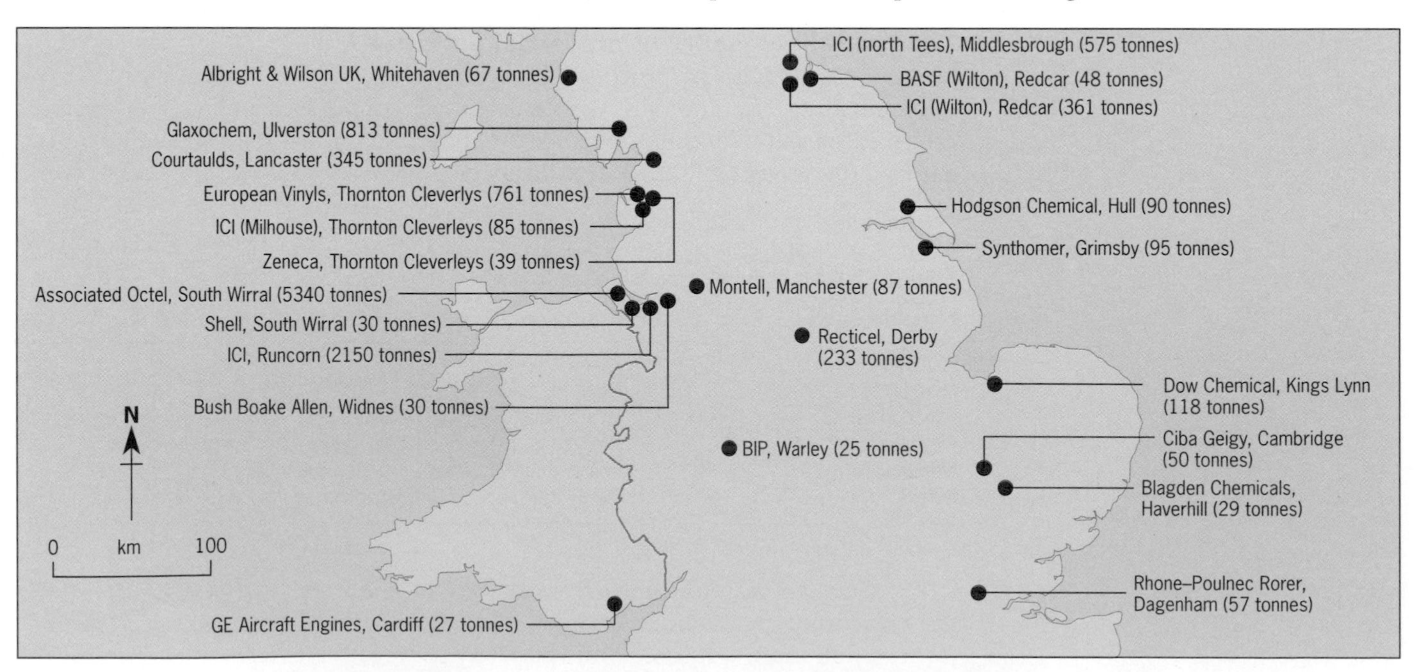

Figure 12.5 Reported emissions of carcinogens into the atmosphere in 1996 in England and Wales

?

Read the newspaper article on the chemical leak at the ICI Tioxide plant on Teesside (Fig. 12.6).

7 Describe the negative externalities created by the plant.

8 Describe the attitudes and beliefs of (a) local residents towards the plants; (b) ICI.

9 Should the plant be closed? Should local people be rehoused? Should local residents put up with pollution because the plant provides jobs? Outline and justify your own attitude towards the issues raised in the newspaper articles.

White smoke leak forces ICI plant closure

Peter Hetherington on fears raised by repeated spills

WHEN the dense white cloud drifted over northern Teesside, people started coughing and wheezing before police warned them to stay indoors.

In their newsagents' shop in the conservation village of Greatham, half a mile from the big chemical plant, Ronnie and Julie Westmoreland remember a "horrible smell".

"My mother phoned from [nearby] Billingham and said 'what's the weather like with you because it's gone all dark here,' recalls Julie. "She was worried."

In the fields beside Greatham, Ronnie said a farmworker went bright red and started spluttering while a woman, dependent on a kidney dialysis machine, needed emergency treatment. Others complained of feeling unwell.

Twelve days after a leak of the potentially dangerous chemical, titanium tetrachloride, part of the ICI-owned Tioxide plant, near Hartlepool, Cleveland, remains closed on Environment Agency orders.

By taking the unusual step of serving a prohibition notice under the Environment Protection Act – after slapping less severe enforcement notices on Tioxide following a string of leaks – the agency is signalling a tougher approach against the chemical giant.

Its senior officials are alarmed that the latest Tioxide escape came shortly after a meeting with ICI nationally when the company agreed to improve the management of plants in an attempt to prevent further serious incidents in the North of England.

Concern was heightened hours afterwards when oil leaked into the Tees from a sister ICI plant at Wilton, prompting the agency's operations director, Archie Robertson, to say: "It is outrageous that within weeks of ICI being called to a meeting with the agency where it promised to clean up its act that its plants have been involved in two further leaks."

Tioxide, part of a huge ICI operation on Teesside, makes titanium dioxide, a white powder pigment used in paint, plastics and paper. Councillors complained less than a month ago when 20 gallons escaped from the plant, creating a dense white cloud.

Then came a bigger leak two weeks later, when water apparently seeped into a cooling circuit. "It reacts very violently with water," said Don Ridley, an senior Environment Agency official called in to investigate. "It's an irritant, very corrosive, causes a dense cloud and people cough and itch."

'ICI realise they have to communicate. People are frightened.'

On Teesside, which contains one of Europe's largest chemical complexes, GPs often report a high level of respiratory ailments. Some schools are concerned about the level of asthma among pupils.

A 1995 study by Newcastle university's department of epidemiology and public health found that women living beside one ICI plant were four times more likely to contract lung cancer than the national average.

By rebuking ICI, the agency appears determined to show it means business after a series of incidents in the North over the past year. The most serious include:

- July 1996: Company fined £15,000 by magistrates at Widnes, Cheshire, after a spill of ethylene dichloride from ICI Runcorn.
- October 1996: Company facing further prosecution by the agency after a leak of vinylidene chloride from Runcorn into a canal.
- May 1997: Fifty tonnes of trichloroethylene escape into the Runcorn canal.
- May 1997: Several hundred tonnes of the liquid petroleum product Naptha spilt at another North Tees ICI site generating a large gas plume. Roads closed and people told to stay indoors.

Next month a special agency team is due to complete a review of management systems at ICI Runcorn while officials at Teesside monitor progress at Tioxide. A company spokesman said reopening would take some time because it still had to pinpoint the cause of the problem.

ICI disputes the claims that its plants pose a health hazard. As a responsible employer, it says, it is in contact with community groups to inform them of its plans. Although escaping chemicals sometimes created a dense cloud, the misty conditions made a leak appear worse.

Tioxide spokesman, Bill Beattie, insisted they were updating and improving the plant regardless of the Environment Agency. The affected section would be reopened when it was operating effectively and safely, he said.

But in Greatham, butcher, Peter Stonehouse, complained they had not been told about the latest incident because Tioxide said the wind was unlikely to push the cloud towards their village. "That was not good enough," he said. "Round here, by the sea, the wind can change at a moment's notice."

But Ronnie Westmoreland, a parish councillor, is now in regular touch with the company. "They realise they have to communicate," he said. "People are frightened. they are well aware of the harm this could do."

ICI's Tioxide plant near Hartlepool, Cleveland, part of it stays closed while an investigation into chemical leaks is carried out.

Figure 12.6 Newspaper report on the chemical leak from ICI's Tioxide plant near Hartlepool, Cleveland (Figure 12.7) (*Source: The Guardian*, 16 June 1997)

Figure 12.7 Teeside: a large concentration of heavy processing industries has created one of the most polluted and potentially one of the most hazardous urban environments in the UK.

polluting factories. Higher income groups avoid polluted areas. They have sufficient wealth to afford property in more salubrious areas. In contrast, demand for housing in heavily polluted neighbourhoods is low. Unable to compete in the housing market, poor families have little choice but to purchase or rent low-cost housing in areas blighted by pollution.

Higher incidences of lung cancer, childhood cancers, respiratory diseases and morbidity are found near polluting factories. However, we cannot say for sure that industrial pollution is the cause of poor health. This is because so many other factors influence death rates and morbidity. For example, the severity of pollution may depend on local influences such as prevailing winds, weather conditions and topography. Equally, ill health and premature death may result from lifestyles (drinking and smoking), housing quality, unemployment, poverty and access to medical services.

?

10 Study the distribution of the emission of carcinogens from factories in England and Wales in 1996 (Fig.12.5).

a Describe the distribution of emissions in Fig.12.5.

b Suggest two possible reasons for the concentration of polluting factories around the coast.

Industrial pollution and human health on Teesside

Teesside has the largest concentration of heavy polluting industries in the UK (Fig. 12.8). The Teesside industrial agglomeration includes iron and steel, chemicals, oil refining, pharmaceuticals and electricity generation. Within the Teesside conurbation these industries are particularly focused at the mouth of the River Tees. Here a single post-code district (TS2) has no fewer than 17 major polluting factories. The average household income in TS2 is just £6,200 a year – 64 per cent below the national figure.

?

11 Study Table 12.3.
a Suggest three possible reasons why the geographical distribution of polluting factories might be disproportionately concentrated in low income neighbourhoods.
b Apply the Chi-squared statistic (appendix A2) to the data in Table 12.3 and test the hypothesis that polluting factories are disproportionately concentrated in low-income post-code sectors. Calculate expected values for the Chi-squared formula by multiplying the total number of polluting factories (i.e. 1,320) by the proportion of post-code sectors for each income class. Thus for post-code sectors with incomes below £10,000 a year, the expected number of factories is 1,320 × 0.0512 = 67.6.
c Calculate the significance of the Chi-squared statistic and comment on the validity of the hypothesis.

12 Log into the Friends of the Earth website on industrial pollution (http://www.foe.co.uk/factorywatch).
a Make a list of polluting factories (i.e. those with IPC registration).
b Download the map that shows their distribution in your locality.
c Select one factory and compile a table to show the types and amounts of pollutants emitted.
d Compare this table with the list of pollutants known to be injurious to human health (available on the website). Comment on the nature of the pollutants emitted.
e Locate your factory on a 1:50,000 OS map of the area. Describe (i) the distribution of residential areas in proximity to the factory, (ii) the relief of the area, (iii) the likely pattern of fall-out in relation to the prevailing wind.
f Taking account of the factory's emissions and the geography of the surrounding area, assess the possible impact of pollution from the factory on local residents.

Two of the largest polluters are the petrochemical works at Wilton, operated by BASF and du Pont, and the Redcar–Lackenby steelworks, which belongs to the Corus group. These two factories release large amounts of sulphur dioxide, nitrous oxide and iron oxide, much of which falls on Teesside. This negative externality is greatest in Grangetown, a residential area immediately adjacent to the factories. Fall-out in this district amounts to 70 tonnes of pollutants each year. Local people believe that industrial pollution is to blame for poor levels of health in the community. Rates of mortality from bronchitis and asthma are three times higher than the national average, and life expectancy is 10 years below the national average. However, pollution is only one possible cause of ill health, which is also linked to bad housing, poor diet and smoking. What is undeniable is that some of the poorest people on Teesside suffer a greater than average share of the environmental costs of industrial production.

A recent investigation into the health of female residents on 27 estates on Teesside and in Sunderland isolated the effects of industrial pollution. After taking poverty into account, it found levels of death from lung cancer in women under 65 years significantly above the national average. This finding provides strong evidence of a causal link between ill health and industrial pollution.

Pollution control

Heightened public awareness of pollution, and a greater political will to tackle the problem, have put pressure on industry to improve its environmental performance. Increasingly, industries are being forced to examine the environmental effects of the entire production process, from obtaining materials, and the use of energy, to the disposal of waste. All transnational chemical companies have greatly increased their expenditure on pollution control in the last ten years. For instance, ICI doubled its environmental budget for the period 1990–95 to $1 billion worldwide. This will amount to 20 per cent of the group's total capital expenditure. Hazardous manufacturing processes and products (including CFCs) are being phased out. In the pulp and paper industry there has been a switch from the use of chlorine to more environmentally friendly bleaching agents.

The usual way to curb pollution has been by command and control. National and international agreements, and national legislation, aim to control emissions and fine companies which break the law. In Western Europe the tightening of pollution regulations has been driven by EU legislation, and the need to harmonise environmental policies following the Single European Act of 1992.

Towards the end of the 1980s an alternative to pollution control began to emerge. This relied on market forces. Green taxes, such as a carbon tax (levied on fossil fuels according to how much carbon dioxide they emit) were promoted to encourage industry to reduce pollution and adopt cleaner production. Such charges are a way of imposing on polluters the external costs they otherwise unload on to the environment. However, any tax aimed at reducing atmospheric pollution can only succeed through widespread international agreement. There is little point in any one country acting alone. A country which shifted to low-energy manufacturing would simply import the high-energy goods it no longer made, with no overall benefit in pollution reduction.

12.3 Steel closures and the social and economic environment

The steel industry underwent massive transformation between the mid-1970s and early 1990s. These changes were a response to powerful economic pressures (see the Case Study on page 158). One result was the wholesale closure of many small steelworks. Because steel is a labour-intensive industry, which often dominates employment in steel-making towns, the effects of closure had serious social and economic effects.

Steel closures in the UK since 1966

In 1967 the British Steel Corporation (BSC) had a workforce of 250 000 and produced 24.7 million tonnes of steel at 23 sites. In 1992 British Steel employed 55 000 workers, and produced 13.8 million tonnes of steel at just four major steel-making sites (Fig. 12.8). This remarkable change was not unique to the UK. Similar changes occurred elsewhere in the EU (notably in Germany, France, Italy and Belgium) and in the USA.

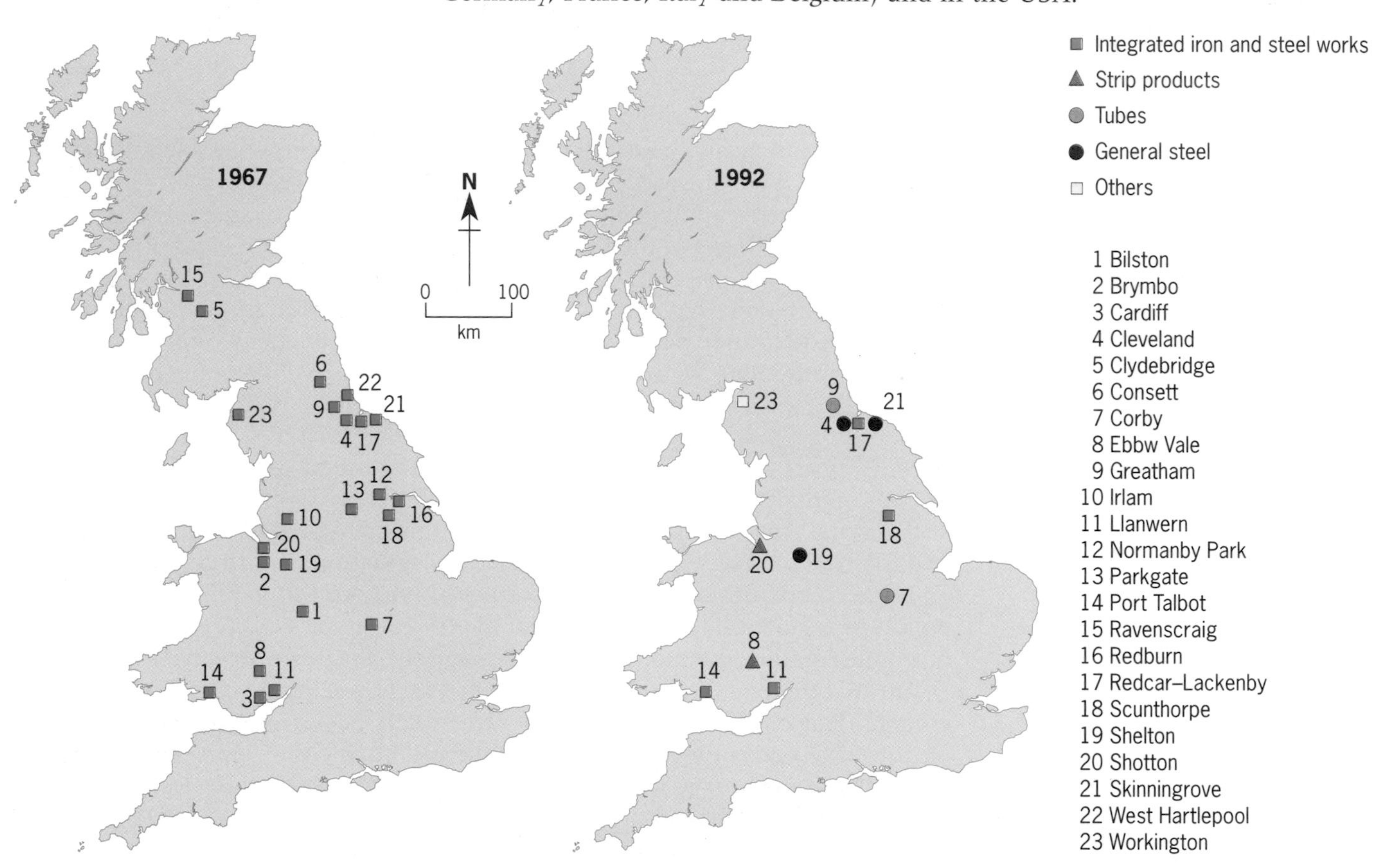

Figure 12.8 The UK's steel industry: the changing geography of production, 1967–92

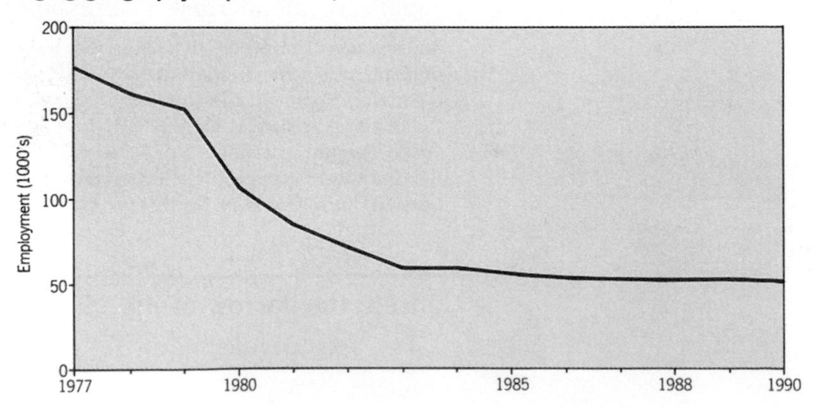

Figure 12.9 British Steel: employment changes, 1977–90

Figure 12.10 The location of Consett in North-East England

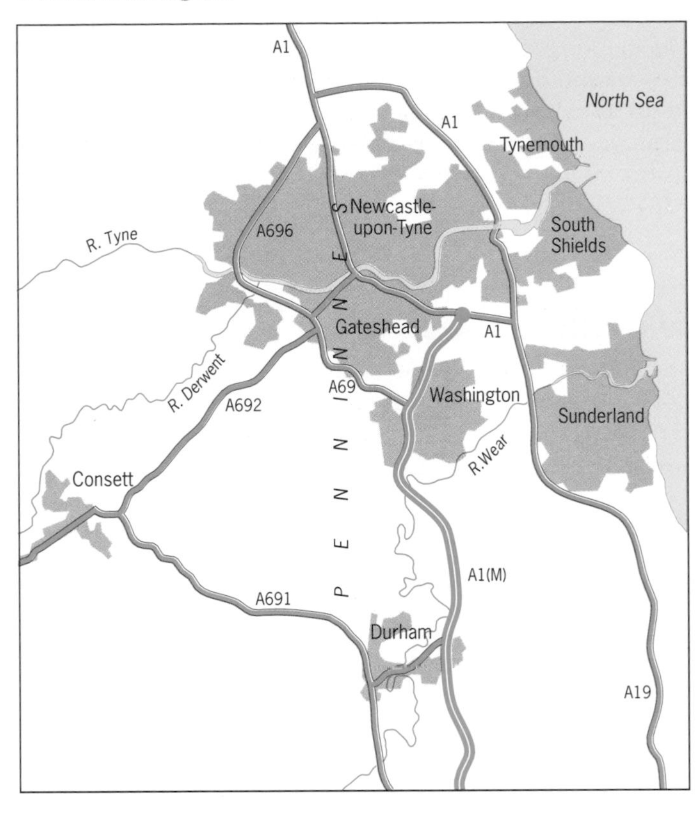

The post-1973 world recession left the steelmaking countries of the EU and North America with huge overcapacity. In many countries the crisis had been made worse by heavy investment in new plant just before the slump. For example, BSC forecast a rise in demand from 22 million in 1973 to 38 million tonnes in 1980. In reality, demand in 1980 was only just over 15 million tonnes.

The response in the EU was **rationalisation**, a policy which brought drastic changes in the geography of production. Many high-cost inland works were closed. In some cases, such as Bilston in the West Midlands and Consett in North-East England (Fig. 12.11), closure ended a tradition of steelmaking stretching back over one hundred years. In the UK, only the biggest integrated works survived the cuts.

Rationalisation brought considerable savings through economies of scale and improvements in productivity. Indeed, by the end of the 1980s, the UK's steel industry was achieving productivity levels comparable with those of Japan and Germany. However, the social and economic price of closures in former steelmaking areas was high. Heavy-industrial regions on the continent, such as South Belgium, Lorraine, Nord–Pas-de-Calais and the Saarland were hard hit. But the most catastrophic effects were felt in one-industry steel towns, like Consett, and Denain in North-East France.

Figure 12.11 Consett steelworks in 1980
Figure 12.12 (Inset) High-tech industrial units on the site of the old Consett steelworks

Steel-making started at Consett in 1840. Local deposits of coking coal and blackband ore attracted the Consett Iron and Steel Company to this remote and windswept corner of North-West Durham. In 1851 the Consett works was described as 'the largest in the kingdom', and by the late 1880s it was producing one-tenth of the nation's steel. The town grew up around the steel-works. It was a company town, with the company owning all the shops and most of the houses its employees lived in. It was also a one-industry town, and remained so until the closure of its steelworks.

Consett: the Jarrow of the 1980s?

ANGER AND BITTERNESS AS A COMMUNITY'S FEARS BECOME REALITY

Steel closure leaves Consett without hope

BY A SPECIAL CORRESPONDENT

FOR THE second time in a generation, the County Durham steel town of Consett faces a future without hope.

The steelworks, the town's major employer for more than 130 years, closes in autumn under the British Steel Corporation's 52,000 jobs cut.

The closure of the works, with the loss of 3,750 jobs, is expected to push male unemployment levels in the town up to at least 30 per cent, and may be nearer 40 per cent.

Already one man in eight is on the dole because of redundancies in the steel industry over the last few years and earlier pit closures.

Mr David Watkins, Consett's Labour MP, says 'We are facing nothing less than a return to the depression. Three out of four people in Consett are directly or indirectly dependent on the steel industry. Consett could become the Jarrow of the 1980s, there is no other industry in the town to speak of.'

For generations, Consett has depended on coal and steel for prosperity. Over the last 20 years 15,000 jobs have been lost in the pits, leaving only one small colliery employing around 200.

The steel industry, which was the remaining pillar of the community, has also shed over 3,000 jobs in the past five years. The local authority and trade unions say Consett has never fully recovered from the pit closures, and with such total dependence on steel, now faces the second major crisis within a generation.

As news of the closure spread through the community of 36,000 people, the reaction was one of anger and bitterness with more than a measure of resignation.

Most of the anger is directed at the BSC which right up to last week insisted long-standing fears about the future of the works were groundless. Some of the apprentices who now face an uncertain future were only taken on last month.

The works, which have a capacity of 1.2m tonnes of steel a year, lost £15.2m last year but only £3.1m in the six months ending in September. In the last three months, the works has even made a modest profit.

'The whole town has been betrayed. We were told that we had to become profitable by March 1980 if we were to have a future. We have done it, but the works is still to close,' said Mr Watkins.

Consett steelworkers have a proud tradition. They produced the steel for Windscale and for Britain's nuclear submarines and are today still acknowledged as the producers of some of the highest quality steel in BSC. Like their MP, they feel betrayed.

Derek Saul, the managing director of BSC's Teesside division – which includes Consett – admits the works have not fallen victim to any inadequacy in its own performance but the general problems of the steel industry.

The town, which suffers from a relatively isolated location in the Derwent Valley, has been bequeathed a legacy of serious environmental problems by the coal and steel industry and neither the local authority nor Durham county council is optimistic about the prospects of attracting new jobs.

The steelworkers face journeys of up to an hour to find work in towns like Newcastle or Sunderland which have considerable unemployment problems of their own.

Consett's communications are already poor, and with the closure of the steelworks, British Rail is expected to shut the railway line.

A measure of the huge problems facing the town is that its major employer will now be Celluware, which although a highly successful business manufacturing table mats, mainly employs girls.

The only two large factories within a few miles of the steel town are the Ransome, Hoffman and Pollard ball-bearing plant at Annfield Plain, and the Ever Ready battery factory at Tanfield Lea, but neither could hope to provide more than a nominal number of new jobs.

The steel corporation's jobhunting organisation BSC (Industry) has been active in Consett since the beginning of the year and has so far managed to attract five small firms employing 150 in total.

BSC (Industry) estimates the numbers employed by the five companies will rise to around 400 by 1982. Other companies will also be attracted to the town in the coming months, but there seems little possibility of success on the scale required.

'I can't see any way that we are going to be able to find even a fraction of the jobs we are going to need. Our experience in Consett over the years has been that for every five jobs we lose only one is replaced,' said Mr Watkins.

'Shops are going to close, local business go bust and houses become virtually worthless. It won't happen immediately but you can't tear the heart out of a town and think things can go on just as before,' said one steel union official.

The closure, which will increase the depopulation which has been a feature of West Durham for many years, has come at a particularly embarrassing time for the local authority.

Work has just started on a multi-million pound redevelopment of the town centre which will now be difficult to finance with the loss of the £1.5m rate income from BSC.

The end of Consett steelworks will create further problems for the Durham coalfield, which has been hard hit by BSC's increasing coke and coal imports. No pit closures will be necessary, according to the National Coal Board, but the loss of a market for 200,000 tons of coking coal a year will have some effect on jobs in some pits.

Figure 12.13 Newspaper report on the closure of the Consett steelworks (*Source: Financial Times*, 13 Dec. 1979)

Nick Garnett reports on efforts to breathe life back into industry in the North-East

A balance sheet on job creation

CONSETT was dubbed 'town for sale' after its integrated steel plant was shut down at the end of 1980 with the loss of almost 4,000 jobs.

By the middle of the following year other closures in Consett, Co. Durham, and the nearby town of Stanley had pushed unemployment in Derwentside up to 28 per cent.

Since then British Steel (Industry), private business, government agencies and the local authorities have been beavering away to try to bring jobs back to the area. The Derwentside Industrial Development Agency, which grew out of these efforts, has just completed its first year. And some conclusions are beginning to emerge about the impact of job creation work.

About 1,800 jobs have been created in the past three years. Business plans for these companies suggest a further 1,800 could be in the pipeline, though this is unlikely to be achieved.

At a time of rising unemployment elsewhere, Derwentside's jobless rate has declined from 28 per cent to 24.4 per cent.

Consett stood out more than three years ago as a social disaster, but unemployment in some other areas of the North-East and elsewhere are likely to soon overtake Derwentside. This is a reflection of the speed at which industrial contraction is hitting other communities.

National calculations indicate that for every 1,000 new jobs, the number of registered unemployed locally probably falls by little more than a third. A number of factors contribute to this: jobs are taken by women who were not on the unemployment register; and the influx of people into the area as new jobs are created.

Unemployment among ex-steel workers and those once employed by roller-bearing manufacturer Ransome Hoffman Pollard, and other companies which have closed sites locally, has not been reduced to anywhere near the extent job creation might suggest.

Companies like Derwent Valley Foods underline the size of incentives on offer on Derwentside. It was started by four partners and began manufacturing snack foods in August 1982. It now employs 55. The partners put up only a ninth of the £500,000 that has gone into the company.

The Derwentside Industrial Development Agency, under Mr Laurie Haveron, its chief executive, worked for BSC (Industry), has a strategy committee largely made up of representatives from government agencies, local authorities and BSC (Industry), and a task force of a few officers providing a 'one-stop' advice centre.

The agency spends considerable time trying to persuade companies which come in from outside to base their management on Derwentside and to source components locally.

Mr Mark I'Anson, a partner in Integrated Micro Products, one of the companies on the Consett number 1 industrial estate, expresses the view of some others when he says: 'If you look at the numbers it's impossible. There's got to be an immense recovery just to bring the area back to where it was.'

The success or otherwise of job creation on Derwentside is likely to be proved only within the next five to ten years.

Figure 12.14 Newspaper report on job creation in Consett (*Source: Financial Times*, 6 Mar. 1984)

The decision to close the works in 1980 was made solely on economic grounds. Raw material assembly costs were high. Local raw materials had long since been exhausted, and the survival of the works (in an obsolete location) was due to inertia. Other disadvantages included the works' relatively small capacity (1.2 million tonnes a year) and the loss of its markets in the North-East with the decline of shipbuilding, coal mining and heavy engineering. In 1979 Consett made a loss of £15 million. Interestingly, BSC justified closure not on loss-making grounds (Consett actually made a small profit in the six months before closure), but on the grounds that it could make higher profits by transferring production to its more cost-effective coastal works (Redcar–Lackenby and Port Talbot).

Closure resulted in the immediate loss of 4,000 full-time jobs, nearly all of them for men. At a stroke, Consett lost any reason for its existence, and the economic and social impact threatened to make the town 'the Jarrow of the 1980s'.

?

Study the newspaper articles (Figs 12.13–12.16) which describe the closure of the Consett steelworks, and its impact on the town.

13 Outline the predicted impact of closure on Consett, as viewed in 1980.

14 Summarise the actual impact (economic, social, demographic and environmental) in the mid-1980s and later 1980s.

15 What attempts were made by BSC (later British Steel) and the Derwentside Industrial Development Agency (DIDA) to attract employment to the town? How successful were these initiatives?

16 Why was Consett's economy more robust in the early 1990s than in 1980?

17 Essay: 'The Jarrow of the 1980s.' Was this a fair description of Consett? (NB: Jarrow is an industrial town on South Tyneside. In the 1930s' slump it was devastated by unemployment, which led to a famous hunger march to London to protest to the government.)

Consett in the 1990s

In the years immediately following the closure of its steelworks, Consett's future looked bleak. To many people there seemed little prospect that the town's economy could ever recover. But from the mid-1980s to the early 1990s the mood changed. Generous financial assistance from the UK government, the EU and other organisations, and the promotional success of the Derwentside Industrial Development Agency secured important inward investments. Meanwhile an enterprise culture fostered many small business start-ups. One SME – Derwent Valley Foods – achieved phenomenal success. New jobs appeared. Most were in service activities, which increased their share of employment from 45 per cent to 70 per cent by 1989. Many jobs were part-timc, employed women rather than men, and had limited security. But the flexibility of Consett's workforce became an attractive selling point to potential investors. By the early 1990s the area seemed to be on the brink of a remarkable recovery.

The prospect in 2000

Economy

Twenty years after the closure of the Consett steelworks, a more sober assessment of the town's economic progress can be made. First, the total number of jobs in the town is still below the 1980 level. In October 1999 unemployment stood at 7.5 per cent, significantly higher than the 5.9 per cent average for County Durham. Only two other districts in the county have higher unemployment rates.

The continuing fragility of the area's economy is reflected in the wide range of financial assistance still on offer. Derwentside is a development area, eligible for the highest levels of assistance under the government's regional policies. The area also has Objective 2 status. This defines it as one of the poorest regions in the EU, with a GDP less than 75 per cent of the EU average. Additional financial and training packages are available from Durham County Council, Derwentside District Council and British Steel (Industry) Ltd.

A major regeneration initiative known as Project Genesis was launched in 1993. Its aim was to develop a 285-ha site of the former steelworks and to attract £100 million of investment in manufacturing, commerce, leisure, educational facilities and conservation. The centrepiece is the 20-ha Hownsgill Industrial Park with purpose-built factory units. However, progress has been slow. By the end of the century Hownsgill had attracted just £16 million of investment. Its biggest employer – International Cuisine – makes chilled and ready meals for several major supermarket chains, and employs 350 people.

Population change

Population changes in Derwentside tell a different story. The contraction of mining after 1950 led to out-migration and depopulation (Fig. 12.17). The closure of the Consett steelworks merely reinforced this trend. However, between 1991 and 1996 the district recorded a small increase in population for the first time for over 50 years. This population turn-round did not reflect a resurgent local economy, but was due to an influx of commuters travelling Tyneside for work, attracted by the area's high-quality environment (the North Pennines is an AONB) and its relatively low-cost housing.

The future

Although it is more than 20 years since Consett lost its steel industry, it remains one of the poorest places in England. Small and isolated within the

Strategy that saved Consett

The Consett story need contain only one dismal note, and that can be disposed of in a couple of sentences. In 1951 there were 17,000 miners in the district; now there are none. In 1975 there were 5,500 steelworkers; now there are none.

When the worst really did happen, Derwentside Council had a strategy in place. The first part of that strategy was improved communications. What was once a mess of potholed and bottlenecked roads to and through the surrounding towns and pit villages is now a fast, first-rate network bringing commercial and business traffic to and from Consett.

Next on the list was environmental rehabilitation. By the time they stopped making steel and mining coal, Derwentside was ravaged: there had never been any impetus to do much more than to keep those industries going as they had for 150 years.

Consett was an unlovely place, with no civic or commercial aspirations beyond the servicing of those polluting sources of livelihood, and when they did go the town was drab and derelict in every respect.

Today it is thriving, with a new shopping centre and supermarket, newsagents that sell the *FT*, the *New Scientist* and *Cosmopolitan* alongside the *Consett and Stanley Trader*, the *Northern Echo* and the *Newcastle Journal* – and grocers who stock Brie and Beaujolais on their shelves.

These changes reflect a cleaned environment – the colliery tips are all but gone and greened over, and there is barely a trace of the vast steelmaking site – but also the renaissance which sprang from two other key elements in Derwentside's recovery plan – the provision of space and premises (notably on land developed with English Estates North) for new businesses; and support in the form of grants and soft loans.

The prime goal was to generate replacement jobs, but not any kind of job. The council was looking for a long-term recovery in which the district would not be as vulnerable to national economic and political oscillations as in the past. Absorption of above-average youth employment remains a problem beyond the resources of the authority on its own, but in the other two areas – replacement and indigenous new business – the efforts have been a dramatic success.

Everyone talks *ad nauseam* of Derwent Valley Foods, makers of savoury snacks and crisps under the Phileas Fogg label and for own-brand clients including Sainsbury and Marks & Spencer. The firm has gone from nothing to 80 employees in less than eight years. And Blue Ridge Care, manufacturer of disposable nappies, has built its considerable sales and marketing successs within the Derwentside strategy.

Yet these are but individual – if shining – jewels in a large and lustrous crown. More than 200 companies have established themselves in and around Consett, creating 3,500 jobs and generating a completely fresh sub-regional economy.

Mr John Carney, director of Derwentside Industrial Development Agency (DIDA), notes that the top 20 local firms – most of which did not exist in 1980 – have a combined annual turnover of £110 million, which is 50 per cent greater than the steelworks at its peak. The next 20 small companies, all of which are new, have a total turnover of £15 million.

Though DIDA has been a prime instrument in bringing this about, particularly through venture-capital assistance, Mr Carney acknowledges the outstanding value for money that the council has secured through its own funding. A total of £1.8 million was spent between 1978 and 1987 – roughly £600 per job – about a third of a new town's advertising budget in total and far less than the capital-per-workplace figures common to major industrial assistance like that for Ford or Nissan.

'We're on our way to the point where deindustrialisation will be a notion of the past and when 1950s levels of unemployment will not be unthinkable. The mood is one of confidence and excitement – and a certainty that we've been doing the right thing,' says Mr Carney.

Figure 12.15
Newspaper report on the economic effects of the regeneration of Consett (*Source: Financial Times*, 29 Mar. 1988)

LIFE AFTER DEATH

Consett was a one-industry steel town until the steel-works and their coal-mine were closed down in 1980.

Consett is a semi-rural market town situated in the beautiful and unspoilt Durham countryside, where the air is the cleanest in Britain. The labour market is buoyant: only one school-leaver is registered as unemployed; more than 4,000 new jobs have been attracted to the community over the last nine years. DOOMED STEEL TOWN (I quote a *Daily Mail* headline from September last year) IS BOOMING AGAIN.

Or alternatively: Consett, a town where 'skivvying' has to pass for work these days (I am talking to Ray Thompson, 34 years at the steelworks and then redundant in 1980, who is now in the Printing Department at the Further Education College); where all the young people are taken off the unemployment register by 'Mickey Mouse dodges' dressed up as YTS schemes. Thompson calls it the 'coolie-ization' of Consett: the bondage of a new light-industrial peasantry, if you like, to badly paid, menial jobs at fly-by-night firms that can up and relocate overnight once their development grants and rates relief run out. Hardly the new El Dorado.

Next year will be the 10th anniversary of the closure of Consett steelworks, as well as the 150th of their opening. If you want The Town That Came Back From The Dead or The Town That Died, I could do you more of either – though Consett is in fact neither a boomtown nor a doomtown.

There are the natty, brightly-coloured foil bags of Phileas Fogg's 'Punjab Puri' crisps, made in Consett by Derwent Valley Foods, whose cool, wry marketing of high-class snacks speaks a language eminently comprehensible to a London business analyst. And the Sylhet Tandoori Restaurant in Front Street, where by 9.30 pm I was still the only customer, and which, like another restaurant a few doors down, where I also ate alone, is up for sale. Consett is a quiet town: quiet in the middle of the day, quiet after dark.

'This used to be the town that never slept,' says John Lee, the steelworkers' union leader at the time of the closure. 'With seven-day round-the-clock production at the Works, the shift patterns giving you as many days off during the week as at weekends, it was like Saturday every day.' The Works (as everyone refers to it) took care of all its employees' social life – was big enough to sustain sports clubs, fishing clubs, works outings, and simply to introduce everyone in the town to each other, because someone in virtually every family would have worked there.

Consett (by which I mean the town centre) is now quiet because its industry is discrete, dispersed. The steelworks was at the bottom of Front Street, a couple of minutes walk from the market-place; the showpiece No. 1 Industrial Estate is a mile out of town along the Medomsley Road; the Leadgate Trading Estate is way back towards Durham. It is a collection of tiny to modest new companies, each employing from 10 to 200 people, and each insulated in its own industrial unit, too small, and too busy consolidating and expanding the business, to promote much communal socialising. 'Consett has,' Ray Thompson weighs the word, 'dis-integrated.'

But from people who used to be at the Works, and have now found other occupations, you also get a strong sense of 'never again' – a wariness of putting all your eggs in one basket, of giving your whole life over to one company. Indeed, this transition from the benevolent despotism of the Works to a minutely compartmentalised economy is official development policy in Consett: 10 new jobs when a computer firm sets up may not do much on their own for the town's employment prospects – but nor will their loss if the firm closes.

Figure 12.16 Newspaper report on the social consequences of the regeneration of Consett (*Source: The Independent* magazine, 11 Nov. 1989)

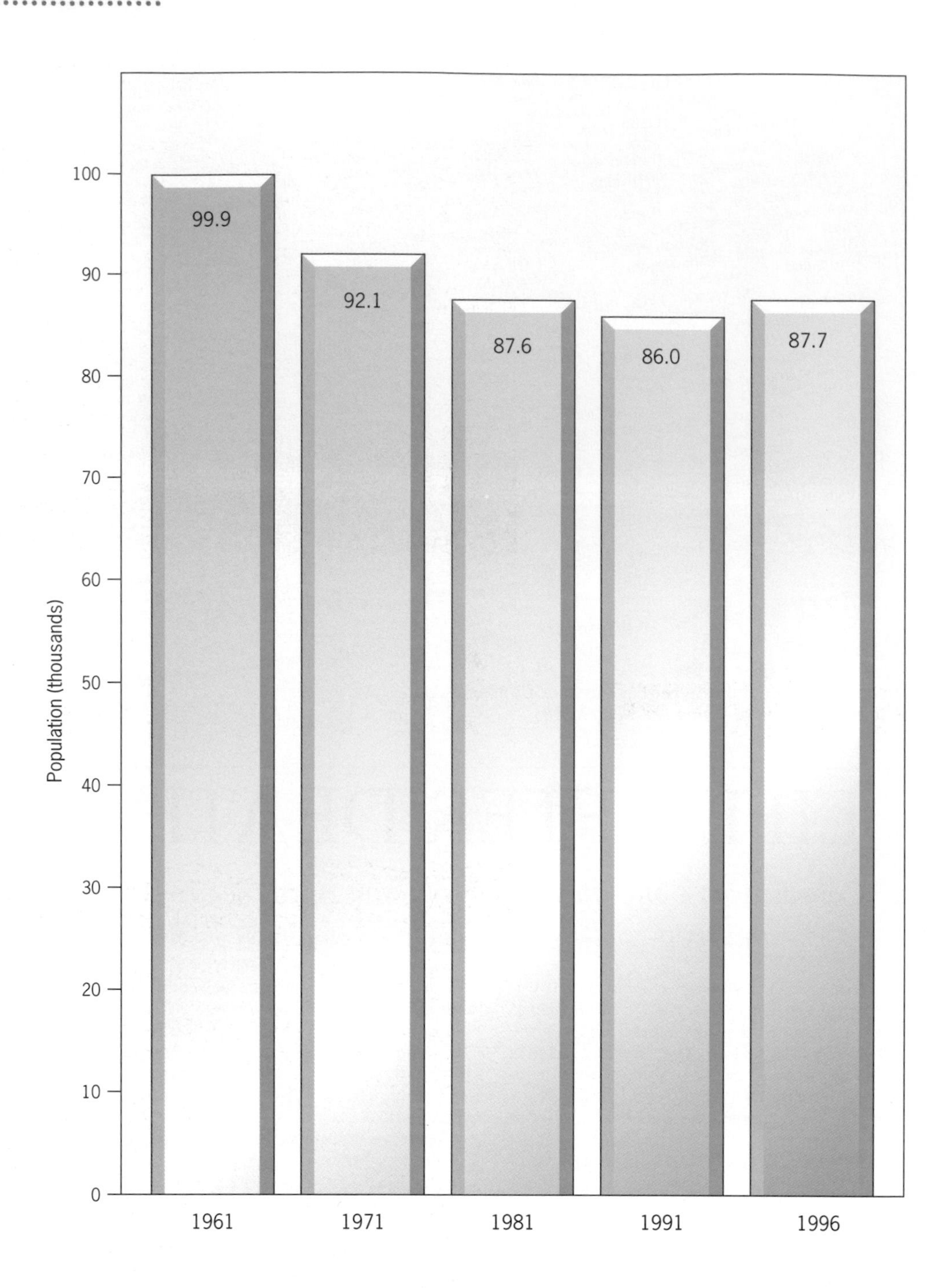

Figure 12.17 Population change in Derwentside, 1961–96

peripheral North-East region, Consett faces major obstacles to regeneration. Comparisons with Corby, a similar-sized steel town in the East Midlands, which experienced deindustrialisation in the 1980s, are instructive. Regeneration at Corby has been more successful. Although many factors have contributed to Corby's revival, its location close to the economic core of South-East England, gives it a significant advantage.

Nonetheless, economic regeneration is a reality at Consett. The pessimists in the early 1980s were wrong. Consett has survived, and its more diversified economy is all the stronger. But progress has been slow and it is now clear that a miraculous recovery forecast in the late 1980s was exaggerated. The fact is that the town's rehabilitation will continue, slowly and unspectacularly. The scars of deindustrialisation are slow to heal in peripheral regions. And it is geography more than anything else that determines the pace of recovery.

Figure 12.18 Hownsgill Industrial Estate, Consett, a project partly funded by the ERDF

Summary

- Manufacturing activities impact on the social and physical as well as the economic environment.
- Manufacturing creates benefits (e.g. jobs, wealth) known as positive externalities for a community, and disbenefits (e.g. pollution, traffic congestion) or negative externalities.
- Externalities have a geographical dimension, such as the impact on a space around a factory. The local space which is affected is called an externality field.
- Large-scale processing industries are the principal manufacturing industries responsible for the pollution of air and water resources.
- Pollution impacts at a global, regional and local scale.
- In MEDCs, manufacturing firms are becoming more environmentally aware. Many are taking steps to control polluting emissions.
- In MEDCs, the steel industry suffered massive decline between 1975 and 1990. This caused considerable social and economic hardship in steelmaking communities.
- Specialised one-industry towns like Consett were hardest hit by steel closures.
- Significant efforts have been made to promote economic regeneration in places most badly affected by deindustrialisation.

13 Manufacturing and land-use conflict

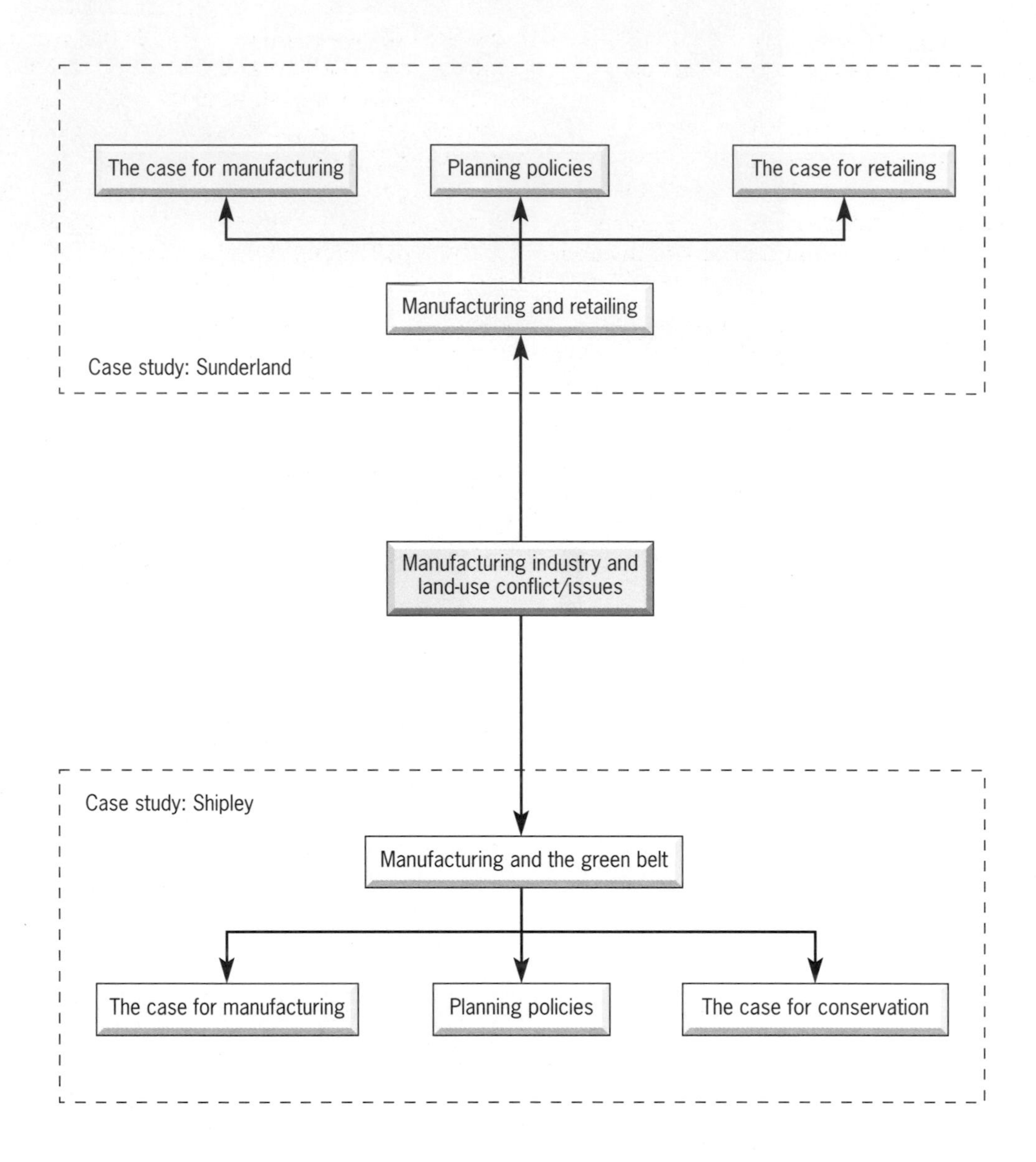

13.1 Introduction

In a free-market economy manufacturing has to compete for land with other users. Often suitable sites are in short supply, especially in urban areas where there is intense competition from retailing, office and recreational activities. Moreover, when suitable sites are found, there is often strong opposition from local residents and conservationist groups who regard industry as an undesirable neighbour. The inevitable outcome is conflict. Such conflicts are normally resolved through the planning process.

In this final chapter we shall look at two case studies of conflict between manufacturing and other activities. Your task is to study the issues, assume the role of a decision-maker and resolve the conflict. Although both case studies are based on real planning issues, in neither case is the actual outcome given.

Land-use conflict between manufacturing and retailing in Sunderland

The first case study concerns the confiict between manufacturing industry and retailing for a site in Sunderland, in North-East England. You are to assume the role of a government inspector to consider an appeal on behalf of Tesco plc for an out-of-centre superstore. Opposed to the development is the local planning council. Your task is to assess the arguments for and against the development. The evidence includes statements by Tesco and Sunderland Borough Council, local newspaper cuttings, an OS map, photographs and census data.

Background to the proposed development

The retail giant Tesco plc plans to build a large food-based superstore (6,500 m^2) at Pennywell in West Sunderland (Figs 13.1–13.3). The development also includes two small shop units (a chemist and a post office), a filling station and a health centre. On-site parking will be provided for 715 cars.

The Pennywell site is 5 km west of Sunderland's city centre and forms part of a small industrial estate. Currently, the site is occupied by a modern single storey factory built for a firm (Bard Ltd) manufacturing

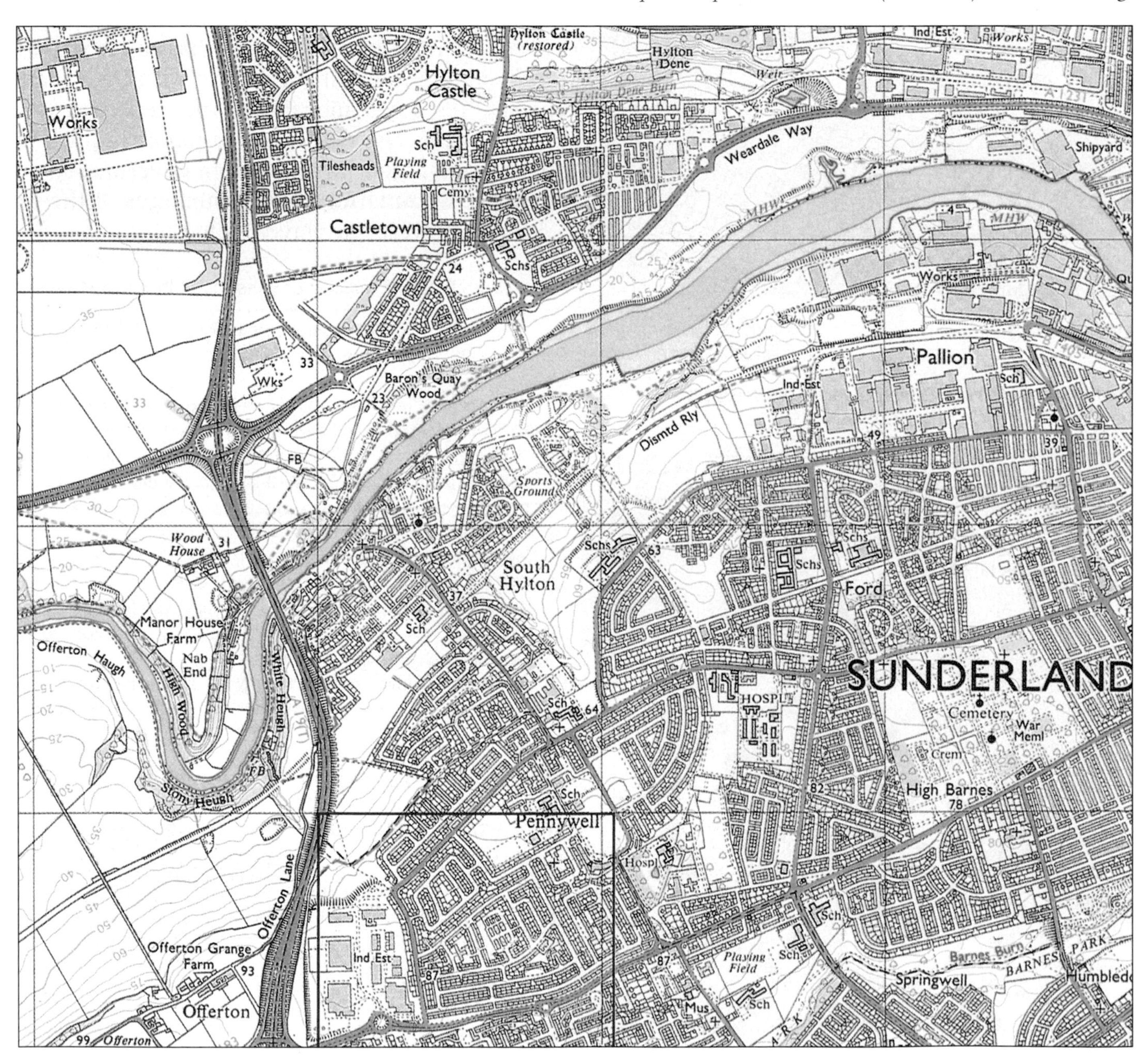

Figure 13.1 The situation of Pennywell (© Crown Copyright)

Proposed site for Tesco superstore

Portsmouth Road district shopping centre

Figure 13.2 Pennywell and the site of the proposed Tesco superstore (© Crown Copyright)

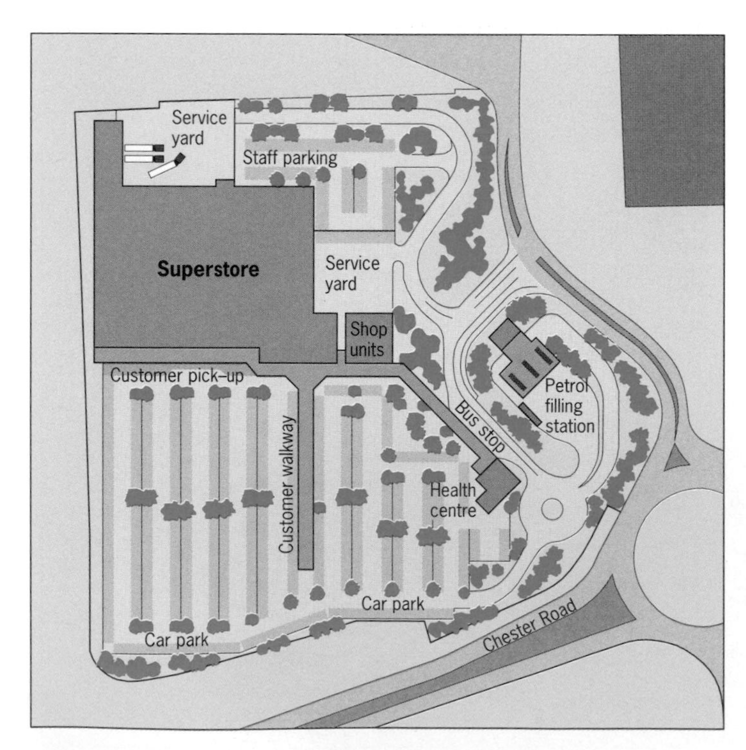

Figure 13.3 Site layout: the Tesco plans for a superstore and petrol filling station at Pennywell

?

1 Describe the site and situation of the Pennywell industrial estate, and its advantages for both manufacturing and retailing.

2 Give a brief description of the main features of the social, economic and built environment of the residential areas within one or two kilometres of the Pennywell site.

3 Summarise the main arguments relating to the impact of the proposal on industry, employment and shopping.

4 As a government inspector, write a report in which you evaluate the arguments presented by Tesco and Sunderland Borough Council. Make a final decision on the appeal, clearly indicating your priorities and justifying your choice.

medical equipment. Bard is a small US transnational employing 7,500 workers worldwide. In a recent rationalisation programme it closed its Sunderland plant and transferred production to Essex.

Sunderland Borough Council has rejected Tesco's proposal, arguing that it conflicts with their planning policies. Tesco has appealed to the Department of the Environment. The outcome is a public inquiry where both parties argue their case to a government inspector. Summaries of the statements of the Borough Council and Tesco are given below.

Table 13.1 Planning policies of Sunderland Borough Council

1	There should be sufficient surplus of industrial land in the borough to allow a wide range of choice of size, type and location for manufacturing firms.
2	Land allocated for industry should be well located in relation to the existing highway network, and have a good public transport service available.
3	A wide range of new economic activities should be created by (*a*) the assembly and preparation of industrial land, (*b*) the development of advance factories and workshop units, and (*c*) provision of suitable access to the highway network.
4	New shopping facilities should normally be located in existing shopping centres, or in areas where there is a lack of service.
5	The development of major shopping facilities outside established centres should be discouraged where they are on a scale which is likely to have too great an impact on an existing centre.
6	Measures should be promoted to upgrade existing centres, with priority given to those centres which provide the worst conditions for shoppers.

Statements presented at the public inquiry

The case for Sunderland Borough Council

Tesco's proposal must be seen in the context of the council's planning policies and the Tyne and Wear Development Plan (DP). The council believes that the Tesco superstore will have an adverse effect in three main areas: industry, employment and shopping.

Industry

The existing DP reserves the Pennywell site for industry. Its development for retailing would mean the loss of a first-class industrial site and building. The site is particularly attractive because of its closeness to the

A19, access to the A1(M), its out-of-centre location well away from congested inner city areas and its pleasant landscape setting. There is an acute shortage of industrial sites in Sunderland which meet these requirements. Good-quality industrial premises are also in short supply. The Bard factory is modern (Fig. 13.4), and if necessary could be subdivided into several smaller units. The council wants to diversify the borough's industrial structure and attract modern industries. Essential to this strategy is the retention of first-class sites and buildings (like those at Pennywell) for manufacturing industry.

Figure 13.4
The Bard factory site at Pennywell

Employment
Tesco's proposal will result in a net loss of employment in Sunderland. The new jobs it creates will not make up for redundancies arising from competition and closure of existing stores. Moreover, most of the new jobs will be part-time, unskilled, poorly paid and likely to be taken by women. The loss of Sunderland's shipbuilding industry in the 1980s has given the city one of the highest male unemployment rates in the UK. There is an urgent need to attract new manufacturing firms which offer full-time employment for men.

Figure 13.5
Portsmouth Road shopping centre

Shopping
The council is not opposed to new retail developments in principle. Indeed, several major schemes have been completed in the borough in the last 10 years. However, the scale of the Tesco superstore is likely to affect adversely the existing shopping hierarchy. Trade would be drawn from a wide area, but the local Portsmouth Road centre (Fig. 13.5), less than one kilometre away, would be especially badly hit. Portsmouth Road is conveniently situated for the surrounding residential areas of South Hylton, Pennywell, Grindon and Hastings Hill (Fig. 13.1). In contrast, the Tesco site is peripheral, and beyond reasonable walking distance for most residents. Those residents without cars would be particularly disadvantaged (Fig. 13.6).

None the less, it is acknowledged that the Portsmouth Road centre is run down and requires redevelopment (Fig. 13.5). A refurbishment scheme is currently under consideration and the council has already approached a number of developers. Tesco's superstore would undermine this initiative and divert investment away from the Portsmouth Road. This would force shop closures and cause further decline. The council believes this would be against the interests of most local residents.

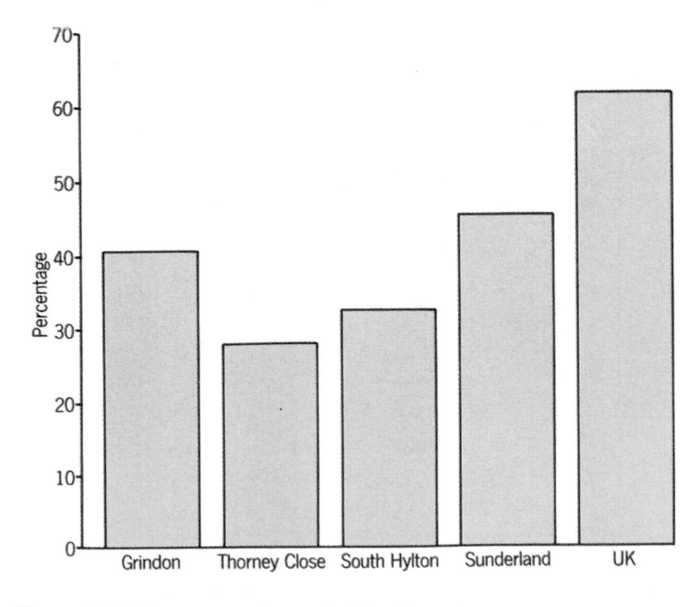

Figure 13.6 Percentage households with a car

The case for Tesco plc

Tesco rejects the argument that industry, employment and shopping will be adversely affected by its proposals. It believes that the council's decision is inconsistent with government guidelines.

Industry
There is no immediate prospect of any manufacturing enterprise occupying the Pennywell site and premises. The most likely alternative use to retailing would be warehousing, which would generate little employment. Otherwise, the site is likely to remain empty for some considerable time. Tesco believes that retailing is preferable to keeping the site vacant. Although Tesco accepts that there is some shortage of industrial land in Sunderland, it does not believe that the situation is critical. The loss of the Pennywell site would still leave the borough with more than sufficient industrial land to satisfy its needs in the foreseeable future.

Employment
The proposed development will provide jobs in commerce and services and contribute significantly to employment in the borough. This is in line with the borough's own planning policies. Sunderland needs real jobs now, not potential jobs in manufacturing in the future. The new superstore will create jobs for 360 people. Even allowing for the closure of its town centre store (if the appeal is successful), this will still leave a net gain of 228 jobs. Furthermore, these jobs will go mainly to people in West Sunderland, where rates of unemployment are the highest in the borough (Fig. 13.7). An extensive range of jobs and training is available at modern Tesco stores, providing opportunities for the unskilled as well as those with specialist skills.

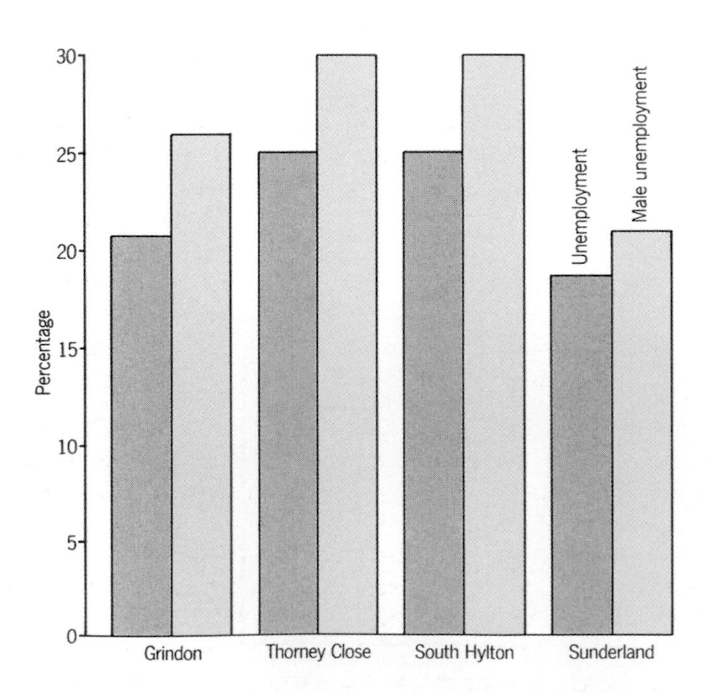

Figure 13.7 General unemployment and male unemployment

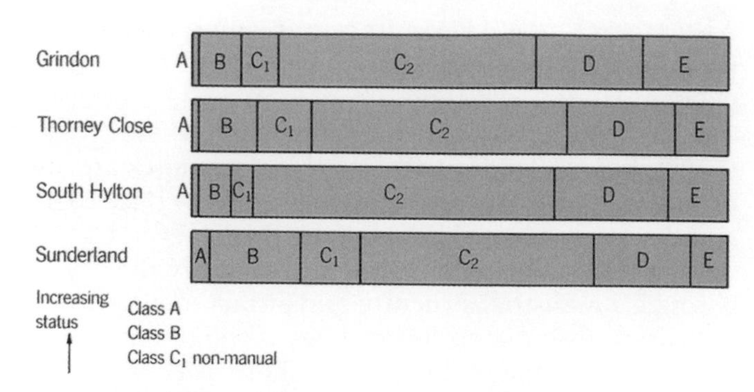

Figure 13.8 Housing tenure

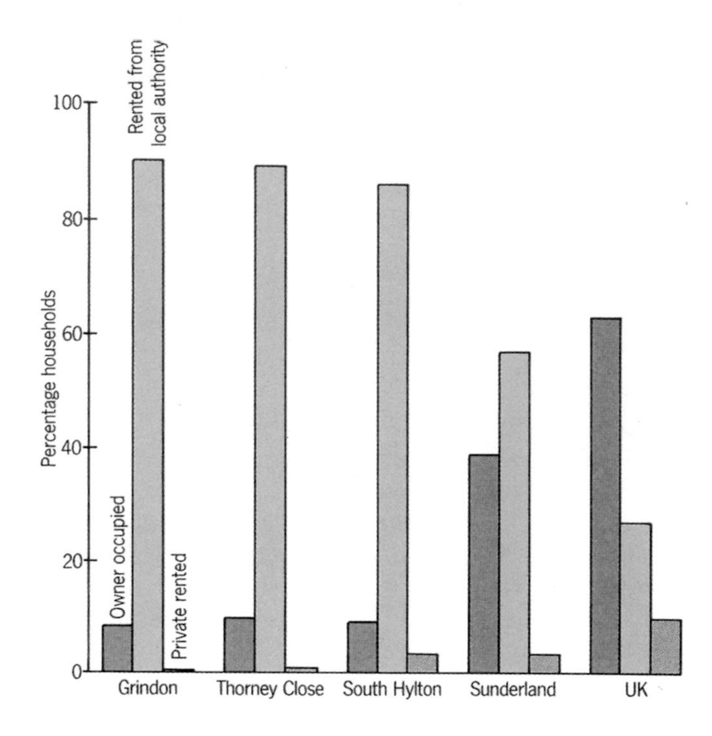

Figure 13.9 Socio-economic groups

Shopping
Tesco's proposal is supported by statutory government guidelines (Table 13.2), which encourage a positive view of new forms of retail development. The council's objection to the superstore on the grounds of its adverse effect on the Portsmouth Road centre goes against this advice.

Tesco does not accept that the new development will have a major impact on the Portsmouth Road centre. Portsmouth Road functions largely as a local convenience centre for 'top-up' shopping and the daily needs of residents. Most residents' major weekly shopping trips already take place outside the neighbourhood, either in Sunderland or Washington town centres, or superstores elsewhere in the borough. Thus, the Tesco store and Portsmouth Road will cater for different kinds of shopping and will not be in direct competition.

The new superstore will bring the following benefits to the shoppers of West Sunderland:

1 It will reduce the need for shoppers to travel outside the area for major convenience shopping.

2 It will provide a high level of convenience for food shoppers travelling by either bus or car.

3 Its size will generate economies of scale which can be passed on to customers in lower prices.

4 It will provide a modern shopping environment, making food shopping a pleasant experience.

5 It will offer a wide choice of products.

In addition, the two shop units alongside the superstore will provide a comprehensive shopping area, as well as a much-needed health centre for the community, which will be built and paid for by Tesco.

Table 13.2 Government guidelines on major new retail developments

1	New forms of retailing which extend range and choice, are more efficient, provide a better service to the public and make shopping more pleasant are to be encouraged.
2	Retailers selling food and convenience goods need large stores and car parking close at hand. The planning system should cater for them.
3	Out-of-centre superstores are to be encouraged because they remove heavy car-borne shopping from town centres and help to relieve traffic congestion.
4	Increases in car ownership have led to changes in shopping habits and retailing methods. However, it is acknowledged that not everyone has a car. Where possible, large new stores should be located to serve both car-borne shoppers and those relying on other forms of transport.
5	The impact of major new retail developments on existing retailers is not a relevant factor in making planning decisions.

Figure 13.10 Newspaper report of an interview with Tesco's managing director (*Source: Sunderland Echo*, 5 Jan. 1989)

Tesco fights for right to move out of town centre

STORE BOSS DEFENDS RELOCATION DECISION

TESCO'S supermarket chain stunned Sunderland this year by announcing its intention to pull out of the town centre and eventually close its shop in The Bridges shopping complex.

The company attracted even more headlines with its plans to build a multi-million pound superstore on Pennywell Industrial Estate, creating hundreds of jobs. Sunderland Council has opposed the Pennywell proposals and the plan is now the subject of a public inquiry.

Reporter GRAEME ANDERSON talks to Tesco's managing director David Malpas, who gave evidence at the public inquiry, about the issues behind the proposals and Tesco's hopes for the future.

TESCO's boss David Malpas makes no secret of the fact he can't wait to start building his firm's next giant superstore on Sunderland's Pennywell Industrial Estate.

Nor does he mind admitting he is hopping mad that Sunderland Council won't let him.

The 48-year-old managing director believes it is in the best interests of his company and the people of Sunderland for Tesco's to move out of the town and up to Pennywell. His links with Sunderland stretch back to 1972 when the present town centre store opened. At the time he was in control of the North-East section of the retail giants.

Now Tesco wants to build a 41,400 square feet superstore on the Pennywell Industrial Estate with a service yard, large car park, a health centre, its own bus terminal and a petrol filling station.

'We have been trading in Sunderland since 1972 and have always been successful – I think at the moment we are the second most profitable store in Sunderland,' said Mr Malpas.

'The store is wholly satisfactory but it's not a useful shop for modern shopping. The building makes our business inefficient and desperately difficult for delivery vehicles to service.'

'Nor can you do a week's bulk shopping there. Have you tried parking in Sunderland? The multi-storey car park is one of the worst I've seen and I've driven into a lot of them.'

'These are fatal disadvantages for the store, and eventually we will inevitably lose out. Our competitors have already relocated.'

'It is not fair to ask 1990s customers to shop in a shop designed in the 1960s,' he said.

Wouldn't the closure of such an important store deal a serious blow to the prestige of The Bridges and affect confidence in the town centre?

'I don't see why it should,' said Mr Malpas, 'Sunderland is uniquely fortunate in being served well by substantial food stores.'

'When we leave we are certain to leave behind a market which will make the remaining stores even more viable and should ensure they stay firmly within the town centre.'

'Let's face it, The Bridges is far more suitable for businesses like clothing department stores. And I am sure if we move out there won't be a gap for too long.'

But why does it have to be the former Bard site on Pennywell Industrial Estate for the building of a new superstore?

'The new site is the best site in Sunderland for us to build a store. At the moment a lot of the town's trade is leaking away to the Gateshead MetroCentre and the Galleries in Washington. This move will help stop that drift. No one will suffer from the move, except perhaps Sainsbury's and Asda outside the town – and I've never known one of their large stores close down yet.'

Mr Malpas rejects arguments that the site should be used for industrial purposes – an argument put forward by Sunderland Council.

Mr Malpas attacked the Labour-controlled authority for blocking the proposals.

'The only word I can think of is "insane". Planners seem to be saying "we don't want these jobs". It is beyond our comprehension. We're astonished.'

'Here we are prepared to invest £12m to £15m building a new superstore in the area which will provide somewhere in the region of 350 jobs. We should have been allowed planning permission. An inquiry is daft.'

The public inquiry is due to resume this month.

• The Echo offered Sunderland Council the opportunity to put its case on the proposed Tesco superstore in a separate article. But a Council spokesman declined, saying it would be inappropriate for it to comment on the matter while the public inquiry was proceeding.

JOB PROSPECTS HIT BY FACTORY CRISIS

By BILL O'SULLIVAN

SUNDERLAND is turning firms and vital jobs away because of a factory space shortage, a borough planner warned today.

The shortage of large buildings for firms wanting to expand or move into the town is set to continue at least until 1991.

Council assistant director of planning, Richard Arkell, said inquiries for large factories had increased markedly.

But he told a public inquiry: 'It is likely that it will be two years before the shortage of larger premises begins to ease.'

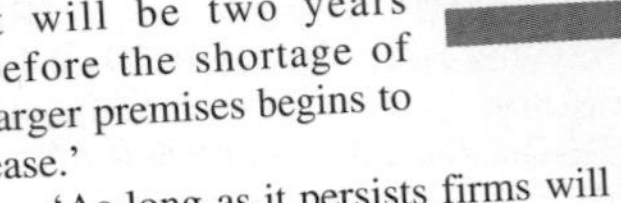

'As long as it persists firms will have to be turned away from the area and existing firms may not be able to expand.'

Mr Arkell was giving evidence at a resumed inquiry to hear an appeal by Tesco Stores against Sunderland Borough Council's failure to take a decision on an application by the company and Bard Limited to build a 67,300 sq. ft. supermarket at Pennywell Industrial Estate.

The store, with 715 parking spaces and a petrol station, would be built on the Bard site.

Mr Arkell said demolishing Bard would hinder the attraction of jobs to the town.

He added: 'The closure of the shipyards has only served to reinforce the strategic importance of the existing larger factories and sites capable of attracting investment.'

'The industrial value of the Bard site is made all the greater, given the strength of the current industrial market.'

The council has also said that the proposed development on the Bard site would endanger the revitalisation of the Portsmouth Road shopping area.

The council's case against the proposed Tesco development is supported by the Tyne and Wear Urban Development Corporation, residents of the Hastings Hill Estate and Pennywell traders.

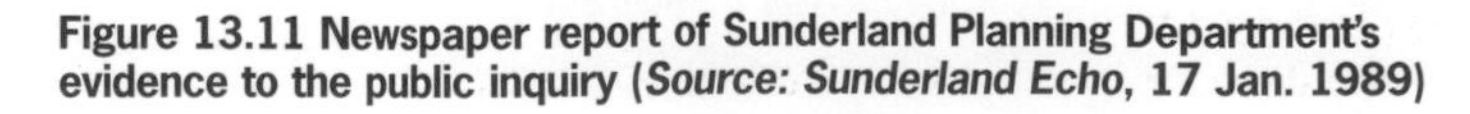

Figure 13.11 Newspaper report of Sunderland Planning Department's evidence to the public inquiry (*Source: Sunderland Echo*, 17 Jan. 1989)

STORE PLAN MAY SPARK JOB BLOW

BY JOHN GELSON

TESCO PUBLIC INQUIRY

A MAJOR revitalisation of Pennywell's rundown Portsmouth Road shopping arcade would benefit the community more than a huge Tesco store nearby, it has been claimed.

And the massive superstore earmarked for the former Bard Chemicals site could do 'demonstrable harm' to hopes of attracting new jobs to Wearside in the wake of the shipyards closure.

The claims were made by David Mole, representing Sunderland Borough Council, as the inquiry into Tesco's superstore scheme for the Pennywell Industrial Estate entered its final day yesterday.

But in his summing-up, Sir Graham Ayre, representing Tesco, claimed that town consumers were 'entitled to the choice, comfort and quality of service' provided by a major superstore.

Mr Mole said a £2.7 million re-development of the Portsmouth Road precinct, put forward by arcade leaseholders Twitchel Investments Limited, 'would serve the urban population much better that the proposed development on the Bard site'.

INVESTMENT

But he warned: 'The chances of Portsmouth Road going ahead if Tesco wins this appeal are minimal, as the investment needed would be deterred.'

Portsmouth Road trader John McMullan claimed 79 jobs on the precinct could be lost if the Tesco store is built.

But Sir Graham claimed that Tesco was facing competition from foodstore rivals already on out-of-town sites, including Asda, Sainsburys and Morrisons.

'After one of the most extensive surveys that has ever been conducted, Tesco identified the Bard site as ideally located without giving rise to legitimate objections on highway or environmental grounds,' he said

DEMAND

Earlier the inquiry heard Norman Batchelor, managing director of property company Washington Developments Limited, reveal that two major international electronic and light engineering companies are currently seeking factory space on Wearside.

'There are no large modern factory or warehouse units available for early occupation in the North-East,' he said.

'There is, however, a pressing demand for such property.'

Figure 13.12 Newspaper report of job forecasts given at the public inquiry (*Source: Sunderland Echo*, 20 Jan. 1989)

Land-use conflict between manufacturing and the green belt in Shipley, West Yorkshire

If you have read Chapter 11 you will know that at present the most attractive locations for manufacturing industry in urban areas are greenfield sites in the outer suburbs. In the UK, this preference often brings industry into conflict with green belt policies. Difficult decisions, which balance the loss of countryside against the creation of jobs, then have to be made.

Study the proposal for a factory extension into green belt land in Shipley in West Yorkshire. The proposal has the support of the District Council. However, because it involves a change to the green belt, approval is required from the Department of the Environment (DoE).

You are a geography graduate working for the DoE and it is your job to advise on the proposal. The following information is provided: the attitudes of Bradford District Council and local residents; industrial land availability in the Bradford area; the agricultural value of the site; government guidelines on green belt land; newspaper articles which present the views of both the chairman of the local residents' association and the managing director of ND Marston; a map of the site.

?

5 Outline briefly the key features of the conflict.

6 Assess the strength of the objectors' arguments, with particular reference to the disbenefits caused by: loss of green belt land; loss of farmland; damage to the viability of a local farm; pollution; traffic increases; and the impact on tourism. You should also consider the objectors' ideas on the availability of alternative sites.

7 Make a similar assessment of the arguments of ND Marston and the benefits of the factory extension.

8 Give a statement of your own priorities with regard to economic–environmental conflicts of this type.

9 Make a recommendation, which should be fully justified (and with due regard to your priorities given in 8 above), to resolve the conflict.

Background to the proposed development

ND Marston, a subsidiary of the Japanese Nippondenso Company (the world's second-largest component supplier for the motor vehicle industry), manufactures car radiators at a site in Shipley, near Bradford. It currently employs 670 people at Shipley and another 330 at its other plant in Leeds. It is one of the area's largest manufacturing employers.

Recently the firm won a contract to supply radiators to Japanese car assembly plants at Derby (Toyota) and Swindon (Honda). This requires an expansion of output which cannot be met by the existing factory. ND Marston are therefore seeking planning permission to extend their Shipley plant into an adjacent field of 10.3 ha, which lies within the green belt (Figs 13.13–13.15).

A single-storey extension to the existing plant is proposed. It will create an extra 150 jobs and safeguard existing employment. The new factory will occupy 3.95 ha of the green belt site. The firm is aware of the sensitivity of the site, which will be properly landscaped and screened by trees. In addition, ND Marston plans to create a 2-ha nature reserve along the banks of the River Aire, and help clean up an unsightly area along the river bank (Fig. 13.16). The public will have access to this area for walking and fishing.

Figure 13.13
Site of the ND Marston proposed extension

Figure 13.14
View down the valley from the site of the extension

If ND Marston's application is turned down, the firm may be forced to relocate outside West Yorkshire, putting a total of 1,000 jobs at risk. There is no possibility of expansion at its Leeds site, which is a crowded inner-city location. Moreover, there is no suitable site in the Bradford area which is big enough to accommodate the existing plant and its proposed extension. The possibility of two geographically separate sites is ruled out on grounds of cost and efficiency.

The agricultural value of the site

The site covers 3.95 ha of land on the flood plain of the River Aire. Most of it comprises an active farmholding and is classified as green belt. The Ministry of Agriculture, Fisheries and Food (MAFF) has allocated an agricultural land classification of grade 2 (on a scale 1–5, where 1 is highest-quality land). Arguments for retaining the land in agricultural use are:

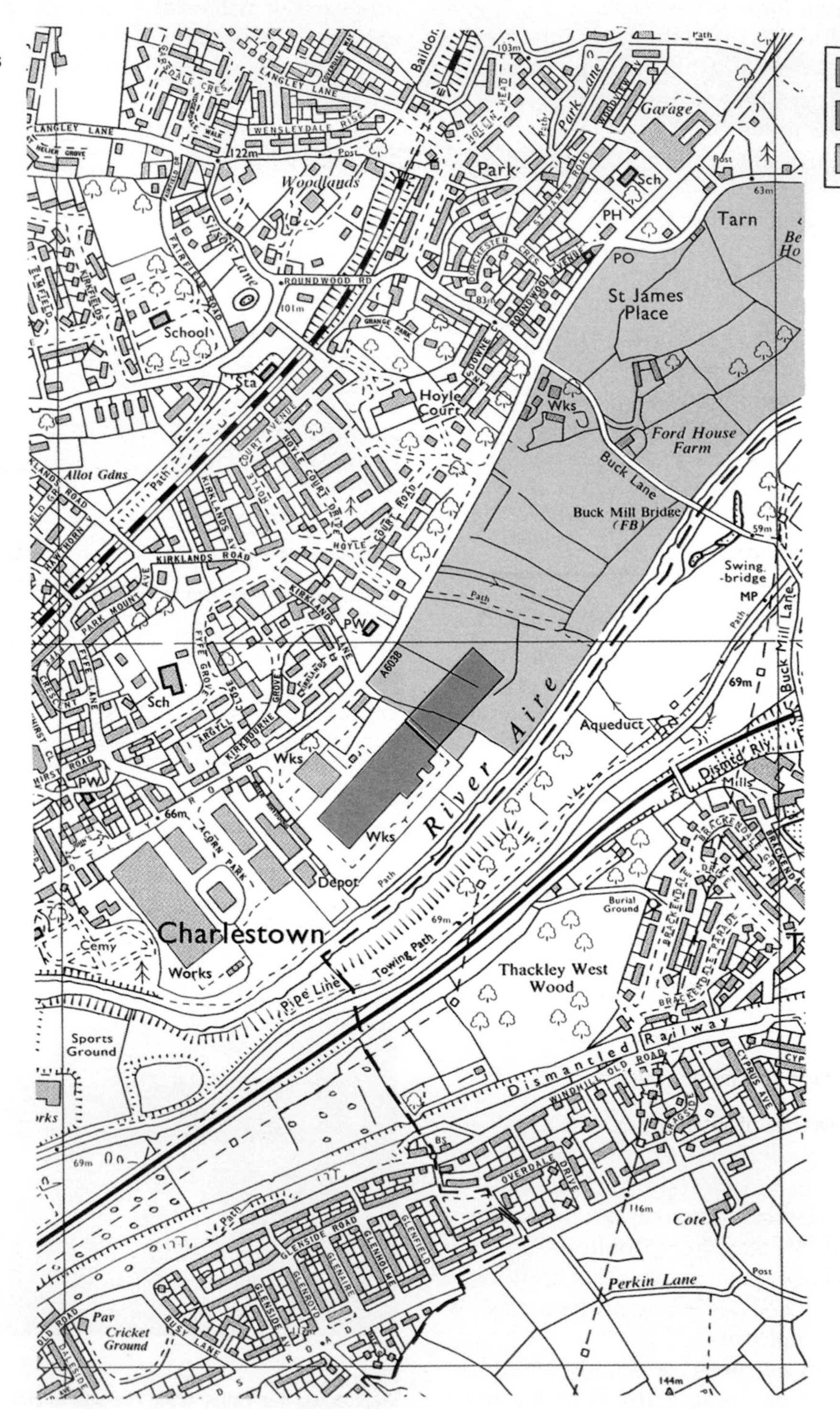

Figure 13.15 Site and situation of ND Marston's proposed factory extension (© Crown Copyright)

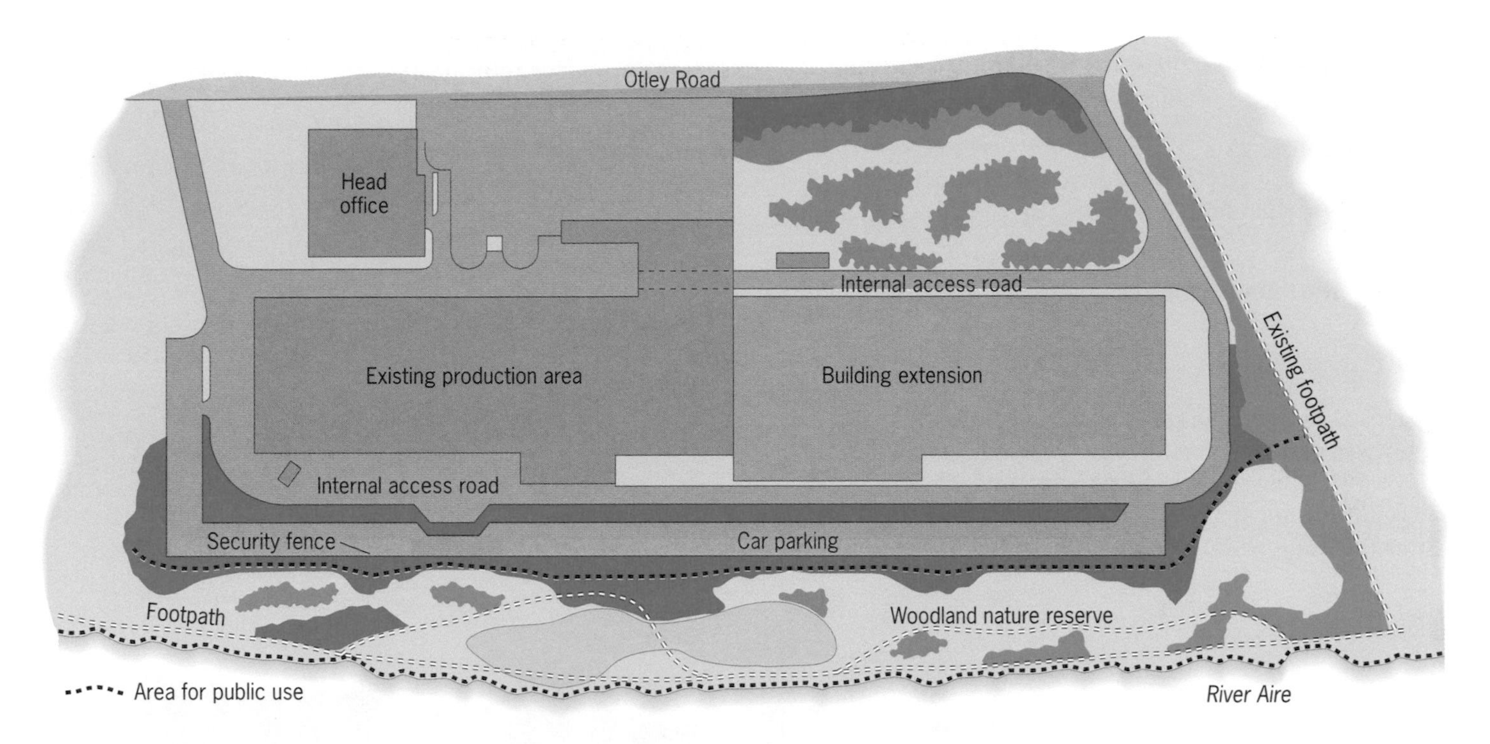

Figure 13.16 Layout of ND Marston's proposed factory extension

1 The quality of the land is the highest found in the Bradford area and has few limitations for agricultural use.

2 The land is part of a productive dairy farm and is used for silage to feed dairy herds in the winter months. The loss of 10.3 ha would undermine the farm's viability.

3 The factory extension would mean that the land would be lost to agriculture for all time.

4 Other areas of high-value agricultural land on the flood plain could also be at risk if planning approval is given for the factory.

Table 13.3 Government guidelines for safeguarding green belt land

The purposes of the green belt are:

1 To check the unrestricted sprawl of large urban areas.

2 To safeguard the surrounding countryside from further encroachment.

3 To prevent neighbouring towns from merging together.

4 To preserve the special character of historic towns.

5 To assist in urban regeneration.

NB: The government does not provide statutory protection to small pieces of farmland. Where the loss of high-quality farmland to non-agricultural use is less than 20 ha, MAFF does not have to be consulted. Government guidelines recommend that additional weight should only be given to the agricultural factor where the loss of high-quality farmland is more than 20 ha.

Availability of industrial land in the Bradford area

The council estimated that between 1984 and 1996 249 ha of land for industry would be needed in the district. The current supply of industrial land is very limited (Fig. 13.17). In Bradford in 1989 the amount of land immediately available for development was only 33.6 ha (14.6 per cent of the total). Within this category no site was larger than 3.1 ha, and only two

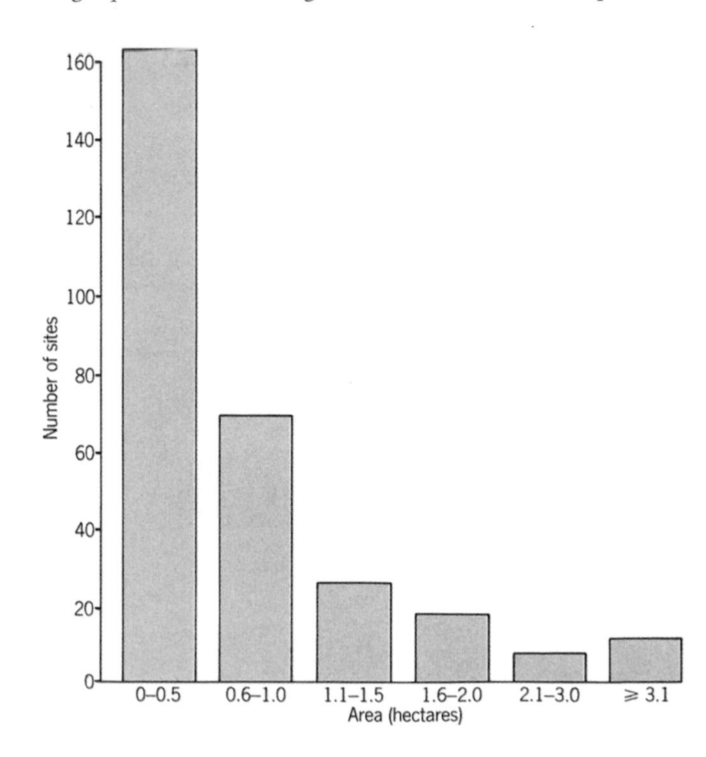

Figure 13.17 Distribution of industrial sites in Bradford by size

sites were above 2 ha. There is an acute lack of medium and large flat sites. Factors which limit the supply of industrial land are:

1 Policy constraints: although new greenfield sites on the edge of the city are popular with industrial developers, the council favours the use of land within the built-up area.

2 Steepness: many undeveloped sites are unsuitable for modern industrial buildings owing to the steepness of slopes.

This shortage of large sites reduces the district's ability to cater for medium to large companies which have outgrown their existing premises and wish to extend or relocate. The council agrees that there are no sites in Bradford immediately available to meet ND Marston's needs, either for the proposed factory extension or for relocation of the entire operation.

The view of Bradford District Council

The council supports the application for the factory extension for the following reasons:

1 There is a critical shortage of medium and large sites for industry in Bradford.

2 If the proposal is not supported, there would be a loss of jobs if ND Marston decided to relocate the firm outside the region.

3 As no suitable alternative sites exist, exceptional circumstances justify the release of land from the green belt.

Green Belt plan firm says: We may leave

by STEVEN TEALE

A COMPANY employing nearly 700 people says it could be forced to move out of the district if it cannot expand on to Green Belt land.

Baildon-based radiator firm ND Marston has been accused of ignoring residents' views about the environment with its plans for a 10,000 square metre extension alongside its premises in Otley Road, Charlestown.

Protesters say about 3,000 people have signed a petition objecting to the plans and many more have sent protest letters to the council.

But Elfed Lewis, managing director of ND Marston, says the company could be forced to move to another part of the country if the development is turned down.

Insensitive

The Company – formerly known as IMI Radiators until it was taken over by Japanese car components group Nippondenso last year – wants to develop the plant to create a new style of radiator to supply two Japanese-owned car assembly factories in Britain.

About 150 jobs would be created, says the company, which employs 680 people at Baildon.

The firm plans to spend between £10 million and £16 million on the new premises and another £300,000 on a nature reserve on adjacent land next to the River Aire.

But the proposals – which are due to go before Bradford Council planners on July 25 – have provoked opposition from villagers who want to preserve the Green Belt.

A public meeting in the village's Ian Clough Hall, called by the firm last night, was attended by about 80 people – less than forecast by some protesters.

But one of the protest leaders, Bernard Stone, said ND Marston had under-estimated the strength of feeling about its plans.

Dr Stone, who is chairman of Bradford and District Planning Association, Baildon Residents Association and Thackley and Baildon Action Group, said: 'The firm was a little insensitive to the enormous opposition to the planned development.

'The plans fly in the face of a public inquiry which ruled that the land should stay Green Belt.'

He said the poor turnout at the meeting was because some people felt it would simply be a public relations exercise by the company.

The firm sent three directors, including Mr Lewis, to the meeting. The firm's planning consultant, Jan Toulson, also attended.

Colin Harris, of Shipley Green Party, praised the nature reserve plans, but said he feared approval for the complex could be a precedent for other firms and the entire valley could be developed.

Figure 13.18 Newspaper report of a public meeting held to discuss the ND Marston planning application
(*Source:* Courtesy of the *Telegraph and Argus*, Bradford, 13 July 1990)

Extension is vital to our firm's future

by ELFED LEWIS
Managing Director
ND Marston Ltd

SINCE it began its operations in Baildon in 1976, ND Marston has always placed great importance on being a good and responsible member of the local community.

As one of the largest employers in the Bradford area, with 670 employees at our Otley Road plant, we spend some £5m a year with local suppliers and services, and our employees' spending power amounts to about £9m a year.

We are committed to supporting Bradford's economic survival and regeneration.

We are also committed to establishing and maintaining high standards of environmental responsibility.

Our premises are carefully landscaped and planted: we are a corporate member of the Yorkshire Wildlife Trust, and in close touch with the National Rivers Authority.

We also maintain contact with the Kirkstall Valley Campaign to clean up the Aire.

We want to stay here as fully participating members of the community, playing our part in providing work and prosperity and helping to make the area an even better place to live.

At the same time, economic circumstances never stand still. We are one of Europe's leading manufacturers of heat exchangers – radiators and charge air coolers – for high-performance vehicles.

Since last November, we have been part of the Japanese Nippondenso group, the largest manufacturer of vehicle radiators and automotive components in the world.

We now have the opportunity to win orders for building a new design of copper radiator, the New Single Row or NSR, for the new Toyota car plant being built near Derby, for the Honda factory at Swindon, and for other manufacturers.

That is why we need to extend our Otley Road plant by about 10,000 square metres, providing, ultimately, a further 150 new jobs.

Production would need to start in late 1992. That means building the extension early in 1991.

It has been asked why we cannot use other production routes. Our existing Otley Road plant is already working close to capacity.

Our other factory, which employs 330 people in Armley Road, Leeds, physically cannot expand.

We cannot expand westwards from the Otley Road plant, and building between the plant and the river would be prohibitively expensive and environmentally unattractive.

Relocating elsewhere in the Bradford area is ruled out through shortage of suitable sites.

We are very well aware of the sensitive nature of this planning application, and of the thoroughly proper regard local residents have for the preservation of amenities of the Aire Valley.

That is why, if this extension is approved, we would develop – and this is a promise, not just a proposal – a two-hectare (five-acre) nature reserve along the banks of the Aire.

This would involve spending £300,000 on careful landscaping and tree and shrub planting, and on cleaning up the river to make an attractive habitat for wildlife, which would also give better access, in pleasanter conditions, to walkers, anglers and others.

This is not an irresponsible attempt on the part of rapacious industrialists to grab a greenfield site.

It is an attempt by a responsible company with a proven track record to strike a balance between the need for economic development in this area (if the extension is refused Nippondenso would have to look elsewhere and up to 1,000 jobs could be at risk) and the desire we all share to preserve and improve the environment we all live in.

Figure 13.19 Statement by the ND Marston managing director (*Source: Aire Valley Target*, 19 July 1990)

Views expressed by local residents at a public meeting

1 The existing factory is noisy and causes some air pollution. This will get worse, increasing the annoyance already felt by residents in the immediate neighbourhood of the factory.

2 The extension will generate increased traffic flows through additional journeys to and from work, and heavy commerical vehicles servicing the site. The area already suffers badly from traffic congestion at peak times. The proposal will therefore worsen an already bad situation.

3 If approved, the proposal will represent further erosion of the green belt. It will set a precedent and open the way for industrial development of the whole flood plain.

4 The creation of a nature reserve is just a sop to environmentalists and planners. If the firm were really interested in the countryside, it would not be seeking to expand into the green belt.

5 The visual impact of the extension would be unsightly, particularly when viewed from Thackley on the south side of the valley (Fig. 13.15). The existing factory is already an eyesore when seen from this direction.

6 Local residents feel very strongly about preserving the fields next to ND Marston's factory. Thousands of signatures have been obtained from people angered by this attempt to encroach into the green belt.

7 There are lots of derelict sites in the inner city. It makes more sense to develop one of these rather than destroy valuable countryside.

8 Many local residents have deliberately chosen to live near to the green belt to escape the worst effects of the urban environment (factories, noise, traffic congestion, etc.).

Don't give in to the old jobs 'blackmail'

by DR BERNARD STONE, Chairman, Baildon Residents' Association

ONLY a few months after a Government Inspector recommended that no industrial development take place on these fields, Japanese-owned Nippondenso Marston applies to extend its factory into this green belt!

Unless this application is rejected, who can have any faith in the processes of public inquiries and consultation?

Concern about the environment is growing fast and any political party that ignores this will be viewed in a jaundiced light in future local elections throughout the District.

ND Marston threatens that 'it may leave' if its application is rejected. This is the same old 'blackmail' theme that is always put forward in these circumstances.

The council should not be panicked by such threats, nor take them at their full face value.

If ND Marston left, its present factory would be rapidly filled by another company, one which might well produce more and higher quality jobs than at present.

In the meantime, ND Marston should put much more imaginative thought into alternative layouts which do not consume green belt land, possibly making use of the council's under-utilised yard adjacent and the spare industrial land in Acorn Park.

The council should fully investigate the various possibilities which arise with it moving off this site, which is in any case unsuitable for the bulk storage of road salt as it is polluting the river.

The council should reject the current application because:

1. The Inquiry Inspector has recommended that industry be not allowed here.
2. Once the green belt is breached, a serious precedent is established.
3. The people of Thackley will also suffer visual disaster if the development goes ahead, as they would look down on the huge roof area of the new factory and Marston, on July 12, admitted they had no solution to this problem.
4. The land is top-grade agricultural land of a quality rare in West Yorkshire and supports a viable farm. The Ministry of Agriculture has opposed industrial expansion on to these fields.
5. Derelict industrial land should be 'reclaimed and recycled'. Bradford Area Planning and Development Association proposals were sent to the council two years ago.
6. Proposals are under way to further develop the tourist potential of this area, including Esholt up to Hawksworth. The Nippondenso factory will diminish this potential. Tourism also means jobs.
7. Thousands of people on both sides of the Lower Airedale Valley (Idle, Thackley, Baildon) have already signified their opposition to building on this area.

In spite of being told that Nippondenso Marston is very environmentally concerned, all we have gathered to date is that they will plant trees around their proposed factory extension and collect up the plastic bags which drift to the edge of the river adjacent to their site!

The 'thumbs down' should be given to this proposal.

Figure 13.20 Statement by the chairman of the residents' association (*Source: Aire Valley Target*, 19 July 1990)

Summary

- In urban areas manufacturing has to compete for sites with other users.
- Suitable manufacturing sites within urban areas are often in short supply.
- Conflicts arise between manufacturing and other users in their competition for sites.
- Conflicts are normally resolved by the planning process at a local level.
- Where conflicts are not resolved at the local level, they may go to an inquiry, with a final decision taken by the Department of the Environment.
- Land-use conflicts highlights differences in values, attitudes and beliefs between planners, private enterprise and private individuals.

Appendices

Al Spearman rank correlation

Correlation is the statistical technique for measuring the relationship or degree of association between two variables. Geographers are often interested in how change in one variable is affected by change in another – i.e. we assume that one variable is a causal factor, bringing about change in another. Causal factors are designated *independent variables* (x); non-causal factors are *dependent variables* (y).

In order to measure the strength of relationships we use a *coefficient* of *correlation* which varies in value from –1.0 to +1.0. These extremes represent perfect correlations, while a value close to zero indicates the absence of any statistical correlation. A positive coefficient tells us that as x increases, y increases. A negative coefficient means that an increase in x causes a decrease in y, or vice versa. Perfect correlations are rare in geography. This is because:

1 There are many random factors (e.g. human free will) which affect geographical phenomena.

2 Dependent variables are normally affected by several causal factors.

3 There is often some degree of error in sampling methods and measurements.

Where a relationship between variables exists, geographers are usually satisfied if this is demonstrated with a coefficient ranging between +0.5 and +1.0, and –0.5 and –1.0.

The Spearman rank correlation coefficient uses data measured on an ordinal or rank scale, and is particularly useful in surveys of attitudes or decision-making, where respondents are often asked to rank their preferences. The equation for calculating the coefficient is:

$$\text{Spearman rank correlation coefficient } (r_s) = 1 - \left[\frac{6\Sigma d^2}{n^3 - n}\right]$$

where d = difference in rank order of each pair of values and n = number of pairs of values.

Example

Manufacturing industry and regional unemployment in Italy

	Manufacturing (%)	**Unemployment (%)**	**Rank**			
	x	***y***	***x***	***y***	***d***	***d*²**
Nord-Ovest	36.7	7.6	3.5	8	4.5	20.25
Lombardia	43.4	4.7	1	11	10.0	100.00
Nord-Est	36.7	6.1	3.5	9	5.5	30.25
Emilia–Romagna	36.5	5.7	5	10	5.0	25.00
Centro	37.0	8.1	2	7	5.0	25.00
Lazio	19.0	10.6	11	5	6.0	36.00
Abruzzi–Molise	28.3	10.0	6	6	0.0	0.00
Campania	24.1	22.9	7	1	6.0	36.00
Sud	22.0	18.4	9	3	6.0	36.00
Sicilia	21.4	18.6	10	2	8.0	64.00
Sardegna	23.9	18.2	8	4	4.0	16.00

$$d^2 = 388.5$$
$$6\Sigma d^2 = 2331.0$$
$$n^3 - n = 1320$$

$$\text{Spearman rank correlation coefficient } (r_s) = 1 - \left[\frac{6\Sigma d^2}{n^3 - n}\right]$$

$$1 - \left[\frac{2331}{1320}\right] = 0.76$$

The correlation coefficient of –0.76 suggests a strong negative relationship between regional unemployment and the importance of manufacturing industry. However, the correlation is far from perfect: this is because manufacturing is only one of several independent variables which control unemployment levels.

Because the correlation between manufacturing and unemployment could be due to chance, it is necessary to establish its statistical significance using the t-distribution,

$$t = r_s . \sqrt{\frac{n-2}{1-r_s^2}}$$

n = number of pairs of data
degrees of freedom (n – 2)

$$t = -0.76 . \sqrt{\frac{11-2}{1-(-0.76^2)}} = 4.65$$

If we refer to t-tables we find that the critical value at the 0.01 level with 9 degrees of freedom is 3.25. As our t-value is greater than this critical value, we conclude that the correlation coefficient of 0.76 is statistically significant, and that there is a strong negative association between unemployment and manufacturing in Italy.

Critical values on student's *t*-distribution

	Confidence limits				
	0.90	**0.95**	**0.98**	**0.99**	**0.999**
	Two-tailed significance levels (one-tailed levels in parentheses)				
	0.10 (0.05)	0.05 (0.025)	0.02 (0.01)	0.01 (0.005)	0.001 (0.0005)
V					
1	6.31	12.71	31.81	63.66	636.60
2	2.92	4.30	6.97	9.93	31.60
3	2.35	3.18	4.54	5.84	12.92
4	2.13	2.78	3.75	4.60	8.61
5	2.02	2.57	3.37	4.03	6.86
6	1.94	2.45	3.14	3.71	5.96
7	1.90	2.37	3.00	3.50	5.41
8	1.86	2.31	2.90	3.36	5.04
9	1.83	2.26	2.82	3.25	4.78
10	1.81	2.23	2.76	3.17	4.59
11	1.80	2.20	2.72	3.11	4.44
12	1.78	2.18	2.68	3.06	4.32
13	1.77	2.16	2.65	3.01	4.23
14	1.76	2.15	2.62	2.98	4.14
15	1.75	2.13	2.60	2.95	4.07
16	1.75	2.12	2.58	2.92	4.02
17	1.74	2.11	2.57	2.90	3.97
18	1.73	2.10	2.55	2.88	3.92

	Confidence limits				
	0.90	0.95	0.98	0.99	0.999
19	1.73	2.09	2.54	2.86	3.88
20	1.73	2.09	2.53	2.85	3.85
21	1.72	2.08	2.52	2.83	3.82
22	1.72	2.07	2.51	2.82	3.79
23	1.71	2.07	2.50	2.81	3.77
24	1.71	2.06	2.49	2.80	3.75
25	1.71	2.06	2.49	2.79	3.73
26	1.71	2.06	2.48	2.78	3.71
27	1.70	2.05	2.47	2.77	3.69
28	1.70	2.05	2.47	2.76	3.67
29	1.70	2.05	2.46	2.76	3.66
30	1.70	2.04	2.46	2.75	3.65
40	1.68	2.02	2.42	2.70	3.55
60	1.67	2.00	2.39	2.66	3.46
over 60		approximates to the normal distribution			
z	1.64	1.96	2.33	2.58	3.29

A2 Chi-squared test

The chi-squared statistic tests whether the observed frequencies of a given phenomenon differ significantly from a hypothesised random frequency. Chi-squared is only applicable where the following conditions are met:

1 Data must be in frequencies.

2 Frequencies must be in absolute values, not percentages or proportions.

3 There should not be too many categories for which expected frequencies are small. For example, with just two categories the expected value in each cell must not be less than 5. With more than two categories, no more than one-fifth of expected frequencies should be less than 5.

Two-sample chi-squared

In this version of chi-squared, two samples of observations are compared with each other rather than with a hypothesised population. The null hypothesis suggests that the two samples were drawn from the same population, and therefore any differences are due to random variation.

The calculation of expected frequencies and degrees of freedom is shown in the example below. This example analyses differences in the size of multi-plant and single-plant firms at Washington in Tyne and Wear. The null hypothesis (H_0) states that there is no difference between the two samples other than that due to random sampling variations from a common population.

Table 1 Employment and multi-plant and single-plant firms

	Employment			
	0–49	50–99	>100	row total *(r)*
Multi-plant firms	15	5	9	29
Single-plant firms	52	6	5	63
Column total *(k)*	67	11	14	total 92

The expected values under the null hypothesis are provided by the marginal totals of rows and columns, so that:

$$\text{Expected cell frequency} = \frac{\text{row total} \times \text{column total}}{\text{grand total}}$$

Table 2 Calculation of expected frequencies

	Employment			
	0–49	**50–99**	**>100**	**row total (*r*)**
Multi-plant firms	(67 × 29)/192	(11 × 29)/92	(14 × 29)/92	29
Single-plant firms	(67 × 63)/192	(11 × 63)/92	(14 × 63)/92	63
Column total (*K*)	67	11	14	total 92

The logic of this procedure is that the probability of a value falling in the first row/first column cell is equal to (67/92) × (29/92) = 0.23.

The chi-squared statistic is calculated from the following formula:

$$\text{chi-squared} = \sum^{r}\sum^{k}\left[\frac{(O-\mathrm{E})^2}{\mathrm{E}}\right]$$

where O = observed frequencies, E = expected frequencies and Σ = summation sign.

The double summation sign means that the additions take place over both rows (r) and columns (k):

$$\begin{aligned}\text{chi-squared} &= 1.77 + 0.67 + 4.78 + 0.82 + 0.31 + 2.20\\ &= 10.55\end{aligned}$$

Degrees of freedom are determined by the product of the number of rows (r) and columns (k), each less one. Using the 0.01 significance level and two degrees of freedom, the critical value is 9.21 (see chi-squared statistical tables). As the chi-squared statistic (10.55) exceeds the critical value, the difference between the two samples cannot be explained by random variation, and the null hypothesis is rejected. Thus we can assume that multi-plant firms are larger than single-plant firms.

Table 3 Critical values on the chi-square distribution

	Significance level				
	0.10	**0.05**	**0.01**	**0.005**	**0.001**
v					
1	2.71	3.84	6.64	7.88	10.83
2	4.60	5.99	9.21	10.60	13.82
3	6.25	7.82	11.34	12.84	16.27
4	7.78	9.49	13.28	14.86	18.46
5	9.24	11.07	15.09	16.75	20.52
6	10.64	12.59	16.81	18.55	22.46
7	12.02	14.07	18.48	20.28	24.32
8	13.36	15.51	20.29	21.96	26.12
9	14.68	16.92	21.67	23.59	27.88
10	15.99	18.31	23.21	25.19	29.59
11	17.28	19.68	24.72	26.76	31.26
12	18.55	21.03	26.22	28.30	32.91
13	19.81	22.36	27.69	30.82	34.55
14	21.06	23.68	29.14	31.32	36.12
15	22.31	25.00	30.58	32.80	37.70
16	23.54	26.30	32.00	34.27	39.29
17	24.77	27.59	33.41	35.72	40.75
18	25.99	28.87	34.80	37.16	42.31
19	27.20	30.14	36.19	38.58	43.82
20	28.41	31.41	37.57	40.00	45.32
21	29.62	32.67	38.93	41.40	46.80
22	30.81	33.92	40.29	42.80	48.27
23	32.01	35.17	41.64	44.18	49.73
24	33.20	36.42	42.98	45.56	51.18
25	34.38	37.65	44.31	46.93	52.62

	Significance level				
	0.10	0.05	0.01	0.005	0.001
26	35.56	35.88	45.64	48.29	54.05
27	36.74	40.11	46.96	49.65	55.48
28	37.92	41.34	48.28	50.99	56.89
29	39.09	42.56	49.59	52.34	58.30
30	40.26	43.77	50.89	53.67	59.70
40	51.81	55.76	63.69	66.77	73.40
50	63.17	67.51	76.15	79.49	86.66
60	74.40	79.08	88.38	91.95	99.61
70	85.53	90.53	100.43	104.22	112.32
80	96.58	101.88	112.33	116.32	124.84
90	105.57	113.15	124.12	128.30	137.21
100	118.50	124.34	135.81	140.17	149.45

A3 Hypotheses

General

1 The output of most manufacturing plants is destined for other manufacturing firms, and not wholesalers and retailers.

Branch plants

2 A significant proportion of multi-plant firms are owned by foreign TNCs.

3 Plants operated by multi-plant firms in peripheral regions are largely concerned with routine production functions.

4 Plants operated by multi-plant firms in peripheral regions employ a large proportion of women.

5 Plants operated by multi-plant firms in peripheral regions employ: (*a*) few highly skilled technical/scientific workers; (*b*) large numbers of unskilled and semi-skilled workers.

6 Most plants operated by multi-plant firms are controlled from outside peripheral regions.

7 Few plants operated by multi-plant firms in peripheral regions have any R&D function.

8 R&D functions of multi-plant firms are mainly found in South-East England and overseas.

9 A large proportion of output from plants operated by multi-plant firms goes to other plants in the company (intra-firm trade).

10 Branch plants generate little local employment through backward linkages in the local region.

11 The output from branch plants is destined for national and international markets.

Single-plant firms

12 Most single-plant firms are relatively new businesses.

13 Most single-plant firms originated in the local region.

14 The present location of most single-plant firms is their first location.

15 Single-plant firms which have relocated have transferred production from conurbations and other large urban areas.

A4 Survey of manufacturing industry

1 Name of firm: ______________________

2 Type of manufacturing activity on site: ______________________

3 Is/are product(s) manufactured?
- finished goods sold to retailer/wholesalers ☐
- semi-finished goods ☐
- finished components for assembly ☐

4 Is the firm UK or foreign-owned?

UK ☐ Foreign ☐ (nationality) ______________

5 How many people are employed in the plant? males ________ females ________

6 Please indicate the approximate numbers of workers in each of the categories below:

- highly skilled non-manual (technicians/scientists/engineers/managers) ________
- clerical ________
- skilled manual semi-skilled ________
- unskilled ________

7 Is the firm a single-plant or multi-plant enterprise?

single-plant ☐ multi-plant ☐ no. of plants ________

8 How long has the firm occupied this site?
- less than 1 year ☐
- 1–2 years ☐
- 3–5 years ☐
- 6–10 years ☐
- more than 10 years ☐

If the firm is a single-plant enterprise, omit Questions 9, 10 and 11

---

9 Which of the following functions are found on this site?

HQ/control ☐ R&D/design ☐ Production ☐

10 If the firm's HQ and R&D are not on this site, where are they located (town/city and country)?

HQ ______________________

R&D ______________________

11 Approximately what proportion of production goes to other plants in the company?

none ☐ quarter ☐ half ☐ three-quarters ☐ all ☐

12 Please indicate (with ticks) the sources of materials used in the plant, and the destinations of products from the plant.

	Materials	Markets
Estate and local area	☐	☐
Elsewhere in the region	☐	☐
Elsewhere in the UK	☐	☐
European Community	☐	☐
Other overseas	☐	☐

13 If the firm is a single-plant enterprise, please place ticks as appropriate in the boxes below.

Firm originated in the local area ☐

Firm located on this site since its foundation ☐

Firm relocated to this site from elsewhere ☐

Previous location (town/city and district/suburb) ________________

Thank you for your co-operation

A5 Questionnaire: externality fields

1 What in your opinion are the problems/disadvantages (if any) caused by factory X?

a air pollution ☐
b water pollution ☐
c noise ☐
d smell ☐
e traffic/congestion ☐
f others ☐

2 How strongly do you feel about these problems/disadvantages?

	a	**b**	**c**	**d**	**e**	**f**
very strongly	☐	☐	☐	☐	☐	☐
quite strongly	☐	☐	☐	☐	☐	☐
not very strongly	☐	☐	☐	☐	☐	☐

3 If the only solution to the problems/disadvantages were to close the factory, would this be acceptable to you?

Yes ☐
No ☐

4 Where do you live? (name of locality/postcode) ________________

5 Estimate the age and sex of the respondent.

	male	female
15–25 years	☐	☐
26–40 years	☐	☐
41–60 years	☐	☐
>60 years	☐	☐

References

Angel, D P (1989), 'The labour market for engineers in the US semi-conductor industry', *Economic Geography, 65,* 2.

Bale, J (1978), 'Externality gradients', *Area,* 10, 5.

Birch, A (1967), *An economic history of the British iron and steel industry, 1784–1879,* Frank Cass.

Black Country Partners (1997), *A Black Country regeneration framework.*

Burn, D (1961), *An economic history of steelmaking,* Cambridge.

Castree, N (1992), 'Industrial location: in need of flexibility?', *Geography Review,* 5, 4.

Chapman, K and Walker, D (1990), *Industrial location* (second edition), Blackwell.

Clark, G (1983), *Industrial location,* Macmillan.

Cowlard, K (1987), 'Inner cities: now the good news', *Geographical Magazine, Analysis,* 2.

Daniels, P W (1990), 'Producer services and economic development', in Pinder (ed.), *Western Europe: challenge and change,* Belhaven Press.

Dicken, P (1986), *Global shift: industrial change in a turbulent world,* Harper & Row.

Dicken, P (1990a), 'European industry and global competition', in Pinder (ed.), *Western Europe: challenge and change,* Belhaven Press.

Dicken, P (1990b), 'Japanese industrial investments in the UK', *Geography,* 75, 4, The Geographical Association.

Divelly, D (1990), 'Toyota comes to Derby', *Worldwide,* 1, 1.

Foley, P, Hutchinson, J and Herbane, B (1996), 'The impact of Toyota on Derbyshire's local economy and labour market', *Tijdschrift voor Econ. en Soc. Geografie,* 87, 1.

Fothergill, S et al. (1985), *Urban industrial change,* HMSO.

Gripaios, P et al. (1989), 'High-tech industry in a peripheral area: the case of Plymouth', *Regional Studies,* 23, 2.

Hahn, R and Wellens, C (1989), 'Where does high-tech grow? An analysis of a developing high-tech region: the Washington–Baltimore corridor', *Tijdschrift voor Econ. en Soc. Geografie,* 80, 4.

Hoare, A (1983), *The location of industry in Britain,* Cambridge.

Hoover, E M (1948), *The location of economic activity,* McGraw-Hill.

Hudson, R (1988), 'Regional uneven development in the EEC', *Geography Review,* 3.

Hudson, R (1992), 'The post-steel economy at Consett-1', *Geography Review,* 6, 1.

Hudson, R (1992), 'The post-steel economy at Consett-2', *Geography Review,* 6, 2.

Hudson, R (1997), 'Changing gear? The automobile industry in Europe in the 1990s', *Tijdschrift voor Econ. en Soc. Geografie,* 88, 5

Hudson, R (1998), 'Globalisation and employment in Europe', *Geography Review,* 11, 5.

Hudson, R (1999), 'Regenerating England's inner cities', *Geography Review,* 12, 3.

Hudson, R and Sadler, D (1989), *The international steel industry,* Routledge.

Humphrys, G (1988), 'Changing places', *Geography,* 73, 4, The Geographical Association.

Keeble, D (1987), 'Industrial change in the United Kingdom', in Lever (ed.), *Industrial change in the United Kingdom,* Longman.

Keeble, D (1989), 'Core-periphery disparities, recession and new regional dynamism in the European Community', *Geography,* 74, 1, The Geographical Association.

Keeble, D (1990a), 'Small firms, new firms and uneven regional development in the United Kingdom', *Area,* 22, 3.

Lever, W F (1987), *Industrial change in the United Kingdom,* Longman.

Lovering, I and Boddy, M (1988), 'The geography of military industry in Britain', *Area,* 20, 1.

Mason, C (1990), 'Venture capital in the United Kingdom: a geographical perspective', *Nat. West Quarterly Review,* May.

Morris, I (1989), 'Japanese inward investment and the importation of subcontracting complexes: three case studies', *Area,* 21, 1.

Mountjoy, A B (1982), *The Mezzogiorno,* Oxford.

Muntendam, I (1989), 'Transnational companies are going global', *Tijdschrift voor Econ. en Soc. Geografle,* 80, 3.

Oakey, R P (1984), 'High-technology industry', *Geography,* 69, 2, The Geographical Association.

Oberhauser, A M (1990), 'State policy, employment and the spatial organisation of production: the establishment of an automobile plant in the north of France', *Tijdschrift voor Econ. en Soc. Geografie,* 81, 1.

Osleeb, J P and Cromley, R G (1978), 'The location of plants of a uniform delivered price: a case study of Coca Cola Ltd', *Economic Geography, 54,* 1.

Peck, F (1990), 'Nissan in the North-East: the multiplier effects', *Geography, 75,* 4, The Geographical Association.

Scott, A J and Mattingly, D J (1989), 'The aircraft and aircraft parts industry in southern California', *Economic Geography,* 65, 1.

Smallbone, D and North, D (19??), *Employment generation in SMEs in contrasting external environments,* Middlesex University.

Smith, D M (1987), 'Neoclassical location theory', in Lever (ed.), *Industrial change in the United Kingdom,* Longman.

Sømme, A (1968), *A Geography of Norden,* Heinemann.

South, R B (1990), 'Transnational *maquiladora* location', *AAAG,* 80, 4.

Taylor, J and Wren, C (1997), 'UK regional policy: an evaluation', *Regional Studies,* 31, 9.

Watts, H D (1987), *Industrial geography,* Longman.

Watts, H D (1990a), 'Manufacturing trends, corporate restructuring and spatial change', in Pinder (ed.), *Western Europe: challenge and change,* Belhaven Press.

Watts, H D (1990b), 'Manufacturing, the corporate sector and locational change', *Geography,* 75, 4, The Geographical Association.

Watts, H D (1991), 'Understanding plant closures: the UK brewing industry', *Geography,* 76, 4, The Geographical Association.

Wood, P A (1987a), 'How industry vanished from the inner cities', *Geographical Magazine, Analysis,* 1, September.

Wood, P A (1987b), 'Behavioural approaches to industrial location studies', in Lever (ed.), *Industrial change in the United Kingdom,* Longman.

Wren, C (1990), 'Regional policy in the 1980s', *Nat. West Quarterly Review,* November.

Glossary

Acquired advantages Advantages such as a skilled workforce and external economies built up over many years in long-established industrial regions. *See* Initial advantages.

Agglomeration economies Savings which arise from the concentration of industries in urban areas and their location close to linked activities. Also known as external economies.

Assisted areas Peripheral or semi-peripheral areas which suffer a range of economic problems (e.g. low growth, declining industries, high unemployment, etc.). They are defined and targeted by government policy for financial assistance (e.g. tax concessions, development grants, etc.).

Backward linkages Functional linkages between manufacturing firms in the direction of raw materials (e.g. automobile–steel industry linkage).

Backwash effects Initially, according to Myrdal's model of cumulative growth, the core develops at the expense of the periphery. At this stage capital, materials, people, etc. move from the periphery to the core. These movements are backwash effects.

Basing-point pricing A discriminatory pricing policy determined by adding the transport costs to production costs from some chosen point (base point).

Behavioural matrix Probabilistic model of decision-making in which the quality of a decision is determined by an individual's ability, level of knowledge, and chance.

Bounded knowledge The limited knowledge possessed by a decision-maker owing to the nature of perception, searching and other behavioural factors. *See* Satisficers.

Branch plant economy A peripheral region where a large proportion of employment in manufacturing is in externally controlled branch plants.

Break-of-bulk location A location where there is a forced transfer of freight from one transpoirt medium to another (e.g. a port).

Brownfield site A site for industry, commerce, housing, etc., previously occupied by urban land use.

Capital-intensive industries Industries with a large ratio of capital investment in plant arid machinery to workers (e.g. oil refining and petrochemicals).

Capitalist system Politico-economic system determined by supply and demand and the operation of the free market.

Command economy A planned, centrally controlled economy found under totalitarian political systems (e.g. China).

Comparative advantage The cost advantages of a region or country for the production of certain manufactured goods.

Core areas Prosperous, centrally situated regions, dominated by tertiary activities, and with many control functions (government, HQ of large firms, etc.).

Constrained location theory The idea that in economically developed countries the urban–rural manufacturing shift results from the lack of space, unsuitable factory buildings, congestion, etc., in large urban areas.

Counter-urbanisation The movement of population in economically developed countries from conurbations and large cities to small towns, commuter belt areas and remote rural areas.

Critical isodapane The isodapane in Weber's theory where the extra cost of transport from the point of minimum transport cost is equal to the saving in some other locational factor (e.g. cheap labour).

Deindustrialisation The contraction of a country's manufacturing base. Most evident in the decline of traditional industries in economically developed countries in the 1970s and 1980s.

Dependency Situation of regions or countries remote from centres of power and decision-making and whose economies depend on external control from the centre.

Deterministic theory Theory or model in which the outcome is fixed (i.e. the same inputs always generate the same output).

Diffuse industrialisation The widespread growth of new industries and small businesses in less industrialised regions like East Anglia and North-East Italy.

Diversification The attempt to strengthen the economy of an area by broadening its range of economic activities. *See* Specialisation.

Economic man The decision-maker in classical economic theory who has perfect knowledge and aims to achieve either maximum profits or minimum costs.

Economic infrastructure Transport networks, gas, electricity, water grids, sewerage systems, etc.

Energy-oriented industry An industry whose location is dominated by the cost (or immobility) of energy (e.g. aluminium smelting).

Enterprise Zones Small, run-down inner-city areas and other areas of industrial decline in the UK where financial incentives are available to encourage investment and regeneration.

Externalities Side-effects of manufacturing felt beyond the factory site. Positive externalities include the employment created by a factory; negative externalities include pollution, traffic congestion, etc.

Financial capital Funds available for investment.

Fixed capital Investment in plant and machinery. Fixed capital is highly immobile and the major cause of industrial inertia.

Flexible industries Industries which make a range of customised products and are able to respond quickly to changes in demand using computer and robotics technology.

Footloose industries Industries which are not constrained in their choice of location by traditional factors such as materials, energy and transport.

Fordist industries Large-scale, mass-production industries making standardised products. Named after the production methods (i.e. assembly lines) used in the early automobile industry.

Forward linkages Functional linkages between manufacturing firms in the direction of the finished product (e.g. the linkage between pulp and paper manufacture).

Free-on-board pricing A pricing system where the customer pays for the full cost of transport from the factory.

Greenfield site An industrial site often located on the edge of town, previously of non-urban land use.

Gross domestic product The value of a country's total production of goods and services.

Growth pole Small areas in less prosperous regions in which investment is concentrated to stimulate self-sustained growth.

Haulage costs That part of transport costs which comprise wages for the crew, fuel. etc. Haulage costs vary with distance. Also known as line costs.

Import-substitution industries Industries developed in the second stage of industrialisation to replace imported products.

Industrial inertia The survival of an industry in an area even though the initial advantages of location are no longer relevant. Industrial inertia is mainly due to the immobility of fixed capital (plant, machinery).

Initial advantages The reasons why an industry first located in a place (e.g. localised resource advantages).

Integrated production strategy Policy adopted by some TNCs where production is organised on a continental-wide basis in order to achieve scale economies and lower unit costs.

Internal economies of scale Savings in unit costs which arise from large-scale production.

Inward investment Investment by foreign-based firms into a country (e.g. Japanese investment in the automobile industry in the UK). *See* Outward investment.

Isodapanes In Weber's theory, isolines of total transport cost which increase in value with distance from the point of minimum transport cost. *See* Critical isodapane.

Labour-intensive industries Industries with a high ratio of labour costs to capital costs (e.g. the clothing industry).

Localisation economies That part of agglomeration economies comprising interlinkages between firms.

Localised materials Materials used by industries which are available only at specific locations.

Location quotient The ratio of the percentage employed in an industry in a region to the national average in that industry. If the quotient exceeds 1 it indicates an above-average concentration of that industry in a region.

Locational triangle In Weber's theory, the triangle formed by the location of the market and two material inputs. The least-cost transport location always lies within the triangle.

Market potential A measure of the aggregated access of a place to the market (based on the gravity concept).

Market-oriented industries Industries attracted to locations at the market (e.g. soft drinks). In Weber's theory, this happens when the transport costs for a finished product exceed the transport costs for raw materials. *See* Material index.

Material index The ratio of the weight of localised materials to the weight ot the finished product in a manufacturing industry. If the index >1 the materials are weight-losing and the industry should be material-oriented. Where the index is <1 the materials gain weight, suggesting a market location.

Material-oriented industries Industries attracted to locations at materials (e.g. brick-making). In Weber's theory, this happens when the transport costs for raw materials exceeds the transport costs for a finished product. *See* Material index.

Minimum efficient scale of production The size of plant which minimises the unit costs of production.

Mobility of labour Either the geographical movement of labour to employment, or the movement of labour between different occupations and industries.

Multi-locational firm Firms which operate plants at more than one location.

Multi-plant firms Firms which operate more than one plant.

Newly industrialising country A country which has undergone rapid and successful industrialisation in the last 30 years (e.g. Taiwan).

Optimal location In classical location theory, the location which either maximises profits or minimises costs.

Optimiser In economic terms, a decision-maker whose goal is either to maximise profits or minimise costs.

Organisational disintegration The devolution of control in a firm from HQ to individual plants.

Outward investment Investment by firms outside their country of origin. *See* Inward investment.

Perfect knowledge In classical location theory, economic man has complete knowledge of all the factors which influence location.

Periphery Areas which are geographically remote from a central core region and/or less prosperous than the core region.

Post-industrial economy The economies of economically developed countries where most employment is in service industries.

Probabilistic theory Theory in which the outcome is not predetermined, but is subject to chance/random influences.

Processing industries Industries based directly on the processing of raw materials (e.g. metal smelting).

Producer-goods Industries Industries like steel which manufacture a basic product for many industries. They are very vulnerable to economic recession.

Producer services Quaternary activities such as auditing, advertising, consultancy, etc., which provide services to other firms, including manufacturing industry.

Product life cycle The idea that every manufactured product has a life cycle with three stages: development, maturity, and standardisation. Each stage makes different demands on production factors (labour, capital, etc.), and in multi-plant firms influences location.

Psychic income Satisfaction derived from non-material rewards (e.g. a small business locating a factory in a high-amenity area).

Rationalisation A reduction in the production capacity of a multi-plant firm by factory closure.

Regional disparities Spatial differences in economic wellbeing within a country.

Reindustrialisation The development of new industries (often based on high-technology) and small businesses in many economically developed countries in the 1980s.

Research and development (R&D) The branch of a manufacturing firm concerned with the design and development of new products. R&D employs highly skilled workers and is often located close to company HQ.

Satisficer Decision-maker who has bounded knowledge, limited ability, imperfect perception and may be motivated by non-economic goals.

Screwdriver industries Industries based on the routine assembly of products manufactured outside a country (e.g. assembly of cars from imported kits).

Sectoral spatial division of labour The concentration of particular labour skills in particular countries or regions.

Semi-periphery Regions of intermediate prosperity between core and periphery (e.g. English East Midlands).

Small manufacturing enterprises (SMEs) Small businesses employing fewer than 100 workers.

Social infrastructure Housing, schools, hospitals and other services available to a workforce.

Spatial margins of profitability In Smith's locational theory, spatial margins delimit the area where a firm can operate profitably.

Specialisation index Measurement of the extent of a region's dependence on a narrow range of industries. The index ranges from I to 100. The higher the score the greater the specialisation.

Standard industrial classification The UK's official classification of industry. Manufacturing covers classes 24.

Structuralism A philosophy which says that patterns of economic activity and change can only be understood with reference to the prevailing politico-economic system (e.g. capitalism) and its so-called hidden structures.

Sub-optimal location A profitable location, but one which neither maximises profits nor minimises costs.

Terminal costs That part of transport costs which comprises the handling of freight. They are fixed regardless of the length of a journey and are responsible for the tapering effect of transport costs with distance.

Tertiarisation The rapid growth of service activities to a position of dominance in MEDCs.

Tidewater sites Sites adjacent to deep-water terminals which give access to imported bulk materials.

Transnational corporation (TNC) Very large firms like IBM and GM which have worldwide manufacturing capability.

Transplant An assembly plant owned and operated by a foreign-based TNC (e.g. Japanese car assembly plant in the EU).

Ubiquitous materials Materials used by industries which are available everywhere (e.g. water, etc.) and therefore have no influence on location.

Uniform delivered pricing A system of pricing whereby prices are the same everywhere regardless of the distance between producer and consumer.

Unit labour costs The ratio of wage rates to output or productivity of a workforce.

Urban Development Corporations Government-appointed bodies in the UK responsible for the revitalisation of rundown inner-city areas.

Urbanisation diseconomies The rise in unit costs as cities increase in size. Costs rise because of greater competition for land and labour, traffic congestion, etc.

Urbanisation economies That part of agglomeration economies arising from the advantages of social and economic infrastructure in large urban areas.

Urban–rural shift The relative movement of manufacturing in MEDCs from conurbations and cities to small towns and rural areas.

Venture capital Funds available for investment in high-risk businesses.

Vertical integration A production system where several stages of manufacture are located on the same site (e.g. modern steelworks which make iron, steel and finished steel products on the same site).

Index

Collins Educational
An imprint of HarperCollins*Publishers*
77–85 Fulham Palace Road
London W6 8JB

First Edition published 1993

ISBN 0 00 3266494

Michael Raw asserts his moral right to be identified as the Author of this work.

Edited by Ron Hawkins
Designed by Jacky Wedgwood
Picture research by Caroline Thompson
Artwork by John Booth, Jerry Fowler, and Contour Publishing

Typeset by Cambridge Publishing Management, Cambridge

Printed and Bound at Scotprint, Haddington

Acknowledgements

Every effort has been made to contact the holders of copyright material, but if any have been inadvertently overlooked the publishers will be pleased to make the necessary arrangements at the first opportunity.

Photographs
The publishers would like to thank the following for permission to reproduce photographs.
ASSI AB, Fig. 3.5;
Aerofilms Ltd. Figs 11.3, 11.12;
Aerophoto Eelde, Fig. 6.3;
Airbus Industrie, Figs 6.11, 6.12;
Aire Valley Target, Fig. 13.13;
Alba Centre, Livingston, Fig. 5.19;
Bradford Heritage Recording Unit, Fig. 6.5;
British Steel plc, Fig. 3.39;
British Sugar plc, Fig. 3.1;
J Allan Cash Photolibrary. Figs 2.14, 8.16;
Corus – Construction & Industrial, Fig. 3.8;
De Beers Centenary A.G., Fig. 2.21;
Derwentside District Council, Fig. 12.14;
Dudley Metropolitan Borough, Planning & Architecture Dept., Fig. 11.12;
Fujitsu Microelectronics Ltd, Fig. 2.6;
Leslie Garland Picture Library, Figs 2.4,3.36.9.3, 9.6, 11.17.12.12;
Alex Gillespie Photography, Fig. 10.3;
Matthew Gloag & Sons Ltd, Fig. 10.4;
Honda of America Manufacturing, Fig. 4.14;
Hulton Deutsch Collection Ltd, Figs 3.20, 3.21;
Hydro Aluminium, Fig. 3.9;
IBM UK Ltd, Figs 5.11, 5.12;
ICI, Fig. 1.3;
IGN Paris © 1988 Fig. 2.7, © 1990 Fig. 10.9;
ILVA (UK) Ltd, Fig. 9.12;
Alain Le Garsmeur/Impact Photos, Fig. 11.10;
London Aerial Photo Library, Figs 1.3 (aerial), 2.8, 5.16, 8.7, 11.11;
LEGO System A/S, Fig. 7,4;
Manchester Public Libraries, Fig. 3.19;
Museum of London, Fig. 11.8;
J Sturrock/Network Photographers, Fig. 5.6;
Nissan Motor Manufacturing (UK) Ltd, Fig. 4.16;
SCOPE/Jacques Guillard, Fig. 10.8;
Chris Sharpe/South American Pictures, Fig 8.16;
Sony Manufacturing Company UK/West Air Photography, Fig. 8.3;
The Stock Market, Figs 3.6, 3.7;
Tony Stone Images. Fig. 2.10, 9.3;
Sunderland, City of, Fig. 13.5;
Syndication International, Fig. 12.3;
Taipei Representative Office in the UK, Fig. 2.12;
Welsh Industrial & Maritime Museum, Fig. 3.35;
Whitbread Beer Company, Fig. 4.1;
White House Studios, Edinburgh, Fig. 2.6.

Cover picture
Tony Stone Images

Maps
Figs 13.1, 13.2, 13.15 – reproduced with the permission of the Controller of Her Majesty's Stationery Office, © Crown Copyright, from: Sunderland 1990 1:25000, Sunderland 1987 1:10000 and Shipley 1982 1:10000 Ordnance Survey Maps.

Case study
The material on pp. 59–60 was supplied by Anders Pellmyr of Perstorp AB.